# Adhesion-GPCRs

# ADVANCES IN EXPERIMENTAL MEDICINE AND BIOLOGY

Editorial Board:
NATHAN BACK, *State University of New York at Buffalo*
IRUN R. COHEN, *The Weizmann Institute of Science*
ABEL LAJTHA, *N.S. Kline Institute for Psychiatric Research*
JOHN D. LAMBRIS, *University of Pennsylvania*
RODOLFO PAOLETTI, *University of Milan*

Recent Volumes in this Series

Volume 698
BIO-FARMS FOR NUTRACEUTICALS: FUNCTIONAL FOOD AND SAFETY
CONTROL BY BIOSENSORS
  Maria Teresa Giardi, Giuseppina Rea and Bruno Berra

Volume 699
MCR 2009: PROCEEDINGS OF THE 4TH INTERNATIONAL CONFERENCE
ON MULTI-COMPONENT REACTIONS AND RELATED CHEMISTRY,
EKATERINBURG, RUSSIA
  Maxim A. Mironov

Volume 700
REGULATION OF MICRORNAS
  Helge Großhans

Volume 701
OXYGEN TRANSPORT TO TISSUE XXXII
  Duane F. Bruley and J.C. LaManna

Volume 702
RNA EXOSOME
  Torben Heick Jensen

Volume 703
INFLAMMATION AND RETINAL DISEASE
  John D. Lambris and Anthony P. Adamis

Volume 704
TRANSIENT RECEPTOR POTENTIAL CHANNELS
  Md. Shahidul Islam

Volume 705
THE MOLECULAR IMMUNOLOGY OF COMPLEX CARBOHYDRATES-3
  Albert M. Wu

Volume 706
ADHESION-GPCRs: STRUCTURE TO FUNCTION
  Simon Yona and Martin Stacey

A Continuation Order Plan is available for this series. A continuation order will bring delivery of each new volume immediately upon publication. Volumes are billed only upon actual shipment. For further information please contact the publisher.

# Adhesion-GPCRs

## Structure to Function

Edited by

**Simon Yona, PhD**
*Department of Immunology, The Weizmann Institute of Science, Rehovot, Israel*

**Martin Stacey, DPhil**
*Institute of Molecular and Cellular Biology, University of Leeds, Leeds, UK*

**Springer Science+Business Media, LLC**

**Landes Bioscience**

Springer Science+Business Media, LLC
Landes Bioscience

Copyright ©2010 Landes Bioscience and Springer Science+Business Media, LLC

All rights reserved.
No part of this book may be reproduced or transmitted in any form or by any means, electronic or mechanical, including photocopy, recording, or any information storage and retrieval system, without permission in writing from the publisher, with the exception of any material supplied specifically for the purpose of being entered and executed on a computer system; for exclusive use by the Purchaser of the work.

Printed in the USA.

Springer Science+Business Media, LLC, 233 Spring Street, New York, New York 10013, USA
http://www.springer.com

Please address all inquiries to the publishers:
Landes Bioscience, 1806 Rio Grande, Austin, Texas 78701, USA
Phone: 512/ 637 6050; FAX: 512/ 637 6079
http://www.landesbioscience.com

The chapters in this book are available in the Madame Curie Bioscience Database.
http://www.landesbioscience.com/curie

*Adhesion-GPCRs: Structure to Function*, edited by Simon Yona and Martin Stacey. Landes Bioscience / Springer Science+Business Media, LLC dual imprint / Springer series: Advances in Experimental Medicine and Biology.

ISBN: 978-1-4419-7912-4

While the authors, editors and publisher believe that drug selection and dosage and the specifications and usage of equipment and devices, as set forth in this book, are in accord with current recommendations and practice at the time of publication, they make no warranty, expressed or implied, with respect to material described in this book. In view of the ongoing research, equipment development, changes in governmental regulations and the rapid accumulation of information relating to the biomedical sciences, the reader is urged to carefully review and evaluate the information provided herein.

**Library of Congress Cataloging-in-Publication Data**

Adhesion-GPCRs structure to function / edited by Simon Yona, Martin Stacey.
   p. ; cm. -- (Advances in experimental medicine and biology ; v. 706)
Includes bibliographical references and index.
ISBN 978-1-4419-7912-4
  1. G proteins. 2. Cell receptors. 3. Cell adhesion molecules. I. Yona, Simon, 1976- II. Stacey, Martin, 1973- III. Series: Advances in experimental medicine and biology ; v. 706. 0065-2598
  [DNLM: 1. Receptors, G-Protein-Coupled--genetics. 2. Antigens, CD--physiology. W1 AD559 v.706 2010 / QU 55.7]
  QP552.G16A34 2010
  572'.69--dc22
                  2010040249

# PREFACE

Upon completion of the human genome project over 800 G protein-coupled receptor (GPCR) genes, subdivided into five categories, were identified.[1] These receptors sense a diverse array of stimuli, including peptides, ions, lipid analogues, light and odour, in a discriminating fashion. Subsequently, they transduce a signal from the ligand–receptor complex into numerous cellular responses. The importance of GPCRs is further reflected in the fact that they constitute the most common target for therapeutic drugs across a wide range of human disorders. Phylogenetic analysis of GPCRs produced the GRAFS[2] classification system, which subdivides GPCRs into five discrete families: glutamate, rhodopsin, adhesion, frizzled/taste2 and secretin receptors. The adhesion-GPCR family can be further subdivided into eight groups.[2]

The field of adhesion-GPCR biology has indeed become large enough to require a volume dedicated solely to this field. The contributors to this book have made a courageous effort to address the key concepts of adhesion-GPCR biology, including the evolution and biochemistry of adhesion-GPCRs; there are extensive discussions on the functional nature of these receptors during development, the immune response and tumourgenesis. Finally, there are chapters dedicated to adhesion-GPCR signalling, an area of intense investigation.

This volume focuses on the recent advances in adhesion-GPCR biology. In Chapter 1, we learn about the evolution of the *adhesion-GPCR* genes in several species including mouse, rat, dog, chicken and the early vertebrate Branchiostoma. In Chapter 2, Formstone continues examining both invertebrate and vertebrate adhesion-GPCRs while discussing Flamingo/Starry Night (*Drosophila*) and Celsr (vertebrate). Both are of particular interest as core components of planar cell polarity during embryonic development. The roles of adhesion-GPCRs regarding embryogenesis and organogenesis are further analyzed in Chapter 3, in which Langenhan and Russ describe their recent observations concerning the adhesion-GPCR lat-1 as a new signalling receptor, which is essential to control long-range tissue polarity in the *C. elegans* embryo.

Structurally, the adhesion-GPCR family is defined by a large extracellular region linked to a TM7 moiety via a GPS (G protein-coupled receptor proteolytic site)-containing stalk region.[3] In Chapter 4, Lin et al explore how this proteolytic cleavage was identified as an intrinsic protein modification process in the majority adhesion-GPCRs and dissect its mechanism and functional consequences. Silva and Ushkaryov further develop the functional value of the GPS site through their description of Latrophilin. This neuronal adhesion-GPCR is the major brain receptor for the black widow spider toxin α-latrotoxin, which stimulates neuronal exocytosis in vertebrates. Chapter 5 presents the latest data regarding the function, signaling and ligands for latrophilin and its related receptors, in addition to dissecting the unusual aspects of post-translational cleavage and signalling by its receptor subunits.

The extended N-terminus region of adhesion-GPCRs often contain common structural domains including epidermal growth factor-like (EGF), thrombospondin repeats, leucine-rich repeats (LRR), lectin-like, immunoglobulin (Ig), cadherins and numerous others. In other proteins many of these domains are involved in protein–protein interactions and cell adhesion; hence the "adhesion-GPCR" nomenclature was conceived reflecting the potential dual roles in cellular adhesion and signaling.[3] In Chapter 6, McMillan and White discuss the very large G protein-coupled receptor 1 (VLGR1) which is most notable for being the largest cell surface receptor in man. The large ectodomain of the protein contains several repeated motifs, including some 35-calcium binding, Calx-β repeats and seven copies of an epitempin repeat thought to be associated with the development of epilepsy. At least two spontaneous and two targeted mutant mouse lines are currently known. Mutant mice are sensitive to audiogenic seizures, have cochlear defects and significant, progressive hearing impairment. Mutations in VLGR1 in humans result in one form (2C) of Usher syndrome, the most common genetic cause of combined blindness and deafness.

Mutations in other adhesion-GPCRs, including the receptor GPR56, are also known to cause human disease. In Chapter 7, Strokes and Piao discuss how these mutations cause excess neuronal migration and a malformed cerebral cortex in the CNS in both primates and rodents. With the emerging effort in studying developmental processes, the vital roles in the development and function of the CNS of other members will be described. In Chapter 8, Xu explains other aspects of GPR56 biology, describing its binding to tissue transglutaminase, a major crosslinking enzyme in the extracellular matrix, and how its expression is suppressed in melanoma metastasis. The functions of GPR56 in cancer progression and the signalling pathways it mediates are also discussed. Further support of the potential importance of adhesion-GPCRs in tumorogenesis is discussed in Chapter 9. Aust profiles the expression of adhesion-GPCRs in tumors from databases and primary research articles and discusses their relevant roles in cell-cell communication, cell migration and angiogenesis.

The EGF-TM7 adhesion-GPCR subfamily are predominately expressed by leukocytes and are involved in coordinating both the innate and acquired immune responses. In Chapter 10, Yona et al highlight some recent immunological advances in relation to EGF-TM7 proteins and other members of the adhesion-GPCR family. Hamann et al, in Chapter 11, show how the use of specific antibodies towards the EGF-TM7 adhesion-GPCR CD97 inhibit the accumulation of granulocytes at sites of inflammation, thereby affecting innate immune responses. Spendlove and Sutavani expand on the role of CD97 through its interaction with the complement control protein DAF/CD55 in Chapter 12. The structural aspects of the CD55-CD97 complex are examined and its functional consequences in T-cell activation are also discussed. In Chapter 13, Lin et al review the historical and functional aspect of the macrophage specific adhesion-GPCR,

F4/80. The F4/80 antigen has now been used for over 30 years as an excellent marker for tissue macrophages. More recently, the receptor has been cloned and identified as an EGF-TM7 receptor critical for the induction of efferent CD8[+] regulatory T cells responsible for peripheral immune tolerance.

Until recently, the signaling cascades of almost all adhesion-GPCRs have remained a mystery. In Chapter 14, Mizuno and Itoh review previous reports which suggest G protein-dependent and independent signaling pathways of adhesion-GPCRs and present successful approaches used to investigate the signal transduction of GPR56. In Chapter 15, Park and Ravichandran describe a signaling success story and review the phylogeny, structure, associating proteins, and proposed functions of BAI1. These include its role as a signaling phosphatidylserine receptor in the uptake of apoptotic cells by phagocytes.

Finally, Chapter 16 by Davies and Kirchhoff describes the expression of adhesion-GPCRs within the male reproductive tract and reviews their potential contribution in reproductive competence.

We would like to record our sincere thanks to all our contributors.

*Simon Yona, PhD*
*Department of Immunology, The Weizmann Institute of Science*
*Rehovot, Israel*

*Martin Stacey, DPhil*
*Institute of Molecular and Cellular Biology, University of Leeds*
*Leeds, UK*

## REFERENCES

1. Lander ES, Linton LM, Birren B et al. Initial sequencing and analysis of the human genome. Nature 2001; 409(6822):860-921.
2. Lagerstrom MC, Schioth HB. Structural diversity of G protein-coupled receptors and significance for drug discovery. Nat Rev Drug Discov 2008; 7(4):339-357.
3. Yona S, Lin HH, Siu WO et al. Adhesion-GPCRs: emerging roles for novel receptors. Trends Biochem Sci 2008; 33(10):491-500.

# ABOUT THE EDITORS...

SIMON YONA, PhD, graduated from Kings College, University of London with a BSc in Physiology and a MSc in Pharmacology, before completing a PhD with Prof. R.J. Flower FRS and Prof. M. Perretti, at St. Bartholomew's Hospital, University of London. Following his doctorate Simon took up a Postdoctoral Research position with Prof. S. Gordon FRS, at the Sir William Dunn School of Pathology, University of Oxford, where he investigated the roles and functions of the leukocyte restricted adhesion-GPCRs. Dr. Yona moved from Oxford to join the group of Prof. S. Jung, at the Weizmann Institute of Science, Rehovot, Israel when he was awarded a Federation of European Biochemical Societies, International Fellowship. Currently he is investigating the developmental profile of mononuclear phagocytes in a number of pathologies.

# ABOUT THE EDITORS...

MARTIN STACEY, DPhil (Oxon), graduated from Hertford College, University of Oxford with an MBiochem before completing a DPhil in the laboratory of Prof. S. Gordon FRS. He continued working at the Sir William Dunn School of Pathology for a number of years where he cloned and characterized leukocyte adhesion-GPCRs and demonstrated the existence of their cell surface ligands. More recently, Dr. Stacey has been appointed as a Lecturer of Immunology, at the University of Leeds where his laboratory focuses on adhesion-GPCRs and role of myeloid cells in human disease.

# PARTICIPANTS

Gabriela Aust
Department of Surgery
Research Laboratories
University of Leipzig
Leipzig
Germany

Annemieke M. Boots
Department of Immune Therapeutics
Schering-Plough Research Institute
Molenstraat
The Netherlands

Gin-Wen Chang
Department of Microbiology
 and Immunology
College of Medicine
Chang Gung University
Kwei-San, Tao-Yuan
Taiwan

Ben Davies
Wellcome Trust Centre for Human
 Genetics
University of Oxford
Oxford
UK

Dorien M. de Groot
Department of Immune Therapeutics
Schering-Plough Research Institute
Molenstraat
The Netherlands

Hans van Eenennaam
Department of Immune Therapeutics
Schering-Plough Research Institute
Molenstraat
The Netherlands

Caroline J. Formstone
MRC Centre for Developmental
 Neurobiology
Kings College London
London
UK

Robert Fredriksson
Department of Neuroscience
Biomedical Center
Uppsala University
Uppsala
Sweden

Siamon Gordon
Sir William Dunn School of Pathology
University of Oxford
Oxford
UK

Jörg Hamann
Department of Experimental Immunology
Academic Medical Center
University of Amsterdam
Amsterdam
The Netherlands

Claudia L. Hofstra
Department of Immune Therapeutics
Schering-Plough Research Institute
Molenstraat
The Netherlands

Hiroshi Itoh
Department of Cell Biology
Nara Institute of Science and Technology
Takayama, Ikoma
Japan

Christiane Kirchhoff
Department of Andrology
University Hospital Hamburg Eppendorf
Hamburg
Germany

Jon D. Laman
Department of Immunology
  and MS Centre ErasMS
University Medical Center Rotterdam
Rotterdam
The Netherlands

Tobias Langenhan
Institute of Physiology
and
Rudolf Virchow Center
DFG Research Center for Experimental
  Biomedicine
University of Würzburg
Würzburg
Germany
and
Department of Biochemistry
Laboratory of Genes and Development
University of Oxford
Oxford
UK

Hsi-Hsien Lin
Department of Microbiology
  and Immunology
College of Medicine
Chang Gung University
Kwei-San, Tao-Yuan
Taiwan

D. Randy McMillan
Department of Pediatrics
University of Texas Southwestern
  Medical Center
Dallas, Texas
USA

Norikazu Mizuno
Department of Cell Biology
Nara Institute of Science and Technology
Takayama, Ikoma
Japan

Karl J.V. Nordström
Department of Neuroscience
Biomedical Center
Uppsala University
Uppsala
Sweden

Daeho Park
Center for Cell Clearance
and
Beirne Carter Center for Immunology
  Research
University of Virginia
Charlottesville, Virginia
USA

Xianhua Piao
Division of Newborn Medicine
Children's Hospital Boston
Harvard Medical School
Boston, Massachusetts
USA

Kodi S. Ravichandran
Center for Cell Clearance
and
Beirne Carter Center for Immunology
  Research
and
Department of Microbiology
University of Virginia
Charlottesville, Virginia
USA

## PARTICIPANTS

Andreas P. Russ
Department of Biochemistry
Laboratory of Genes and Development
University of Oxford
Oxford
UK

Helgi B. Schiöth
Department of Neuroscience
Biomedical Center
Uppsala University
Uppsala
Sweden

John-Paul Silva
Division of Cell and Molecular Biology
Imperial College London
London
UK

Ian Spendlove
The City Hospital
Academic Clinical Oncology
University of Nottingham
Nottingham
UK

Martin Stacey
Institute of Molecular and Cellular Biology
University of Leeds
Leeds
UK

Joan Stein-Streilein
Department of Ophthalmology
Schepens Eye Research Institute
Boston, Massachusetts
USA

Natalie Strokes
Division of Newborn Medicine
Children's Hospital Boston
Harvard Medical School
Boston, Massachusetts
USA

Ruhcha Sutavani
The City Hospital
Academic Clinical Oncology
University of Nottingham
Nottingham
UK

Paul P. Tak
Division of Clinical Immunology
 and Rheumatology
Academic Medical Center
University of Amsterdam
Amsterdam
The Netherlands

Yuri A. Ushkaryov
Division of Cell and Molecular Biology
Imperial College London
London
UK

Henrike Veninga
Department of Experimental Immunology
Academic Medical Center
University of Amsterdam
Amsterdam
The Netherlands

Lizette Visser
Department of Immunology
 and MS Centre ErasMS
University Medical Center Rotterdam
Rotterdam
The Netherlands

Perrin C. White
Department of Pediatrics
University of Texas Southwestern
 Medical Center
Dallas, Texas
USA

Lei Xu
Department of Biomedical Genetics
University of Rochester Medical Center
Rochester, New York
USA

Simon Yona
Department of Immunology
The Weizmann Institute of Science
Rehovot
Israel

# CONTENTS

## 1. THE *ADHESION* GPCRs; GENE REPERTOIRE, PHYLOGENY AND EVOLUTION ........................................................................................ 1
Helgi B. Schiöth, Karl J.V. Nordström and Robert Fredriksson

Abstract .................................................................................................................... 1
Introduction ............................................................................................................. 2
The Mouse Gene Repertoire ................................................................................... 2
The Rat Gene Repertoire ........................................................................................ 4
The Dog Gene Repertoire ....................................................................................... 4
The Chicken Gene Repertoire ................................................................................ 5
Fish ........................................................................................................................... 5
The Amphioxus Gene Repertoire ........................................................................... 6
The Nematostella Gene Repertoire ........................................................................ 7
The *Adhesion* Family is an Ancestor of the *Secretin* Family .............................. 7
Conclusion and Perspective ................................................................................... 8

## 2. 7TM-CADHERINS: DEVELOPMENTAL ROLES AND FUTURE CHALLENGES ........................................................................................... 14
Caroline J. Formstone

Abstract .................................................................................................................. 14
The 7TM-Cadherins: A Unique Group of Adhesion-GPCRs .............................. 14
The Family Tree .................................................................................................... 15
Vertebrate 7TM-Cadherins Are Developmentally Regulated ............................ 16
7TM-Cadherins Play Pleiotropic Roles During Embryonic Development ........ 18
Future Directions: Linking 7TM-Cadherin Structure to Function ................... 30
Conclusion ............................................................................................................. 31

## 3. LATROPHILIN SIGNALLING IN TISSUE POLARITY AND MORPHOGENESIS .................................................................. 37

Tobias Langenhan and Andreas P. Russ

Abstract ............................................................................................................. 37
Introduction ....................................................................................................... 37
The Role of Adhesion-GPCRs in Development ................................................ 38
Latrophilins and Tissue Polarity ....................................................................... 42
Conclusion ........................................................................................................ 46

## 4. GPS PROTEOLYTIC CLEAVAGE OF ADHESION-GPCRs .......................... 49

Hsi-Hsien Lin, Martin Stacey, Simon Yona and Gin-Wen Chang

Abstract ............................................................................................................. 49
Introduction ....................................................................................................... 49
The Identification of GPS Proteolytic Modification in Adhesion-GPCRs ....... 50
The Molecular Mechanism of GPS Proteolysis ............................................... 52
The Regulation of GPS Proteolysis .................................................................. 54
The Role of GPS Proteolysis in Receptor Function and Human Disease ...... 54
The Fate and Functional Interaction of the Extracellular and 7TM Subunits
    following GPS Proteolysis ........................................................................ 55
Conclusion ........................................................................................................ 57

## 5. THE LATROPHILINS, "SPLIT-PERSONALITY" RECEPTORS ................... 59

John-Paul Silva and Yuri A. Ushkaryov

Abstract ............................................................................................................. 59
Introduction ....................................................................................................... 59
The Isolation of Latrophilin ............................................................................... 60
The Latrophilin Family ...................................................................................... 61
Expression Patterns of Latrophilins .................................................................. 62
The Structure of Latrophilin .............................................................................. 64
Latrophilin as a GPCR ....................................................................................... 68
Ligands and Interacting Partners of Latrophilins ............................................. 69
Latrophilin Gene Knockouts ............................................................................. 71
Latrophilins in Disease ...................................................................................... 72
Conclusion ........................................................................................................ 72

## 6. STUDIES ON THE VERY LARGE G PROTEIN-COUPLED RECEPTOR: FROM INITIAL DISCOVERY TO DETERMINING ITS ROLE IN SENSORINEURAL DEAFNESS IN HIGHER ANIMALS ............................................................ 76

D. Randy McMillan and Perrin C. White

Abstract ............................................................................................................. 76
Introduction ....................................................................................................... 76
Gene Structure .................................................................................................. 77
Protein Structure ............................................................................................... 79
Gene Expression ............................................................................................... 80
Mice with Mutations in VLGR1 ........................................................................ 80

Expression in the Cochlea and Retina .................................................................................. 81
Conclusion .............................................................................................................................. 83

## 7. ADHESION-GPCRs IN THE CNS .......................................................................... 87

Natalie Strokes and Xianhua Piao

Abstract................................................................................................................................... 87
Introduction............................................................................................................................. 87
Adhesion-GPCRs in Brain Development................................................................................ 89
GPR56 in Brain Development and Malformation ................................................................. 90
Conclusion .............................................................................................................................. 94

## 8. GPR56 INTERACTS WITH EXTRACELLULAR MATRIX AND REGULATES CANCER PROGRESSION ........................................... 98

Lei Xu

Abstract................................................................................................................................... 98
Introduction............................................................................................................................. 98
The Primary Structure of Gpr56 Gene and its Encoded Protein ........................................... 99
Functions of GPR56 in Cancer ............................................................................................. 101
Signaling Pathways Mediated by GPR56 ............................................................................ 103
Conclusion ............................................................................................................................ 105

## 9. ADHESION-GPCRs IN TUMORIGENESIS ........................................................ 109

Gabriela Aust

Abstract................................................................................................................................. 109
Databases ............................................................................................................................. 109
Overview on the Adhesion-GPCR Family ........................................................................... 110
Group I: Latrophilin-Like (Latrophilin 1-3, ETL) ............................................................... 110
Group II: EGF-Like (CD97, EMR1-4) ................................................................................. 111
Group III: IgG-Like (GPR123, GPR124, GPR125) ............................................................ 112
Group IV: CELSR-Like (CELSR 1-3) ................................................................................. 113
Group V: GPR133 and 144................................................................................................... 113
Group VI: GPR110, GPR111, GPR113, GPR115, GPR116............................................... 113
Group VII: BAI-Like (BAI 1-3)............................................................................................ 113
Group VIII: Miscellaneous (GPR56, GPR64, GPR97, GPR112,
    GPR114, GPR126)......................................................................................................... 116
Ungrouped (GPR128, VLGR1)............................................................................................. 117
Conclusion ............................................................................................................................ 117

## 10. IMMUNITY AND ADHESION-GPCRs .............................................................. 121

Simon Yona, Hsi-Hsien Lin and Martin Stacey

Abstract................................................................................................................................. 121
Introduction: Adhesion-GPCRs in Immunology ................................................................. 121
The EGF-TM7 Family .......................................................................................................... 122
Gone Fishing......................................................................................................................... 123
BAI1 ...................................................................................................................................... 125
GPR56 ................................................................................................................................... 125
Conclusion ............................................................................................................................ 126

## 11. CD97 IN LEUKOCYTE TRAFFICKING ............................................................. 128

Jörg Hamann, Henrike Veninga, Dorien M. de Groot, Lizette Visser,
   Claudia L. Hofstra, Paul P. Tak, Jon D. Laman, Annemieke M. Boots
   and Hans van Eenennaam

Abstract ............................................................................................................................. 128
CD97 is a Prototypical EGF-TM7 Receptor .................................................................... 129
CD97 Antibody Treatment Inhibits Granulocyte Trafficking .......................................... 129
CD97 Targeting in Antigen-Driven Disease Models ....................................................... 132
Antibody Treatment versus Gene Targeting ..................................................................... 134
In Vivo Studies Start to Unveil the CD97 Mechanism of Action .................................... 135
Conclusion ........................................................................................................................ 136

## 12. THE ROLE OF CD97 IN REGULATING ADAPTIVE T-CELL RESPONSES ............................................................................ 138

Ian Spendlove and Ruhcha Sutavani

Abstract ............................................................................................................................. 138
Introduction: Structure of CD97 and its Interaction with CD55 ...................................... 138
Costimulation of T Cells .................................................................................................. 140
T-Cell Signalling .............................................................................................................. 141
Effect on Different T-Cell Populations ............................................................................ 142
Effects of Complement on CD55 and T Cells ................................................................. 145
Conclusion ........................................................................................................................ 146

## 13. F4/80: THE MACROPHAGE-SPECIFIC ADHESION-GPCR AND ITS ROLE IN IMMUNOREGULATION ..................................... 149

Hsi-Hsien Lin, Martin Stacey, Joan Stein-Streilein and Siamon Gordon

Abstract ............................................................................................................................. 149
Introduction ...................................................................................................................... 149
Generation and Application of F4/80 mAb: The Phenotypic and Functional
   Characterization of Mouse Macrophage Subpopulations ........................................... 150
Molecular Cloning and Characterization of the F4/80 (Emr1) Gene .............................. 151
Generation and Analysis of F4/80-Deficient Animals ..................................................... 153
The Role of F4/80 in Immunoregulation .......................................................................... 153
Conclusion ........................................................................................................................ 154

## 14. SIGNAL TRANSDUCTION MEDIATED THROUGH ADHESION-GPCRs ........................................................................... 157

Norikazu Mizuno and Hiroshi Itoh

Abstract ............................................................................................................................. 157
Difficulties in Studying the Signal Transduction of Adhesion-GPCRs ........................... 157
G Protein-Dependent Signaling Pathway of Adhesion-GPCRs ....................................... 158
Oligomerization of Adhesion-GPCR for Activation ........................................................ 159
Functional Antibodies against Adhesion-GPCRs ............................................................ 161
Other Studies that Suggest the Signaling Pathway of Adhesion-GPCRs ........................ 162
Complexity of Signal Transduction via Adhesion-GPCRs .............................................. 163
Conclusion ........................................................................................................................ 164

## 15. EMERGING ROLES OF BRAIN-SPECIFIC ANGIOGENESIS INHIBITOR 1 .................................................................................. 167

Daeho Park and Kodi S. Ravichandran

Abstract ........................................................................................................... 167
Introduction .................................................................................................... 167
Initial Identification of BAI1 ......................................................................... 168
Structural and Functional Domains of BAI1 .............................................. 168
BAI1 as an Engulfment Receptor for Apoptotic Cells ................................ 170
Role of BAI1 in Glioblastomas ...................................................................... 173
Other Known Interacting Partners of BAI1 ................................................ 174
Conclusion ...................................................................................................... 176

## 16. ADHESION-GPCRs IN THE MALE REPRODUCTIVE TRACT ............... 179

Ben Davies and Christiane Kirchhoff

Abstract ........................................................................................................... 179
Introduction .................................................................................................... 179
GPR64 .............................................................................................................. 180
GPR124 and GPR125 ..................................................................................... 184
CELSR1-3 ........................................................................................................ 185
Other Adhesion-GPCRs ................................................................................. 185
Conclusion ...................................................................................................... 186

## APPENDIX: MAMMALIAN ADHESION-GPCRs ............................................... 189

Simon Yona and Martin Stacey

**INDEX** ............................................................................................................... 195

CHAPTER 1

# THE *ADHESION* GPCRs; GENE REPERTOIRE, PHYLOGENY AND EVOLUTION

Helgi B. Schiöth,* Karl J.V. Nordström and Robert Fredriksson

**Abstract:** The *Adhesion* family is unique among the GPCR (G protein-coupled receptor) families because of several features including long N-termini with multiple domains. The gene repertoire has recently been mined in great detail in several species including mouse, rat, dog, chicken and the early vertebrate Branchiostoma (*Branchiostoma floridae*) and one of the most primitive animals, the cnidarian Nematostella (*Nematostella vectensis*). There is a one-to-one relationship of the rodent (mouse and rat) and human orthologues with the exception the EMR2 and EMR3 that do not seem to have orthologues in either rat or mouse. All 33 human *Adhesion* GPCR genes are present in the dog genome but the dog genome also contains 5 additional full-length *Adhesion* genes. The dog and human *Adhesion* orthologues have higher average protein sequence identity than the rodent (rat and mouse) and the human sequences. The *Adhesion* family is well-represented in chicken with 21 one-to-one orthologous with humans, while 12 human *Adhesion* GPCRs lack a chicken ortholog. Branchiostoma has rich repertoire of *Adhesion* GPCRs with at least 37 genes. Moreover, the *Adhesion* GPCRs in Branchiostoma have several novel domains their N-termini, like Somatomedin B, Kringle, Lectin C-type, SRCR, LDLa, Immunoglobulin I-set, CUB and TNFR. Nematostella has also *Adhesion* GPCRs that are show domain structure and sequence similarities in the transmembrane regions with different classes of mammalian GPCRs. The Nematostella genome has a unique set of *Adhesion*-like sequences lacking GPS domains. There is considerable evidence showing that the *Adhesion* family is ancestral to the peptide hormone binding *Secretin* family of GPCRs.

*Corresponding Author: Helgi B. Schiöth—Department of Neuroscience, Biomedical Center, Box 593, 75 124 Uppsala, Sweden. Email: helgis@bmc.uu.se

*Adhesion-GPCRs: Structure to Function*, edited by Simon Yona and Martin Stacey.
©2010 Landes Bioscience and Springer Science+Business Media.

## INTRODUCTION

Our interest in the *Adhesion* GPCRs started when we were mining and performing phylogenetic analysis of the entire repertoire of human GPCRs. This work resulted in an overall classification of the GPCRs based on strict phylogenetic critera.[1] During our quest to include as many human GPCRs as possible, we found that there were several GPCRs with long N-termini that were not annotated. The *Adhesion* GPCRs have been notoriously difficult to study because of their large genome size, being up to over 7000 nucleotides, multiple introns (up to 100) which also can be alternatively spliced.[2] We worked with the HUGO nomenclature committee to give these new genes names and 16 of the 33 human *Adhesion* GPCRs were given names in three separate publications.[3-5] Further work of the GPCR repertoire resulted in thorough mining of the gene repertoire of GPCRs in mouse, rat and chickens,[3,6,7] also reviewed in reference 8. The *Adhesion* family, or parts of it, has been called various names including EGF (epidermal growth factor)-TM7, B2, LN (long N-termini) B-TM7, or LN-7TM receptors.[9-13] In the classification of GPCRs[1] we showed that the *Adhesion* family was clearly unique family among GPCRs and we thus named it the *Adhesion* GPCRs and this name has since received broad acceptance. The *Adhesion* GPCRs have also been classified to the pfam class B together with the *Secretin* GPCR family and the *Methuselah* sequences found in insects. These pfam models were based on various such sequences and do thus not distinguish between these groups. Also, at this time, many of the *Adhesion* GPCRs were not included in the original training sets for these models, that are continuously updated and they therefore lack specificity and sensitivity. However, there are many very clear differences between these families. There is striking differences within the N-terminal domain architecture between the *Secretin* and the *Adhesion* family receptors. The *Adhesion* GPCRs display the GPCR proteolytic (GPS) domain while the *Secretin* receptors lack this domain. The *Secretin* GPCRs have only one type of a domain, a hormone binding domain (HBD). The ligands of the different families deviate also as the de-orphanized *Adhesion* family receptors bind extracellular matrix molecules whereas the *Secretin* GPCRs bind peptide hormones. Recent mining of the *Adhesion* GPCRs has shown that they are very well-conserved in evolutionary terms and sequences from one of the most primitive animals Nematostella, as well as the single-celled and colony-forming eukaryotes Monosiga brevicollis and Dictyostelium discoideum contain *Adhesion*-like sequences.[14] Remarkably some of these sequences have domain compositions that resemble that of the human receptors. Moreover, there are additional *Adhesion*-like subgroups in distant species such as Branchiostoma and Nematostella that are not found in mammals. In this chapter we review the genomes where the *Adhesion* family has been mined and assembled to great detail.

## THE MOUSE GENE REPERTOIRE

The mouse genome had early on one of the best assemblies of the mammalian genomes. We characterized the mouse repertoire of *Adhesion* GPCRs using several bioinformatics methods.[15] In this work we identified the two last members of the human *Adhesion* GPCRs resulting in 33 as the final number of human *Adhesion* GPCRs. Two human pseudogenes of the *Adhesion* family, both similar to GPR116, have received official names from HUGO Gene Nomenclature Committee and are now referred to as GPR116P1 and GPR116P2, respectively.[16] The mouse genome contains two fewer

*Adhesion* GPCRs. We found that there is a one-to-one relationship of the mouse and human orthologues, with the exception the EMR2 and EMR3 that do not seem to have orthologues in mouse. In general, the mouse and human repertoire share several features, which suggested that these receptors are likely to have similar functional role in both species. The study also showed that the *Adhesion* family is by no means a coherent group of equally divergent clusters. The overall similarity of the TM region among the *Adhesion* GPCRs is in many cases as low as 18-19% in amino acid identity (between GPR124 and GPR128), as revealed through inspection of a percentage identify chart for all the receptors included in the phylogenetic analysis that is only based on the TM regions. Some of the phylogenetic groups are more highly similar such as the BAI receptor cluster that shares 52-76% amino acid identity. Other clusters are more variable as the sequences in the phylogenetic cluster VIII share 21 to 73% amino acid identity. It is also interesting that there exists a high degree of one-to-one pairing of the mouse and human orthologues although the percent identity between the human and mouse orthologues varies to a large extent. The human and mouse BAI3 receptors in human and mouse share for example 98% amino acid identity while the orthologues pairs of GPR127, GPR112 and GPR113 only share 54, 55 and 58% amino acid identity, respectively, suggesting that these genes are under less evolutionary constraint as compared with the BAI genes. Overall, ten of the human-mouse orthologues pairs share over 90% identity while nine have identity below 70%. It is tempting to speculate that the low degree on conservation of the TM regions among some of the *Adhesion* GPCRs may be due to less specific functional roles of this region as compared with other GPCRs. This should be considered in light of the fact that the *Adhesion* GPCRs, unlike most GPCRs, are not believed to have a ligand binding site within the TM regions. The lack of clarity whether the *Adhesion* GPCRs do couple to second messengers, in either specific or important fashion, has also contributed to speculation about low importance of the TM region, or only has a function or anchoring to the membrane.[17] The uneven degree of conservation between the mouse and human orthologues perhaps indicates that the role of the TM region might be highly variable. In order to search for specific elements, we created consensus sequences for the different phylogenetic groups.[3] These sequences reveal that there exist some highly conserved motifs in the TM regions. These are in particular found in TMIII where there are three residues conserved in all human and mouse *Adhesion* GPCRs including the hydrophilic His and Trp. Moreover, there are high conservation of a Glu and Tyr that add to the notion that TMIII could have an important and common role for these receptors. It is intriguing that studies on *Rhodopsin* GPCRs have revealed that a motif that is only found in *Rhodopsin* GPCRs, namely the DRY in the ascending part of TMIII, has a role in keeping several of the *Rhodopsin* GPCRs in an inactive conformation.[18] Mutagenesis studies have indicated that this region is crucial for the intracellular signalling of many *Rhodopsin* GPCRs. The TMIII is found in the interior of the TM helical bundle. If the overall topology of *Adhesion* GPCRs is similar to that of the bovine rhodopsin,[19] it is interesting to speculate if the rotation of TMIII may be crucial for the transduction of signal to the interior of the cell, which in fact has been suggested to possibly be a common feature of *Rhodopsin* GPCRs.[20] The high degree of conservation of the TMIII region thus points to that the region could be commonly important for the *Adhesion* GPCRs as well, perhaps for coupling to second messengers. There are in general very few motifs in the *Adhesion* GPCRs that show specific similarities to the other families of GPCRs. In TMIV there is a Trp that is found in the consensus sequences of the *Rhodopsin*, *Secretin* and *Frizzled* receptors but not in the *Glutamate* receptors.[1] This conserved residue is followed by a PAL/V motif found in both the *Secretin* and *Frizzled* families. This conserved Pro is interesting as this residue is rather uncommon in TM regions and is believed to form kinking of the α-helix. Another feature that the *Adhesion* GPCRs share with the rest of the GPCRs families is two conserved Cys. One between TMI and TMII and another conserved Cys between TMIII and TMIV. These residues are believed to create a disulphide bridge between these loops and to be important for the structural integrity of the protein. On the whole the seven TM regions in the *Adhesion* GPCRs are fairly divergent from other GPCRs, while these few common motifs provided support for common evolutionary origin of the *Adhesion* GPCRs with the other mammalian families of GPCRs.

## THE RAT GENE REPERTOIRE

Detailed mining of the *Adhesion* GPCRs was a part of a work that compared the overall repertoire of all GPCRs in the human, mouse and the rat genome.[7] The percentage of one-to-one GPCR orthologues is only 58% between rats and humans and only 70% between the rat and mouse, which is much lower than stated for the entire set of all genes. Moreover, the average protein sequence identities of the GPCR orthologue pairs are in general also lower than for the whole genomes. However, the proportions of orthologous and species-specific genes vary significantly between the different GPCR families. The largest diversification is seen for GPCRs that respond to exogenous stimuli. Moreover, the *Adhesion* GPCRs display relatively low sequence conservation (72%) between rats and humans. However, in contrast to the receptors that respond to exogenous stimuli, the *Adhesion* family orthologues repertoire is relatively well-conserved. 100% of the rat and mouse and 91% of the rat and human *Adhesions* make up one-to-one orthologous pairs. The mouse and rat have the same *Adhesion* gene set up and the two genes missing in mice (EMR2 and EMR3) are also missing in the rat genome. It should also be noted that the rat and mouse gene sequences of GPR144 contain stop codons within the transmembrane region and are thus likely to be pseudogenes.

## THE DOG GENE REPERTOIRE

The dog is an important model in biomedical research for several genetic and pharmaceutical reasons and we mined this genome for GPCRs.[21] The *Adhesion* family displays partly unconventional orthology relationships between dog and human. All 33 human *Adhesion* GPCRs are present in the dog genome. But, interestingly, the dog also contains an additional 5 full-length genes; EMR2b, EMR2c, EMR2d, EMR4b and EMR4c; and 1 pseudogene GPR133b. These *Adhesion* GPCR genes seem to be specific for the dog lineage, as they have not been found in the other mammals we have studied. In dog, two EMR2-like GPCRs have been reported previously [22] and we found one additional EMR2 and two EMR4 gene duplicates. The dog and human one-to-one *Adhesion* receptor orthologues have an average protein sequence identity of 83%. This is higher than between the rodents (rat and mouse) in comparison to the human genome, phenomena found also for the other GPCR families. EMR2 and EMR3, which are full length in human but absent in rodents, appear to be functional (are full-length) in dog. The gene sequences of BAI1, EMR2d, EMR4c, GPR123 and GPR124 are incomplete but this could be related to incomplete genome assembly. Moreover, we found additional EMR2 duplicates in cow. We performed a phylogenetic analysis based on the 5 dog-specific EMR receptor sequences together with the dog, human, cow and opossum EMR1-EMR4 and CD97. It has been suggested that EMR2 has a chimeric structure. The seven transmembrane (7TM) segments of EMR2 are most similar to those in EMR3 while the EGF domains in EMR2 are almost identical to those in CD97.[22] Interestingly, in our phylogenetic analysis based on the 7TM segments, EMR2 and EMR3 orthologs did not cluster together and instead the receptor paralogs grouped together. This is in line with the previous hypothesis about chimeric gene structures in this group.[22] We found this pattern to be the same for the human, dog, cow and opossum receptors. The new genes that we found in dog provided additional evidence for the unique evolution of the EMR subfamily of *Adhesion* GPCRs that seem not only shuffle domains within the N-terminal region but also larger segments of the N-termini.

## THE CHICKEN GENE REPERTOIRE

We studied the overall repertoire of the chicken GPCRs.[6] We manually edited and verified, i.e., curated, the coding regions of each of the GPCRs (557 in total) and provided the first high-quality collection of GPCR sequences from the full genome of a nonmammalian species. We found that 259 of the 557 chicken GPCRs have a one-to-one ortholog in the human genome. There are 21 cases of one-to-one orthologous relationships between human and chicken *Adhesion* GPCRs, while 12 human *Adhesion* GPCRs lack a chicken orthologue. We found that the group of that contains the lectomedin receptors (LEC1-LEC3) and the EGF-TM7-latrophilin—related protein (ETL) receptor, are relatively well-conserved in the chicken in comparison with humans and only the LEC1 receptor was missing. The human group, that contains the CD97 and four EGF-like modules containing mucin-like receptor proteins (EMR1-4), does not seem have any chicken orthologues. Since CD97 is present in the teleost, Fugu, *Takifugu rubripes* this receptor appears to be has been lost in the lineage leading to the chicken, while the EMRs have probably expanded in mammals. Looking at the five main GPCR families,[1,23] the *Adhesion* family displays the lowest percentage identity (68.8%) between orthologous pairs and this could be due to the fact that the *Adhesion* GPCRs utilize the TM regions differently than the other families of GPCRs.

## FISH

Our initial survey of GPCRs suggested presence of at least 22 *Adhesion* GPCR in zebrafish (Danio rerio) and 6 in Fugu.[23] Metpally and Sowdhamini[24] took a special look at the Fugu and found 29 *Adhesion* sequences. It should be noted that the assembly of these genomes were quite poor at this time and the identity of the in particular such complex genes as *Adhesion* were difficult to determine. In our latest work, we found 23 unique *Adhesion* GPCRs in *T. nigroviridis*.[14] We can conclude that the fish lineage has members from several of the human *Adhesion* family including CELSR (IV), LECs (I), CD97 (II), BAI (VII) as well as the orphan groups III, V, VI and VIII. These genes have long N-termini with large similarities in the domain composition with the corresponding mammalian group members.[14]

## THE AMPHIOXUS GENE REPERTOIRE

Amphioxus (*Branchiostoma floridae*) is one of the closest now living relatives to vertebrates. Amphioxus shares several features with vertebrates like a dorsal, hollow nerve cord, notochord, segmental muscles and pharyngeal gill slits, while they are missing the pronounced head region of vertebrates as well as not having neural crest cells functioning similar to those in vertebrates, paraxial skeletal tissue and some visceral organs.[25] The amphioxus genome is remarkably rich in various GPCR subtypes and we found evidence for the presence of at least 664 distinct GPCRs distributed among all the main families of GPCRs. However, the main GPCR groups known to sense exogenous substances (such as Taste 2, mammalian olfactory, nematode chemosensory, gustatory, vomeronasal and odorant receptors) in other bilateral species are absent. Interestingly, we found a very rich repertoire of *Adhesion* GPCRs, in total 37, in amphioxus.[26] Some of the mammalian groups are missing, such as the groups I (lectomedin receptors) and II (EGF-containing genes). Three of the human *Adhesion* groups have both mammalian and amphioxus members. These are group III (orphans, expressed in CNS), V (orphans) and VIII (orphans with highly variable N-terminal length) and also one orthologous receptor to the very long G protein-coupled receptor 1 (VLGR1) is also present. The remaining missing groups are group IV (CELSR with multiple cadherin domains), VI (GPR110, GPR111, GPR113, GPR115 and GPR116) and VII (BAI). Many of the amphioxus *Adhesion* GPCRs have multiple domains in the N-termini but rather surprisingly, we found several novel domains in these N-termini like Somatomedin B, Kringle, Lectin C-type and SRCR (for more details see ref. 26) which were at that point unique for amphioxus among the GPCRs. Also the domains LDLa, Immunoglobulin I-set, CUB and TNFR cannot be found in mammalian *Adhesion* GPCRs. The Kringle and Somatomedin B domains are of a special importance because they are found in sequences from a previously uncharacterized *Adhesion* expansion of ten genes. The Kringle domain is a protein-binding domain[27] is present in urokinase-type plasminogen activator (uPA) while the Somatomedin B domain can be found in vitronectin.[28] These two proteins interact and the Somatomedin B domain helps in the localization of uPA to focal adhesion in microvessel endothelial cells. Interestingly, all of these domains not previously identified in *Adhesion* GPCRs, have a large number of conserved cysteines which is a feature consistent with many other common N-terminal domains found in this family. This similarity could suggest that these domains are inserted through domain shuffling of similar stretches of DNA. Moreover, many of these new domains have recognized cell adhesion properties and participate in cell guidance. We found that some of the new genes share a resemblance to invertebrate genes according to the top five hits in BLAST searches against the NCBIs nonredundant (nr) database. These invertebrate genes are primarily from either *Strongylocentrotus purpuratus* which is closely related to amphioxus[29] or from the more distantly related cnidarian Nematostella. These findings suggested that the *Adhesion* family, with its unique structural and functional characteristics, has a very long evolutionary history and that these are likely to be present in most vertebrates. Moreover, the functions of the *Adhesion* GPCRs seem to have undergone further diversification within the lineage leading to amphioxus and are likely to have gained additional roles in cell guidance.

## THE NEMATOSTELLA GENE REPERTOIRE

The starlet sea anemone, *Nematostella vectensis*, is becoming an increasingly important model system in genomics and other disciplines. Nematostella belongs to cnidarians or the animal phylum that includes sea anemones, corals, jellyfish and hydra which occupy a critical position in the tree of life. They are the most primitive animals with epithelial cells, neurons, stem cells, complex extracellular matrix, muscle fibers and a fixed axis of symmetry in their body organization. Interestingly, the Nematostella has much richer overall gene repertoire as compared with the artopods (flies, drosophilia and mosquito)[30] and this is also the case for the *Adhesion* GPCRs and therefore we focus on this genome. Overall, the Nematostella genome has a remarkably rich set of *Adhesion* GPCRs with many of the domains found in mammalian GPCRs. Both the phylogeny[26] and the domain composition show that Nematostella has members, which clearly belong to group III (GPR123, GPR124 and GPR125) and groups IV (Celsr) and V (GPR133 and GRP144) of the mammalian *Adhesion* GPCRs. Nematostella has a Celsr like gene that has multiple copies of the CA domain as well as LAMG, EGF, GPS and a HBD showing that the complex multidomain structure of the *Adhesion* GPCRs is extremely well-conserved though the evolutions. Nematostella has also GRR133/GRP144 like sequences that share the CLECT domain with the mammalian homologs as well as the GPS domain. Moreover, Nematostella has four genes that are homologs to GPR123/GPR124/GPR125 where at least two have a Ig domain. Nematostella has also a copy of the VLGR with multiple Calx-beta domains and a GPS domain. Taken together it is clear that the *Adhesion* family is one of the most ancient of the five main families of GPCRs and it is likely to be found in most animal genomes.

Looking at the hierarchy among the *Adhesion* subgroups we found it likely that groups I, IV and V share a common ancestor because they group together in the phylogenetic analysis. This hypothesis is also strengthened by the fact that three of their common conserved splice sites (css2, css5 and css6) are missing in the other groups present in Nematostella. We also find it likely that group VII, which we only found in vertebrates, arose from this branch of the evolutionary tree, most likely from group I or IV according to the domain composition. It is also evident that group II originates from group I. Group III is likely to have branched from the branch that contains groups I, IV and V, which in turn gives rise to groups VI and VIII.

The Nematostella genome provided further interesting finding as we found a set (total of 13) of unique *Adhesion*-like GPCRs in this genome. These genes do not have any GPS domain but have a TM domain that can readily be aligned with *Adhesion* GPCRs (amino acid identities range from 21% to 30%). These sequences do not only have long N-termini but also they contain one Somatomedin_B domain each. This Somatomedin_B domain is not found in any mammalian *Adhesion* GPCR. However, these are found in a set of *Adhesion*-like GPCRs found in amphioxus.[26] This unique N-terminal composition with no GPS domains and their relatively high amino acid identities (21–38%) suggests that these two groups could be related.

## THE *ADHESION* FAMILY IS AN ANCESTOR OF THE *SECRETIN* FAMILY

Surprisingly, we found that the Nematostella genome does not have any *Secretin* genes.[26] Moreover, we did not find any *Secretin* genes in *M. brevicollis* or *D. discoideum* although both these genomes have *Adhesion*-like GPCRs. We can thus draw the conclusion that the *Adhesion* GPCRs are a more ancient GPCR family than the *Secretin* GPCRs. This work[26] showed for the first evolutionary hierarchy among the five main families of vertebrate GPCRs. Subsequently, we searched for evidence that the *Secretin* GPCRs, which are found in both *D. melanogaster* and *C. elegans*, might have originated from one branches of *Adhesion* GPCRs. We performed repeated iteractive phylogenetic analysis and found that group V (GPR133/GPR144) is the closest relative to the *Secretin* family among the groups of the *Adhesion* family. It further strengthened this hypothesis that the group V sequences in Nematostella (Nv_201898 and Nv_204814) both shared the same splice site setup as the *Secretin* GPCRs. This splice site setup is not shared by any of the other ancient groups. One of the most conserved motifs in the whole *Secretin* family is only found in group V of the *Adhesion* family. In conclusion, these results provided strong evidence that the *Secretin* family of GPCRs could have originated from group V of *Adhesion* GPCRs. It is however clear that the *Secretin* group has taken over more specific functions and does not share the domain shuffling properties which is so uniquely dominant in the evolution of the *Adhesion* family.

## CONCLUSION AND PERSPECTIVE

Our studies suggest that the *Adhesion* family is likely to be present in most, if not all animal species and this finding is summarized in Table 1. It is thus conceivable that the *Adhesion* GPCRs serve some essential functions that are shared by many animal species. However it is also clear that there has been a large variation of the evolutionary pressure among the different families. While some of the families are highly conserved with high percentage amino acid identity between orthologues others have very low such similarities even among orthologues of closely related species. We also found that there is a good correlation between the compositon of the N-terminal domains and the results from the phylogenetic analysis of the TM regions (see Figs. 1 and 2). While we

**Table 1.** Number of sequences in the different *Adhesion* subgroups I-VIII in eight different species. Data for *C. familiaris* was taken from reference 21, for *B. floridae* data is from Figure 2 and from reference 26 and for the remaining species data came from reference 14.

| Group | H. sapiens | C. familiaris | G. gallus | T. nigroviridis | B. floridae | D. melanogaster | C. elegans | N. vectensis |
|---|---|---|---|---|---|---|---|---|
| I | 4 | 4 | 3 | 4 | - | 1 | 2 | 1 |
| II | 5 | 10 | - | 2 | - | - | - | - |
| III | 3 | 3 | 2 | 3 | 1 | 1 | - | 4 |
| IV | 3 | 3 | 2 | 2 | - | 1 | 1 | 1 |
| V | 2 | 2 | 2 | 1 | 2 | - | - | 3 |
| VI | 5 | 5 | 2 | 2 | - | - | - | - |
| VII | 3 | 3 | 2 | 3 | - | - | - | - |
| VIII | 6 | 6 | 2 | 6 | 3 | - | - | - |
| Other | 2 | 2 | 2 | . | 31 | 2 | 1 | 28 |

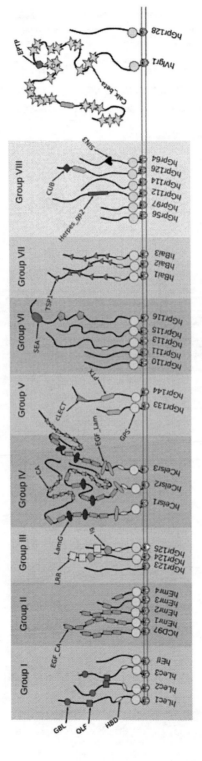

**Figure 1.** Schematic presentation of the domains in N-termini of human *Adhesion* GPCRs. The domains were identified with rps-blast and an e-value cutoff at 0.1. The following domains were found: GPS (GPCR proteolytic site), HBD (hormone binding domain), OLF (olfactomedin domain), GBL (galactose binding lectin domain), EGF_CA (calcium binding epidermal growth factor-like domain), LRR (leucine rich repeat), CA (cadherin repeats), TSP1 (thrombospondin repeats, Type 1), PTX (Pentraxin domain), LamG (laminin G domain), EGF_Lam (laminin type epidermal growth factor domain), Calx-beta domain, CUB (C1r/C1s urinary epidermal growth factor and bone morphogenetic domain), Ig (immunoglobulin domain), SEA (sea urchin sperm protein domain), EPTP (epitempin protein domain), cLECT (C-type lectin-like domain), Herpes_gp2 (Equine herpesvirus glycoprotein gp2 domain), SIN3 (Histone deacetylase complex).

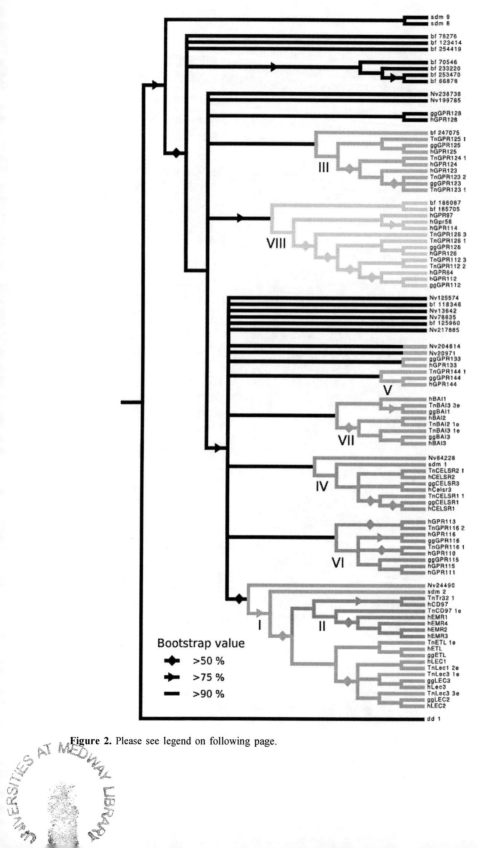

**Figure 2.** Please see legend on following page.

**Figure 2, viewed on previous page.** A phylogenetic tree of *Adhesion* GPCRs from *Dictyostelium discoideum* (dd), *Nematostella vectensis* (Nv), *Drosophila melanogaster* (sdm), *Branchiostoma floridae* (bf), *Tetraodon nigroviridis* (Tn), *Gallus gallus* (gg) and *Homo sapiens* (h). The tree was calculated with RA × ML[32] based on a retranslated alignment of the 7TM regions made with Mafft-einsi[33] and is bootstrapped 100 times. Nodes with bootstrap values below 50% are collapsed. Branches to nodes with bootstrap values above 50% are marked with a square and branches to nodes with bootstrap values above 75% are marked with a triangle. Branches to nodes with bootstrap values above 90% are left unmarked. Compared to the tree in[14] this tree lacks resolution for group III and V, which probably is due to the more advanced methods used in that work. Group VI is found in the node holding group I, II, IV, V and VII, which could be a result of group VI:s promiscuity. The group is hard-placed as it only vertebrate members, but in our previous calculations has been located closest to the out-group.

are starting to get the rough picture for the evolution of the *Adhesion* family (Fig. 3) there is much more work left to receive detailed high resolution charts of the molecular events that have shaped this family in animal evolution. New genomes are arriving that will provide important information such as the *Trichoplax adhaerens*, which is one of the most primitive animals studied to date. During the preparation of this work a paper was published that described mining of GPCRs in mined in Xenopus tropicalis[31] identifying 24 *Adhesion* GPCRs members in this species. As more sequences have been assembled the possibility increases to perform more detailed analysis of the structural and splicing

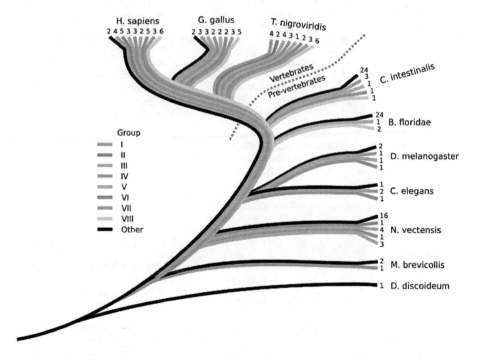

**Figure 3.** The schematic tree presents the number of *Adhesion* GPCRs in different species separated in groups. The numbers for *B. floridae* are taken from reference 26 and modified with respect to Figure [phylogenetic_tree]. The numbers for Ciona intestinalis are taken from reference 34 and for *G. gallus* from reference 6. The information about the remaining species are taken from reference 14. Each group is represented by a color as shown in the legend. A color version of this image is available at www.landesbioscience.com/curie.

elements that are likely to be important for the different subfamilies. It is evident that the *Adhesion* family is among the most ancient branches of GPCRs and the evolution of this family is thus of great interest for resolving the overall evolution of GPCRs including questions about the overall hierarchy among the main families of GPCRs.

## ACKNOWLEDGEMENTS

The studies were supported by The Swedish Research Council, Svenska Läkaresällskapet, Åke Wiberg Foundation, NOVO Nordisk Foundation, the Magnus Bergvall Foundation, Swedish Royal Society of Sciences, Byggmästare Engkvists Foundation and Åhlens Foundation.

## REFERENCES

1. Fredriksson R, Lagerstrom MC, Lundin LG et al. The G-protein-coupled receptors in the human genome form five main families. Phylogenetic analysis, paralogon groups and fingerprints. Mol Pharmacol 2003; 63(6):1256-1272.
2. Bjarnadottir TK, Geirardsdottir K, Ingemansson M et al. Identification of novel splice variants of Adhesion G protein-coupled receptors. Gene 2007; 387(1-2):38-48.
3. Bjarnadottir TK, Fredriksson R, Hoglund PJ et al. The human and mouse repertoire of the adhesion family of G-protein-coupled receptors. Genomics 2004; 84(1):23-33.
4. Fredriksson R, Gloriam DE, Hoglund PJ et al. There exist at least 30 human G-protein-coupled receptors with long Ser/Thr-rich N-termini. Biochem Biophys Res Commun 2003; 301(3):725-734.
5. Fredriksson R, Lagerstrom MC, Hoglund PJ et al. Novel human G protein-coupled receptors with long N-terminals containing GPS domains and Ser/Thr-rich regions. FEBS Lett 2002; 531(3):407-414.
6. Lagerstrom MC, Hellstrom AR, Gloriam DE et al. The G Protein-Coupled Receptor Subset of the Chicken Genome. PLoS Comput Biol 2006; 2(6):e54.
7. Gloriam DE, Fredriksson R, Schioth HB. The G protein-coupled receptor subset of the rat genome. BMC genomics 2007; 8:338.
8. Lagerstrom MC, Schioth HB. Structural diversity of G protein-coupled receptors and significance for drug discovery. Nat Rev Drug Discov 2008; 7(4):339-357.
9. Harmar AJ. Family-B G-protein-coupled receptors. Genome Biol 2001; 2(12):3013.
10. McKnight AJ, Gordon S. The EGF-TM7 family: unusual structures at the leukocyte surface. J Leukoc Biol 1998; 63(3):271-280.
11. Parmentier ML, Galvez T, Acher F et al. Conservation of the ligand recognition site of metabotropic glutamate receptors during evolution. Neuropharmacology 2000; 39(7):1119-1131.
12. Stacey M, Lin HH, Gordon S et al. LNB-TM7, a group of seven-transmembrane proteins related to family-B G-protein-coupled receptors. Trends Biochem Sci 2000; 25(6):284-289.
13. Kwakkenbos MJ, Kop EN, Stacey M et al. The EGF-TM7 family: a postgenomic view. Immunogenetics 2004; 55(10):655-666.
14. Nordstrom KJ, Lagerstrom MC, Waller LM et al. The Secretin GPCRs descended from the family of Adhesion-GPCRs. Mol Biol Evol 2009; 26(1):71-84.
15. Bjarnadottir TK, Fredriksson R, Schioth HB. The gene repertoire and the common evolutionary history of glutamate, pheromone (V2R), taste(1) and other related G protein-coupled receptors. Gene 2005; 362:70-84.
16. Bjarnadottir TK, Gloriam DE, Hellstrand SH et al. Comprehensive repertoire and phylogenetic analysis of the G protein-coupled receptors in human and mouse. Genomics 2006; 88(3):263-273.
17. Bjarnadottir TK, Fredriksson R, Schioth HB. The adhesion-GPCRs: a unique family of G protein-coupled receptors with important roles in both central and peripheral tissues. Cell Mol Life Sci 2007; 64(16):2104-2119.
18. Chung DA, Wade SM, Fowler CB et al. Mutagenesis and peptide analysis of the DRY motif in the alpha2A adrenergic receptor: evidence for alternate mechanisms in G protein-coupled receptors. Biochem Biophys Res Commun 2002; 293(4):1233-1241.
19. Palczewski K, Kumasaka T, Hori T et al. Crystal structure of rhodopsin: A G protein-coupled receptor. Science 2000; 289(5480):739-745.

20. Lagerstrom MC, Klovins J, Fredriksson R et al. High affinity agonistic metal ion binding sites within the melanocortin 4 receptor illustrate conformational change of transmembrane region 3. J Biol Chem 2003; 278(51):51521-51526.
21. Haitina T, Fredriksson R, Foord SM et al. The G protein-coupled receptor subset of the dog genome is more similar to that in humans than rodents. BMC genomics 2009; 10:24.
22. Kwakkenbos MJ, Matmati M, Madsen O et al. An unusual mode of concerted evolution of the EGF-TM7 receptor chimera EMR2. FASEB J 2006; 20(14):2582-2584.
23. Fredriksson R, Schioth HB. The repertoire of G-protein coupled receptors in fully sequenced genomes. Mol Pharmacol 2005.
24. Metpally RP, Sowdhamini R. Genome wide survey of G protein-coupled receptors in Tetraodon nigroviridis. BMC Evol Biol 2005; 5:41.
25. Holland LZ, Laudet V, Schubert M. The chordate amphioxus: an emerging model organism for developmental biology. Cell Mol Life Sci 2004; 61(18):2290-2308.
26. Nordstrom KJ, Fredriksson R, Schioth HB. The amphioxus (Branchiostoma floridae) genome contains a highly diversified set of G protein-coupled receptors. BMC Evol Biol 2008; 8:9.
27. Patthy L, Trexler M, Vali Z et al. Modules specialized for protein binding. Homology of the gelatin-binding region of fibronectin with the kringle structures of proteases. FEBS Lett 1984; 171(1):131-136.
28. Salasznyk RM, Zappala M, Zheng M et al. The uPA receptor and the somatomedin B region of vitronectin direct the localization of uPA to focal adhesions in microvessel endothelial cells. Matrix Biol 2007; 26(5):359-370.
29. Delsuc F, Brinkmann H, Chourrout D et al. Tunicates and not cephalochordates are the closest living relatives of vertebrates. Nature 2006; 439(7079):965-968.
30. Putnam NH, Srivastava M, Hellsten U et al. Sea anemone genome reveals ancestral eumetazoan gene repertoire and genomic organization. Science 2007; 317(5834):86-94.
31. Ji Y, Zhang Z, Hu Y. The repertoire of G-protein-coupled receptors in Xenopus tropicalis. BMC genomics 2009; 10:263.
32. Stamatakis A. RAxML-VI-HPC: maximum likelihood-based phylogenetic analyses with thousands of taxa and mixed models. Bioinformatics 2006; 22(21):2688-2690.
33. Katoh K, Toh H. Recent developments in the MAFFT multiple sequence alignment program. Brief Bioinform 2008; 9(4):286-298.
34. Kamesh N, Aradhyam GK, Manoj N. The repertoire of G protein-coupled receptors in the sea squirt Ciona intestinalis. BMC Evol Biol 2008; 8:129.

# CHAPTER 2

# 7TM-CADHERINS:
## Developmental Roles and Future Challenges

Caroline J. Formstone*

**Abstract:** The 7TM-Cadherins, Celsr/Flamingo/Starry night, represent a unique subgroup of adhesion-GPCRs containing atypical cadherin repeats, capable of homophilic interaction, linked to the archetypal adhesion-GPCR seven-transmembrane domain. Studies in *Drosophila* provided a first glimpse of their functional properties, most notably in the regulation of planar cell polarity (PCP) and in the formation of neural architecture. Many of the developmental functions identified in flies are conserved in vertebrates with PCP predicted to influence the development of multiple organ systems. Details of the molecular and cellular functions of 7TM-Cadherins are slowly emerging but many questions remain unanswered. Here the developmental roles of 7TM-Cadherins are discussed and future challenges in understanding their molecular and cellular roles are explored.

## THE 7TM-CADHERINS: A UNIQUE GROUP OF ADHESION-GPCRs

The 7TM-Cadherins are unique within the adhesion-GPCR family as their extracellular domains comprise a series of nine atypical cadherin repeats linked to a combination of EGF-like and Laminin G-like domains (Fig. 1). Cadherin repeats are repetitive subdomains which contain sequences involved in calcium binding and which are capable of homophilic interaction, for a review see reference 1. The cadherin superfamily consists of a number of subgroups which are of a classic and nonclassic type, for a review see reference 2. Classic cadherins interact with cytoplasmic catenins and are well-defined adhesion molecules whereas nonclassic cadherins, such as the protocadherins and 7TM cadherins, do not bind catenins and generally exhibit weak adhesive activities. The combination of EGF-like[3] and Laminin G-like (LG)[4] domains

*Caroline J. Formstone—MRC Centre for Developmental Neurobiology, New Hunt's House, Guy's Campus, King's College London, SE1 1UL, UK. Email: caroline.formstone@kcl.ac.uk

*Adhesion-GPCRs: Structure to Function*, edited by Simon Yona and Martin Stacey.
©2010 Landes Bioscience and Springer Science+Business Media.

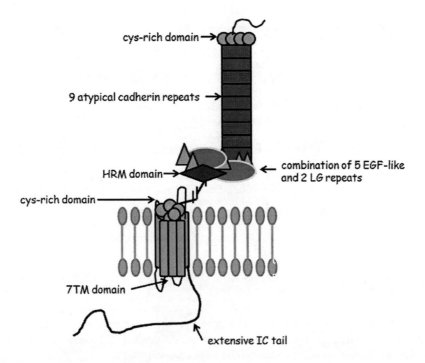

**Figure 1.** Schematic of a typical 7TM-Cadherin family member. 7TM-Cadherins are large (around 400 kDa) glycosylated, multi-domain transmembrane proteins. Cys denotes cysteine-rich domain, HRM denotes hormone-binding domain similar to that found within the family B GPCRs, EGF-like is epidermal growth factor-like domain, LG is Laminin G-like domain, 7TM is seven-transmembrane domain and IC denotes the intracellular tail.

with cadherin repeats found within the extracellular domain of 7TM-Cadherins, is an ancient association[5] and characterises many cadherin-containing proteins in invertebrates. Vertebrate classic cadherins, such as E-cadherin, have lost the EGF/LG region and, strikingly, differ from their invertebrate counterparts in both sub-cellular distribution and adhesive properties. The role of these domains in 7TM-Cadherins and other nonclassic vertebrate cadherins remains unclear.

## THE FAMILY TREE

The 7TM-Cadherins are an evolutionary conserved gene family with homologues identified from ascidians through to mammals (Table 1). Initially identified in mouse and given the name *Celsr*; [C cadherin E EGF L laminin G-like S seven-pass R receptor],[10] two groups then independently identified a gene in *Drosophila melanogaster* (*Drosophila*) which they named *flamingo*[11] or *starry night*.[12] Subsequently a family of 3 genes was identified in mammals; *Celsr1*, *Celsr2* and *Celsr3*.[13-16] In avians only two homologues have been isolated; *c-flamingo1* (*c-fmi1*)[17,18] and *c-flamingo3* (Formstone and Mason, unpublished) whereas in teleosts, a four-member gene family is known to exist including two *Celsr1* homologues.[19,20]

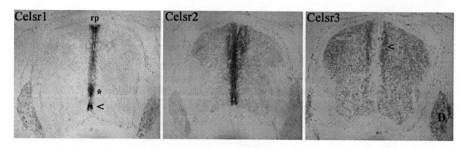

**Figure 2.** Comparative RNA in situ analysis of *Celsr1*, *Celsr2* and *Celsr3* expression in the developing mouse spinal cord at 13.5 days post coitum (dpc). Transverse cryostat sections, dorsal is to the top. Rp is roof plate, D denotes dorsal root ganglion. Arrowheads indicate *Celsr1* enrichment in the floorplate and *Celsr3* enrichment in the subventricular zone of the alar plate. *Denotes a third region of *Celsr1* enrichment within the ventricular zone. Reproduced with permission from: Formstone CJ, Little PFR. Mech Dev 2001; 109:91-94. Copyright ©2001 Elsevier Science Ireland Ltd.

## VERTEBRATE 7TM-CADHERINS ARE DEVELOPMENTALLY REGULATED

Comparative studies on the RNA expression patterns of vertebrate 7TM-Cadherins provide useful insight into their global functions in addition to highlighting distinct roles for individual family members during vertebrate embryogenesis.

The different mammalian *Celsr* genes are characterised by their complementary patterns of expression in different developing tissue and organ systems.[14-16,22] This phenomenon is most marked within the developing nervous system in which all three *Celsr* genes are expressed. In the developing spinal cord for example, the *Celsr* family appear to be expressed in a sequential manner reflecting individual roles in different neural compartments (Fig. 2).[14] In the cerebellum, each *Celsr* occupies a unique cellular territory.[15] Whereas *Celsr3* is predominantly neural-specific,[14] initially decorating newly-born neurons, *Celsr1* and *Celsr2* show contrasting patterns of expression both during gastrulation[14,22] and through later development in nonneural tissues such as the lung, kidney and skin.[15,16] During organogenesis, *Celsr1* and *Celsr2* invariably occupy the different epithelial components of a particular organ. The compartmentalised nature of their individual expression patterns tempts speculation that a function of the *Celsr* family may be to restrict cells to the particular tissue compartment they decorate.

Comparative studies of the early embryonic expression of vertebrate *Celsr1* homologues suggest that they mediate both conserved and species-specific functions (Fig. 3). In mouse and avian embryos, *Celsr1* and the chick homologue, *c-fmi1*, are both expressed within the primitive streak and node and subsequently in the early neural plate at sites where neural tube closure is initiated suggesting a conserved role in this process (Fig. 3B,F).[7,14,17] In zebrafish, expression of a *Celsr1* homologue, *zflamingo1a* (*zfmi1a*), during gastrulation[19] becomes increasingly localised to axial tissues such as shield and notochord (Fig. 3I). During neurula stages in mouse, chick and zebrafish, the rostral limit of *Celsr1/fmi1* expression lies within the presumptive diencephalon (Fig. 3C,G,K). *c-fmi1* and *zfmi1a* expression subsequently enrich within diencephalic tissue (Fig. 3H,M). *c-fmi1* however, exhibits a novel distribution in the developing somites,[17] a pattern which is partially mirrored by *zfmi1a* (Fig. 3H,L).

# 7TM-CADHERINS: DEVELOPMENTAL ROLES AND FUTURE CHALLENGES 17

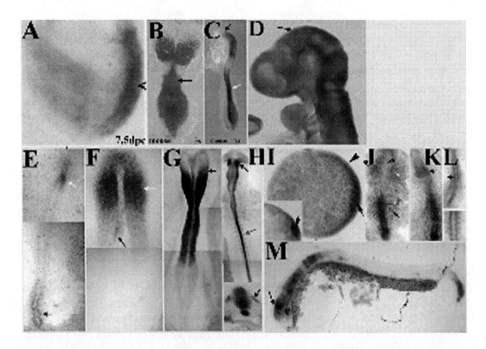

**Figure 3.** Comparative RNA in situ analysis of *Celsr1* expression in mouse (A-D), chick (E-H) and zebrafish (I-M). B-D,F-H,J-L) Anterior is to the top (A) lateral view of a mouse embryo at 7.5 dpc, anterior is to the left. Arrowhead denotes expression of *Celsr1* in the primitive streak (B) Dorsal view of a mouse embryo at the 3 somite stage. Arrow indicates site of initiation of neural tube closure (closure 1) in the mouse embryo. C) Lateral view of a 7-somite mouse embryo. White arrow indicates closed neural tube, black arrow indicates anterior limit of *Celsr1* expression within the forebrain. D) Lateral view of the head of a mouse embryo at 9.5 dpc. Arrow indicates *Celsr1* expression within the diencephalon. E) dorsal view of a Stage 4 chick embryo. White arrow indicates *c-fmi1* expression in the node, black arrow indicates *c-fmi1* expression within the posterior primitive streak. F) Dorsal view of a Stage 5 chick embryo. White arrow denotes *c-fmi1* expression in the presumptive mesencephalon, the site of initiation of neural tube closure in the chick. Black arrow indicates expression of *c-fmi1* in the node. G) Dorsal view of a Stage 8 chick embryo. Black arrow denotes anterior limit of *c-fmi1* expression within the forebrain. H) Dorsal view of a Stage 10 chick embryo. Black arrow indicates enrichment of *c-fmi1* within the diencephalon, grey arrow denotes restricted expression of *c-fmi1* within the medial portion of the developing somites. Inset is a transverse section from a Stage 15 embryo. Black arrow indicates expression of *c-fmi1* within the dermamyotome of the developing somite. I) Dorsal view of a zebrafish embryo at 65% epiboly. Dorsal is to the right. Black arrow indicates *zfmi1a* expression in the shield, black arrowhead indicates *zfmi1a* transcripts within the germ ring. Inset is a lateral view of the same embryo, dorsal is to the right. Black arrow denotes *zfmi1a* expression in the shield. J) Dorsal view of a 3-somite zebrafish embryo. Black arrowhead denotes loss of *zfmi1a* expression within the midline of the developing forebrain, white arrow indicates *zfmi1a* expression within the developing eyefield and black arrow denotes low levels of *zfmi1a* transcripts within lateral tissue. K) Dorsal view of a 5-somite zebrafish embryo. Black arrow indicates enrichment of *zfmi1a* within the diencephalon. L) Dorsal view of the trunk region of a 10-somite zebrafish embryo. Black arrow indicates restriction of *zfmi1a* transcripts to the medial portion of the developing somites. Inset denotes magnified view of *zfmi1a* expression within the medial and lateral regions of developing somites from a 15-somite zebrafish embryo (M) lateral view of 24hr zebrafish embryo, anterior is to the left. Black arrow denotes enrichment of *zfmi1a* expression within the diencephalon, note the absence of *zfmi1a* transcripts within the somites. e indicates the eye. A-C) Reproduced with permission from: Formstone CJ, Little PFR. Mech Dev 2001; 109:91-94. Copyright ©2001 Elsevier Science Ireland Ltd. E-H) Reproduced with permission from: Formstone CJ, Mason I. Dev Dyn 2005; 232:408-413. Copyright ©2005 Wiley-Liss Inc. I-L) Reproduced with permission from: Formstone CJ, Mason I. Dev Biol 2005; 282:320-335. Copyright ©2005 Elsevier Science Ireland Ltd.

Other sites of expression of mammalian *Celsr* genes outside of the nervous system include anterior visceral endoderm,[22] cochlea/vestibular system,[15,16] the limb apical ectodermal ridge[15] (*Celsr1* and *Celsr2*) and testis.[23]

## 7TM-CADHERINS PLAY PLEIOTROPIC ROLES DURING EMBRYONIC DEVELOPMENT

Studies in *Drosophila* provided the first insights into the functional role of 7TM-Cadherins. The elegant analyses possible in this model system have demonstrated that *flamingo/starry night* is essential to *Drosophila* embryonic development,[11] the regulation of tissue polarity[11,12] and to the development of *Drosophila* neuronal architecture.[24-26] In vertebrates, similar functional themes are emerging (Table 1) with Celsr/Flamingo proteins implicated in coordinating both tissue polarity and neuronal outgrowth.[17-20,96-99,107,108]

In *Drosophila,* the developmental functions of Flamingo can be classified under regulation of tissue polarity or nervous system formation, with Flamingo active in distinct cell signalling pathways in each case. In vertebrates however, rationalisation of 7TM-Cadherin function is increasingly complex since a particular morphogenetic event, such as gastrulation, may require input from different Celsr/Flamingo-dependent cell signalling systems.[102] This section covers the known developmental roles of both *Drosophila* and vertebrate 7TM-Cadherins with the description of the latter group organised so as to reflect the greater functional complexity.

### 7TM-Cadherins Regulate Planar Cell Polarity (PCP) Processes

How groups of cells acquire and coordinate cell/tissue polarity either with their neighbours or along particular embryonic axes is the subject of intense study. This phenomenon, termed tissue polarity or planar cell polarity (PCP), was first recognised in insects[27] and has since been extensively studied, through the ease of genetic analysis, in the fruit fly *Drosophila melanogaster*, for a review see references 28-30. Through PCP, *Drosophila* cells make orientated structures such as the hairs and bristles on the wing and abdomen, orientate sensory bristles on the notum via regulation of asymmetric division of sensory organ precursors (SOPs) and specify the cell fate of distinct photoreceptor cells in the ommatidia of the eye through polarised rotation of the photoreceptor cluster. Studies in the *Drosophila* wing and eye have led to the discovery of two independent PCP pathways. The first, acting at a local level between neighbouring cells, is Flamingo/Starry night-dependent and requires a number of other 'core' PCP components (Fig. 4). The second is a more global system, reliant on two other atypical cadherin proteins, Fat and Dachsous.[31] The question of whether the two PCP pathways function in parallel[30] or via a multi-tiered process[32] however, is currently under debate.

Vertebrate counterparts of the 'core' *Drosophila* PCP signalling cascade act together to influence multiple aspects of vertebrate development from regulation of early embryogenesis to organogenesis (Table 2), for a review see reference 33. However, clear differences between PCP signalling pathways in vertebrates and invertebrates are emerging. Some aspects of the *Drosophila* PCP cassette are conserved. For example, mutation of one vertebrate PCP component affects the localisation of others in the same tissue context as has also been observed in *Drosophila*.[49,72] However, in contrast to the

**Table 1.** 7TM-Cadherins are an evolutionary conserved gene family. In the vertebrate families of two or more members, *flamingo* is designated *fmi*. Ascidian is a sea squirt, a urochordate, primitive marine animal of the subphyla Tunicata. C.elegans is the yeast species, *Caenorhabditis elegans*. Drosophila is the insect, *Drosophila melanogaster*. Danio Rerio is the zebrafish.

| Species | Gene | Accession No. | Developmental Function |
|---|---|---|---|
| *Ascidian* | *flamingo*[6] | | Unknown |
| *C. elegans* | *flamingo-*[17,8] | AY314773 | Avon outgrowth, synaptogenesis[9] |
| *Drosophilia* | *flamingo*[11,12] | AB028498 (Fmi) | Orientation of cuticular structures[11,12,84,86] |
| | | NM 165794 (Stan) | Ommatidial rotation/cell fate, specification[85] and axonal and dendritic outgrowth[24-26, 111,112,120,121] |
| | *Celsr/fmi1a*[19] | AY 960152 | Epiboly[102] CE movements[19,102] neuronal migration[20] |
| *Dani Rerio* | *Celsr/fmi1b*[19] | AY960153 | Epiboly[102] CE movements[19,102] neuronal migration[20] |
| | *Celsr/fmi2*[20] | NM 001080577 | Epiboly[102] CE movements[102] neuronal migration[20] |
| | *Celsr/fmi3*[20] | XM 001922677 | Unknown |
| *Xenopus* | *flamingo*[21] | AF 518403 | PCP |
| *Avians* | *c-fmi1*[17] | AY426608 | Neural tube closure[17], inner ear hair patterning[18] |
| | *c-fmi3* | XM 414354 | Unknown |
| *Mammals* | *Celsr1*[7,13] | NM 009886 mouse | Neural tube closure,[96] inner ear hair patterning,[96] hair follicle patterning[49,106] |
| | | AY212290 rat | |
| | | NM 014246 human | Eyelid closure |
| | *Celsr2*[11,13] | NM 001004177 mouse | CNS dendrite morphogenesis[107,108] |
| | | NM 017392 mouse | |
| | | AF177695 rat | |
| | | NM 001408 human | |
| | *Celsr3*[11,14] | NM 080437 mouse | CNS dendrite morphogenesis[108] and axon tract formation,[97,98] forebrain interneuron migration[99] |
| | | NM 031320 rat | |
| | | NM 001407 human | |

*Drosophila* model (Fig. 5B), individual PCP components may distribute to different cellular compartments,[34,72] suggesting that vertebrate PCP may not follow a linear signalling cascade but rather involve multiple signalling systems utilising different PCP components. In addition, new players have emerged onto the vertebrate scene, which have no influence on PCP in *Drosophila*. In both *Caenorhabditis elegans* (*C. elegans*) and vertebrates, strong evidence supports a role for Wnts (small secreted glycoproteins which can act as ligands for Frizzled receptors recruiting the cytoplasmic protein dishevelled) in PCP signalling.[62,73,74] However, despite extensive study[75] no such evidence exists in *Drosophila*. Other new players include Scribble,[67] via the *Circletail* mutant (*Crc*), demonstrated to be an apical-basal polarity determinant in *Drosophila*[76] and PTK-7, a transmembrane protein with tyrosine kinase homology.[70]

**Table 2.** Vertebrate PCP pathway components and their developmental roles. (a) denotes mouse mutant (b) denotes zebrafish mutant (c) denotes single gene mouse knockout (d) denotes double or triple gene mouse knockout. (e) Vangl2 is Vang-like 2, the mouse homologue of Drosophila Vang, and zebrafish strabismus.

| Gene | Developmental Process |
| --- | --- |
| *Frizzled 3 and 6[d]* | Neural tube closure; inner ear sensory hair cell patterning[34] |
| *Frizzled 6[c]* | hair follicle patterning[35,42] |
| *Dishevelled 1[d], 2[c,d] and 3[d]* | Neural tube closure;[36, 37, 39,40] inner ear sensory hair cell patterning;[38,40] heart malformation and cardiomyopathy;[37,40] primary cilia function;[41] angiogenesis[42] |
| *Vangl2[e]; Looptail[a], Strabismus[b]* | CE movements during gastrulation;[43,44] cell intercalation during primitive streak formation in chick;[45] neural tube closure;[46] inner ear sensory hair cell patterning;[47,48] hair follicle patterning;[49] heart malformation and cardiomyopathy;[50] eyelid closure; facial motor neuron migration;[51,61] primary cilia function;[52] asymmetric division and maintenance of cortical progenitors[53] |
| *Prickle* | CE movements during gastrulation[54-56] cell intercalation during primitive streak formation in chick;[45] facial motor neuron migration;[56] angiogenesis[42] |
| *wnt5; pipetail[b]* | CE movements during gastrulation,[57-59, 64] cochlea extension;[59] inner ear sensory hair cell patterning;[59] neural tube closure;[59] elongation of the small intestine;[60] facial motor neuron migration;[61] angiogenesis[42] |
| *wnt11; Silberblick[b]* | CE movements during gastrulation;[62-64] neural crest migration;[65] elongation of early muscle fibres[66] |
| *Scribble; circletail[a]* | Neural tube closure;[67] inner ear sensory hair cell patterning;[47] heart malformation and cardiomyopathy;[68] eyelid closure; facial motor neuron migration[61,69] |
| *PTK-7[c]* | Neural tube closure;[70] inner ear sensory hair cell patterning;[70] neural crest migration[71] |

*Flamingo/Starry Night is a Central Component of the 'Core' Drosophila PCP Pathway*

*Drosophila* genetics uncovered a 'core' cassette of genes (Fig. 4) mediating PCP.[77,78] Frizzled, a 7TM protein distantly related to the 7TM-Cadherin family sits at the top of the hierarchy in this signalling cascade. In PCP, Frizzled acts via a distinct pathway to that involving β-catenin. The latter is a conserved signalling pathway regulating many aspects of growth and differentiation across species (for a review see refs. 79-81). The factor(s) orientating the polarity of Frizzled-PCP signalling[78] remain unclear[75] but elegant mosaic analyses in the fly wing have generated a number of working models for Frizzled-PCP,[82-84] each focussing on a system of local intercellular communication. In one such model system, the 7TM-Cadherin, Flamingo/Starry night (hereafter Flamingo), is a central player.[84]

*Flamingo* was found to be broadly expressed in *Drosophila* epithelia and nervous system.[11] Rescue of embryonic lethality in *flamingo* mutants revealed defects in PCP in

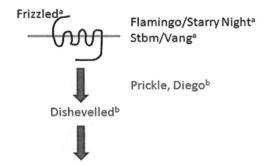

**Figure 4.** Schematic of the Drosophila PCP signalling pathway. a) Indicates transmembrane proteins b) Indicates cytoplasmic proteins. Stbm is strabismus, homologue of Vang.

adult ommatidia of the eye,[85] sensory bristles from the thorax,[86] wing hairs[11] and in hair orientation on the abdomen.[84] This is similar to PCP defects identified in mutants of two other 'core' transmembrane PCP components Frizzled[87] and Vang.[88]

*Flamingo Orientates Cuticular Structures in the Epidermis of Drosophila*

The coordinated orientation of cuticular structures within the epidermis of *Drosophila* has provided an exemplary model for the study of PCP. In the *Drosophila* wing for example, each epidermal cell assembles actin bundles at its distal-most vertex, producing a single prehair that extends away from the cell along the proximo-distal axis (Fig. 5A). Flamingo is one of a group of core PCP components that respond to an extrinsic spatial cue and subsequently define, at the cellular level, the axis of polarity. This response establishes two asymmetric protein complexes distributed to apico-lateral junctions across the proximo-distal hair/bristle cell boundary (Fig. 5B,C). Flamingo protein is found distributed to both proximal and distal cell membranes.[11]

Genetic analyses in the fly wing have placed Flamingo high up in the Frizzled-PCP signalling cascade. It has a key role in localising other polarity components such as Frizzled and Vang to the apico-lateral junctional region of wing hair cells,[11,89,90] a prerequisite step during initiation of PCP signalling in the wing (Fig. 5B). Subsequently Flamingo participates in the promotion of the asymmetric distribution of PCP components (Fig. 5C).

Following observations that (a) Flamingo formed homodimers across neighbouring cell membranes[11] and (b) Flamingo was uniquely distributed along the axis of wing hair cell polarity on both sides of a cell boundary (Fig. 5B,C)[11,89] Lawrence, Casal and Struhl[84] demonstrated that Flamingo was required to both send and receive polarity information between neighbouring cells and subsequently proposed the 'Flamingo bridge' model. In this model, Flamingo senses Frizzled activity between neighbouring cells and acts with either Frizzled or Vang, to send or receive, respectively, polarity information. Using different assay parameters, Chen and colleagues[91] have provided further data in support of the 'Flamingo bridge' model. This study and two others[92,93] are currently fuelling a new debate[94] on whether Flamingo plays a passive or instructive role in propagation of polarising information. Chen et al[91] believe that Flamingo is an active participant in PCP

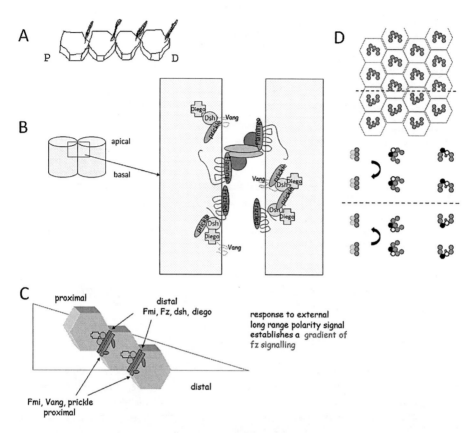

**Figure 5.** Flamingo is a key participant in the establishment of PCP in the Drosophila wing and eye. A) PCP is visualised in the Drosophila wing by the polarised distribution of a single pre-hair to the distal-most vertex of wing hair cells. P is proximal, D is distal. B) A first step in the establishment of PCP in the Drosophila wing is the distribution of 'core' PCP proteins to the apico-lateral membrane of wing hair cells where particular protein complexes are generated. C) Wing hair cells respond to an unknown long-range polarity signal which establishes a gradient of Frizzled signalling along the proximo-distal axis. The second stage of PCP establishment is marked by the asymmetric distribution of PCP proteins across neighbouring cell membranes which mirrors the axis of polarity. Fz; Frizzled, Fmi; Flamingo, dsh; dishevelled. D) Upper panel, polarisation of Drosophila ommatidia. Polarity is reflected in the mirror image arrangement of ommatidia across the dorso-ventral axis or equator (denoted by black hatched line). Lower panel, schematic representation of the establishment of ommatidial polarity. Initially the ommatidial clusters are symmetric. Specification of R3 (black circle) and R4 (white circle with black outline) promotes a 90o rotation towards the equator (black hatched line) breaking the symmetry. The symmetric pre-R3/R4 pair are shown as grey circles.

signalling whereas Wu and Mlodzik[93] propose that Flamingo merely aids direct contact between Frizzled and Vang across neighbouring cells, thus promoting ligand-receptor interaction. Strutt and Strutt[92] suggest a similar model except that in this case Flamingo acts passively to stabilise asymmetric junctional complexes across neighbouring cell membranes. In this third model, Flamingo is proposed to interact preferentially with Frizzled to bias the localisation of a Frizzled/Flamingo complex to distal wing cell edges and a Vang/Flamingo complex to proximal wing cell edges. How Flamingo directs Vang asymmetric distribution is unclear although some data exists to implicate the Flamingo cytoplasmic tail.[92] It is also possible that both Flamingo homodimers and the Frizzled/Vang complex *in trans* can individually propagate polarity information albeit at low efficiency, the notion here being that both complexes act instructively. 'Bridging' between an *in cis* Frizzled/Flamingo complex and Flamingo in a neighbouring cell may be more efficient at promoting Frizzled/Vang interactions in trans thus stabilising the asymmetric complex.

*Flamingo Links Frizzled-PCP and Notch/Delta Signalling in Drosophila Eye Development*

In the *Drosophila* eye, PCP is reflected in the arrangement of ommatidia, which are single eye units (Fig. 5D). Each ommatidium is oriented with respect to both the anterior-posterior (AP) and dorsoventral (DV) axes. The DV arrangement is orchestrated by rotation of ommatidial clusters through 90 degrees toward the DV midline (equator) forming a mirror image alignment (for a review, see ref. 95). Concomitant with rotation, the ommatidia lose their symmetry and opposite chiral forms are established in the dorsal and ventral eye halves (Fig. 5D). Each ommatidium comprises 8 photoreceptor cells. The chirality of each ommatidium following rotation is represented by the asymmetric arrangement of photoreceptors R3 and R4 within the photoreceptor cell cluster. This asymmetry is governed by signalling events which involve both Frizzled-PCP and Notch/Delta pathways. Briefly, the R3/R4 pair are symmetrically arranged with R3 being closer to the equator. A signal from the equator appears to activate higher levels of Frizzled-PCP signalling in the preR3 cell, resulting in higher levels of Delta expression in preR3. Delta then activates Notch signalling in the preR4 cell, thus specifying R4 fate. The specification of R3 and R4 subsequently determines both chirality and direction of ommatidial rotation.

Flamingo plays complex roles in R3/R4 specification.[85] Unlike other 'core' PCP components which are required in either R3 or R4, Flamingo appears to play distinct roles in both R3 and R4. Current data suggests that initially in R3, Flamingo has a positive effect on Frizzled-PCP signalling through maintenance of asymmetric 'core' PCP protein distribution. Subsequently in R4, Flamingo is up-regulated by Notch signalling and acts to suppress Delta expression by dampening Frizzled signalling in this cell. This in turn enforces the initial Frizzled signalling bias across the R3/R4 boundary. The distinct activities of Flamingo in R3 and R4 are echoed by the dynamic expression of Flamingo protein within the apical domains of both cells. Flamingo becomes initially enriched asymmetrically across preR3 on the equatorial borders. As the clusters initiate rotation, Flamingo becomes enriched specifically within R4 in a U-shaped pattern. In this context, Flamingo provides a link between Frizzled-PCP and Notch/Delta signalling in R3/R4 specification.

*Vertebrate 7TM-Cadherins in PCP*

In mammals, two independent ENU-generated mutant alleles of *Celsr1*, *crash* (*crsh*) and *spin-cycle* (*scy*), were identified in mouse and together they provided the first evidence of a role for a 7TM-Cadherin in PCP-related processes in mammals.[96] *crsh* and *spy* are not null mutations; Celsr1 protein expression is evident in mutant tissues,[49] but they may be hypomorphic or dominant mutations. A large body of evidence now exists to suggest that vertebrate 7TM-Cadherins function in many aspects of PCP-related processes during vertebrate embryogenesis (Table 1) and these will be covered in more detail in the next two sections.

Whilst in mammals *Celsr1* is known to regulate PCP, we currently have no evidence of similar roles for *Celsr2* and *Celsr3* in mammalian embryogenesis despite predicted functions from RNA expression studies.[22] Data from *Celsr3* mouse 'knockouts'[97-99] suggest that any early roles for *Celsr3* are either non-essential or redundant with *Celsr1* and *Celsr2*. Generation of double/triple knockouts for *Celsr1*, *Celsr2* and *Celsr3* should resolve this issue.

## Vertebrate 7TM-Cadherins Function in the Morphogenesis of the Early Embryo and of Multiple Organ Systems

### Morphogenetic Movements and Neural Tube Closure

Coordinated cell movement is fundamental to early embryogenesis and formation of the basic body plan. The body axis is established by coordinated and directional movements of cells that include epiboly and convergence and extension (CE), processes requiring cell intercalation along different embryonic axes. Whilst epiboly, a morphogenetic movement specific to amphibian and teleost species, relies on radial cell intercalation enabling the spread of cells along the animal-vegetal axis to cover the yolk and close the blastopore,[100] CE requires medio-lateral cell intercalation which contributes to both a narrowing and elongation of the embryo along the antero-posterior axis (for a review, see ref. 101). Vertebrate 7TM-Cadherins are known to regulate each of these morphogenetic movements.

Gene 'knockdown' strategies have uncovered roles for *Celsr1* homologues in epithelial cell intercalation during avian primitive streak formation[45] and zebrafish CE movements.[19,102] The zebrafish *Celsr1* homologues, *zfmi1a* and *zfmi1b* were found to cooperate to mediate CE and *zfmi1a* was demonstrated to act in a cell autonomous manner to promote cell intercalation along the anterio-posterior axis.[19] Epiboly defects also resulted from gene 'knockdown' of two *Celsr1* homologues in a zebrafish *Celsr2* ENU mutant, *off-road*.[102] Whilst 7TM-cadherins appear to function through a nonPCP pathway to influence radial intercalation during epiboly,[102] their role(s) in medio-lateral cell intercalation require the PCP pathway.[19,45,102]

Neurulation involves an exquisitely coordinated suite of cell movements and changes in cell shape. Briefly, in higher vertebrates, the neural plate narrows medio-laterally and elongates rostrocaudally, its lateral edges elevate and move towards the dorsal midline where they meet and fuse, thereby converting a flat neural plate into a hollow tube (Fig. 6A). Disruption of this process can occur at different levels of the body axis leading to a distinct range of neural tube defects (NTD). Craniorachischisis is the most severe form of NTD and results from failure to initiate neurulation, affecting most of the neural tube. A number of mice mutant for 'core' PCP components, exhibit craniorachischisis in the homozygous state[34,37,39,40,46,67,70,96] implicating PCP processes in neural tube closure. Two of these are *crsh* and *spy* and both contain mutations in *Celsr1*.[96] Mice mutants exhibiting craniorachischisis display medio-lateral expansion of the neural plate which interferes with formation/function of the neural groove, positioning the neural folds too far apart to undergo apposition, preventing closure. Forebrain and rostral midbrain closure occur normally. A recent study by Ybot-Gonzalez et al[103] demonstrated a requirement for CE movements in both axial mesendoderm and neuroepithelium during initiation of mammalian neural tube closure. Defective CE is also linked to delayed neural keel formation in zebrafish.[19,104]

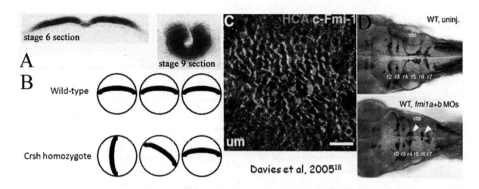

**Figure 6.** Vertebrate planar cell polarity. A) RNA in situ hybridisation of *c-fmi1* within the developing neural tube, transverse sections. At Stage 6, the neural plate is a flat sheet which has rolled up into a tube-shape by Stage 9. Fusion of the neural folds subsequently forms the neural tube. B) Polarity of chevron-shaped outer hair cells (OHC) within the mammalian ear is disrupted in 18.5dpc *scy/crsh* homozygotes compared to their wild-type littermates. C) Distribution of c-Flamingo1 (c-Fmi-1) protein in the sensory epithelium of the chick inner ear. HCA is hair cell antigen. Reproduced with permission from Davies A et al. Dev Dyn 2005; 233:998-1005. Copyright ©2005 Wiley-Liss Inc. D) Facial motor neuron migration (arrowheads) is disrupted in zebrafish embryos injected with morpholinos against *zfmi1a* and *zfmi1b*. Dorsal view of a zebrafish hindbrain stained with an antibody against islet-1. r2-r7 denote rhombomeres. Oto denotes otic vesicle. C. Formstone and A. Chandrasekhar, unpublished data.

## Orientation of Sensory Hair Cells in the Vertebrate Inner Ear

The orientation of sensory hair cells within the developing inner ear has many parallels to the system of oriented hairs/bristles that underpins the study of PCP in *Drosophila*. In the mammalian ear, the oriented alignment of 50-200 stereocilia, a group of modified microvilli that align to form a chevron-shaped pattern with a single large kinocilium located at the apex of the chevron is disrupted in the two independent ENU-generated mutant alleles of *Celsr1* in mouse, *crsh* and *spy*.[96] This also occurs in mice defective in other 'core' PCP components; *vangl2*[47,48] (*looptail* mutant; *lp*) and *frizzled6*[34] (mouse knockout) as well as *scribble*[47] (*circletail* mutant; *crc*) and *PTK7*[70] (mouse knockout). Heterozygote *crsh* and *scy* mice were found to exhibit head shaking behaviour, belly curling and spinning during tail suspension[96] suggestive of some vestibular dysfunction although the mice are not profoundly deaf. Subsequent examination of the adult cochlea of *crsh/scy* heterozygote mice revealed that both mutants exhibit misorientated outer hair cell stereociliary bundles (OHCs). The inner hair cells appear only slightly affected and the supporting cells immediately surrounding the affected hair cells exhibited abnormal shape. In homozygote mice, the OHCs are extensively mis-orientated along the entire length of the cochlear duct consistent with phenotypes from other mutant mice (Fig. 6B). However, there are subtle phenotypic differences between mouse mutants, for example, one group of OHCs, OHC1, are only slightly affected in *crsh/scy* compared to those in *lp* mice which exhibit more severe defects.

The asymmetric distribution of PCP components within the sensory epithelium is a characteristic feature of PCP processes in the inner ear[47,72,105] similar to patterns observed in the wing hair cells of *Drosophila*.[89] In the avian embryo the Celsr1 homologue, c-Fmi1, is distributed in a similar zig-zag pattern[18] to Flamingo in the fly wing,[11] reflecting the axis of hair cell polarity (Fig. 6C).

## Hair Follicle Patterning

Whereas in insects, hairs are uni-directionally oriented within a single cell, in mammals hair follicle alignment is a more complex process characterised by interactions between epidermal epithelial cells and underlying dermal mesenchyme which induce biased outgrowth of a hair follicle that is oriented along the anterior-posterior (A-P) axis of the embryo. Devenport and Fuchs[49] recently demonstrated that hair follicle angling is accompanied by polarised cell shape and cytoskeleton changes in both anterior and posterior cells at the hair follicle-epidermal boundary and subsequent polarised expression within the hair follicle of a number of proteins such as E-cadherin, P-cadherin and Sonic hedgehog. Involvement of the PCP pathway in hair patterning in mammals was first suggested through loss of function studies with *frizzled6*[35] where mosaic analyses of *frizzled6* mutants demonstrated that hair misalignment was due to an influence from the *frizzled6*-expressing epithelium. Analyses of *Celsr1* (*crsh*) and *Vangl2* (*lp*) homozygous mutant mice subsequently revealed a loss of A-P alignment of hair follicles during embryogenesis.[49] In this same study, Celsr1, Vangl2 and Frizzled6 proteins were each demonstrated to be asymmetrically distributed within the basal keratinocyte layer of the developing skin along the A-P axis. The initiation of asymmetric expression of PCP components within the basal layer was coincident with hair follicle formation although the exact nature of their influence on follicular orientation remains unclear. Studies using a *Celsr1* conditional knockout suggest that *Celsr1* may also play a subsequent role in postnatal hair development.[106] During embryogenesis, *crsh* mice display radially orientated hair follicle growth with respect to the skin whereas hair follicles in the *Celsr1* conditional knockout at P1 exhibit a random oblique orientation.

## 7TM-Cadherins Mould the Architecture of the Embryonic Nervous System

Early studies on *Drosophila Flamingo* mutants revealed central nervous system (CNS) defects essential to *Drosophila* development, comprising local disconnection of longitudinal axons.[11] 7TM-Cadherins have subsequently been demonstrated to play conserved roles in axon tract formation, dendritic arbour morphogenesis, axon-axon interactions, axon target selection and the promotion of neuronal migration.

It is intriguing that the function of both invertebrate and vertebrate 7TM-Cadherins in neuronal outgrowth appears independent of the Frizzled-PCP signalling pathway although PCP components such as Frizzled are implicated with vertebrate 7TM-Cadherins in axonal morphogenesis.[109]

### Axon Targeting and Axon Tract Development

A number of elegant studies have recently demonstrated roles for Flamingo in facilitating axon-axon interactions and axon target selection in the *Drosophila* visual system.[25,26]

In wild-type and control mosaic flies, photoreceptor (R) cell axons terminate in smooth topographic arrays in three distinct layers of the optic lobe. R1-6 target to the lamina and R8 terminates in the medulla. Senti et al[26] generated mosaics of *flamingo* mutant photoreceptors revealing disorganisation of R8 axon termination. These analyses additionally demonstrated a requirement for Flamingo activity in R8 axons alone and not in their target cells. Flies with mutations in other 'core' PCP components exhibited wild-type/control R axon projections. Lee et al[25] focussed on interactions between R1-R6 axons and revealed that Flamingo was also required for their proper targeting. Since Flamingo protein was found to be expressed mainly in R axons and not target lamina cells, a further study considered the mechanism by which Flamingo mediated the distinct R1-R6 growth trajectories. Using mosaic analyses with a repressible cell marker (MARCM) to generate elegant single cell manipulations, Chen and Clandinin[111] were able to alter Flamingo levels on individual R1-R6 growth cones. They found that the manipulated growth cones were sensitive to relative differences in Flamingo activity and not the absolute level of Flamingo activity in any single cell. Accordingly, the authors proposed a model where individual photoreceptor growth cones can compare their levels of Flamingo with that of their neighbours and either increase or decrease contact with them in order to balance precisely the Flamingo-mediated interactions on all sides. This balance of opposing forces establishes an appropriate trajectory for each growth cone.

Expanding on observations of defects in axon fasciculation in the *Drosophila* CNS,[11] subsequent studies in the abdominal sensory system of *Drosophila* revealed that Flamingo is also required here for axon advancement.[112] Growth of *Flamingo* mutant sensory axons stalled at reproducible positions corresponding to key intermediate target cells leading to the authors proposal that Flamingo mediates some form of interaction with intermediate cellular targets to ensure axon advancement. Whether the interaction requires Flamingo homophilic activity or interaction with an unknown ligand remains unproven.

The importance of Flamingo function in axonogenesis during *Drosophila* development[11,112] has parallels in mammals since defects in axon tract development are a major phenotype in a *Celsr3* knockout.[97] There are selective anomalies in several major axonal tracts, the most notable being the thalamocortical tract, resulting in complete disconnection of the neocortex from subcortical structures. This phenotype is similar to that observed in a *frizzled-3* knockout, prompting Tissir et al[97] to propose that an analogous pathway to that controlling PCP is operating in axon tract formation. Further study however, has failed to corroborate the existence of such a pathway.[109,113] It is clear however, that the Celsr3/Frizzled3 combination may hold some functional parallels with that of Celsr1/Frizzled6[34,42,96] but the molecular and cellular mechanisms involved are as yet unclear. In the future it will be interesting to test the similarity between the molecular/cellular roles of Celsr1/Frizzled6 in, for example, neural tube closure to those of Celsr3/Frizzled3 in axonal outgrowth.

More recently, conditional inactivation of *Celsr3* specifically in the telencephalon, ventral forebrain and cortex has demonstrated an essential role for *Celsr3* in neurons that project axons to the anterior commissure and sub-cerebral targets as well as in cells that guide axons through the internal capsule (Fig. 7A).[98,114] This latter process is reminiscent of Flamingo function in the *Drosophila* abdominal sensory system.[112]

*Dendrite Morphogenesis*

During development, neurons elaborate specific dendritic branching patterns to cover particular spatial territories known as dendritic fields. Dendritic fields of individual neurons do not overlap and yet these fields can completely fill a tissue area, for example the entire retina. An early study from the lab of Yuh Jung Jan of

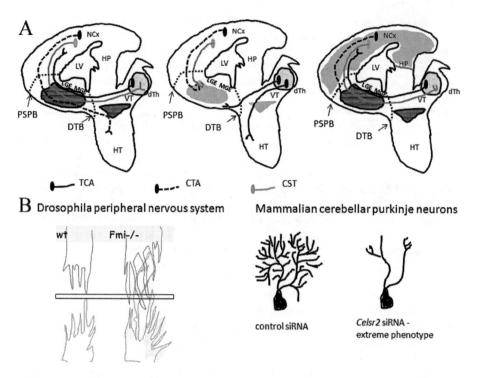

Figure 7. 7TM-Cadherins function in axonal and dendritic morphogenesis. A) Schematic representation of the region-specific inactivation of Celsr3 within the internal capsule (IC). Three tract components of the IC, namely TCAs (black), CTA fibers (black dotted), and subcerebral projections (CST, grey) are observed in wild-type mice (left). In mice where Celsr3 was deleted in the ventral telencephalon using Dlx5/6-Cre mice that express Cre in the ventral telencephalon (middle), Celsr3 is inactivated in the intermediate region of the ventral telencephalon and diencephalon (grey areas). Thalamic fibers fail to turn normally at the diencephalon–telencephalon boundary (DTB), whereas cortical axons are blocked at the external aspect of the basal telencephalon. In Celsr3|Emx1 mice, in which Celsr3 is inactivated early from E10.5, in the cortical anlagen (right, grey area), reciprocal thalamocortical projections form normally, but subcerebral projections fail to cross the pallial subpallial border (PSPB). NCx: neocortex; LV: lateral ventricle; LGE and MGE: lateral and medial ganglionic eminences; dTh: dorsal thalamus; VT: ventral thalamus; HT: hypothalamus; HP: hippocampus. Expression pattern of Dlx5/6 is denoted by stippled area . Drawings adapted from Zhou L, Qu Y, Tissir F et al. Cereb. Cortex 2009; Epub ahead of print. B) Schematic representation of the role of 7TM-Cadherins in dendrite morphogenesis. Drosophila flamingo regulates the limit of dorsal dendrite extension (left). In larvae, definition of the dorsal midline (line) is based on its equal distance to both dorsal clusters of PNS sensory neurons in the two hemisegments. Whereas in wild-type flies, the dorsal dendrites from homologous neurons extend to the dorsal midline and their dendritic fields have minimal overlap, in flamingo mutant embryos dorsal dendrites exhibit an overextension phenotype. Celsr2 promotes dendrite elaboration in mammalian cerebellar purkinje neurons (right). Control neurons exhibited elaborate dendritic trees whereas 37% of Celsr2-siRNA transfected neurons exhibited an extreme class of phenotype with a more simplistic branching pattern. Adapted from: Shima Y, Kengaku M, Hirano T et al. Dev Cell 2004; 7:205-216.

dendritic field formation in Drosophila showed that Flamingo is one gene involved in the regulation of this process.[24] Flamingo was found to promote competition between the dorsal dendrites of homologous peripheral nervous system (PNS) neurons situated within contralateral hemisegments, contributing to their mature dendritic morphology and preventing overlap between dendritic fields across the dorsal midline (Fig. 7B).[24]

As in the experiments of Chen and Clandinin[111] in the retina, levels of Flamingo protein expression appeared to be of crucial importance for Flamingo activity in dendritic field formation suggesting parallels between the two processes. The action of Flamingo in controlling dendritic field formation also appears to occur independently of Frizzled.[24] Studies in mushroom body neurons defective in Flamingo activity also exhibit overextension of dendrites.[121]

Further studies by Gao and colleagues utilised the MARCM technique to allow visualisation of single multiple dendritic (MD) neurons in living *Drosophila* larvae.[114] They found that Flamingo did not affect general dendritic branching patterns in postmitotic neurons but controlled in a cell autonomous manner, the extension of dorsal dendrites (Fig. 7B). When initiation of dorsal dendrite extension was visualised, disruption of Flamingo activity in individual neurons promoted precocious initiation of dorsal dendrite extension suggesting that Flamingo regulates the timing of dendrite extension.

In mammals, Tadashi Uemura and colleagues used siRNA technology to silence *Celsr2* in postnatal day 4 brain-slice cultures in which cortical pyramidal and cerebellar purkinje cell neurons were marked by expression of eGFP. Their studies elegantly demonstrated a role for *Celsr2* in dendrite stabilisation.[107,108] Following an initial phase of outgrowth comparable to that of control neurons, *Celsr2* siRNA-expressing neurons subsequently exhibited shorter and fewer basal dendrites (Fig. 7B). The authors also explored the Celsr2 protein domain required for the retraction phenotype and proposed that neurite outgrowth relied on Celsr2/Celsr2 homophilic interaction.[107,108] Using the same assay system, an opposing phenotype was observed for *Celsr3* shRNA-expressing neurons which extended longer basal dendrites with more branches. The *Celsr3* data however, has not been supported by the studies of Tissir and colleagues[97] who did not observe dendritic abnormalities in their *Celsr3* knockout.

Overall, 7TM-Cadherins appear to play conserved roles in dendrite stabilisation. An explanation for the apparently different consequences of disrupting Flamingo and Celsr2 activity, namely overextension and retraction of dendrite extension, should be explored however, possibly through time-lapse analysis, to understand better how 7TM-Cadherins regulate dendrite morphogenesis.[120]

*Neuronal Migration*

Vertebrate nervous system development depends on migration of immature neurons from their site of origin, often involving tangential migration through different neural layers. Facial (nVII) motor neurons, born within rhombomere 4 (r4), migrate caudally within the hindbrain in both mouse and zebrafish caudally to r6 where they form the facial nucleus (for a review see ref. 110) and this process is halted when activity of a number of vertebrate PCP components are disrupted (Fig. 6D).[20,44,51,56,61] Loss of *Celsr2/zfmi2*[20] in the *off-road* zebrafish ENU mutant[69] similarly disrupts facial neuron migration with most facial neurons remaining in r4. *Celsr1a/zfmi1a* and *Celsr1b/zfmi1b* appear to function redundantly with *Celsr2/zfmi2* to regulate this process.[69]

A GFP-knockin at the *Celsr3* locus has revealed its crucial role in both tangential and radial interneuron migration in the developing mouse forebrain.[99] *Celsr3* inactivation disrupts tangential migration of the calretinin-positive and radial migration of calbindin-positive interneuron classes, the latter being GABAergic neurons. Disruption in the development of these interneuron classes is known to contribute to major neuropsychiatric disorders such as schizophrenia and depression.[122]

## FUTURE DIRECTIONS: LINKING 7TM-CADHERIN STRUCTURE TO FUNCTION

With our understanding of 7TM-Cadherin function in both PCP and nervous system development rapidly unravelling, further insights into their roles during vertebrate organogenesis and nervous system development are expected. The advent of conditional knockouts for mammalian *Celsr* genes[98,106] will significantly extend our knowledge and enable investigation of global *Celsr* function during mammalian embryogenesis. Most importantly, these mice herald an exciting new phase of study into the roles of mammalian 7TM-Cadherins from postnatal development through to the adult.

With regard to the cellular and molecular roles of 7TM-Cadherins however, little is understood. Although pioneering studies in both *Drosophila*[91-93] and mammals[107,115] have provided useful insight into structure/function relationships, many questions remain. These include how posttranslational processing and intracellular trafficking of 7TM-Cadherins regulate protein function, what mechanisms exist for engagement of 7TM-Cadherins in cell signalling other than homophilic interaction and finally whether 7TM-Cadherins directly influence cell signalling through G protein coupling.

The adhesion-GPCR family are characterised by cleavage at the GPS domain, studies on which have provided an intriguing scenario which suggests that the two fragments generated by this process may act independently.[116,117] *Drosophila* Flamingo is a strong candidate for GPS cleavage; western data is supported by immunological and structure/function analyses which hint at separable functions for the extracellular and cytoplasmic domains.[11,111,112,114,115] The hypothesis that the 7TM/cytoplasmic regions of 7TM-Cadherins can act independently of their extracellular domains has gained further credence recently from a study which suggests separable functions for the extracellular domains and intracellular regions of zebrafish Celsr homologues during zebrafish gastrulation movements.[102] The cleavage status of the zebrafish Celsr protein used in these experiments was however, not reported. Published western data for mammalian 7TM-Cadherins suggest that Celsr2[15] and Celsr3[98] exist uncleaved, supporting functional data that the protocadherin and 7TM domains of Celsr2 and Celsr3 must be directly coupled to mediate cell signalling via homophilic interaction.[108] The cleavage status of 7TM-Cadherins needs to be explored in more detail therefore and the tantalising notion that cleavage status and functional specificity are intimately linked should be addressed.

Similarly, how 7TM-Cadherins traffic through the cell is an important question. At present we do not understand the molecular requirements for cell surface expression of these enigmatic proteins nor do we appreciate how 7TM-Cadherins are assigned to different cellular compartments. This would expand on intriguing data from Carreira-Barbosa et al[102] who provide an exciting glimpse at a potential mechanism for regulating the release of functional 7TM-Cadherin protein to the plasma membrane.

Many studies have touched upon the question of whether 7TM-Cadherins are adhesion proteins. Current data favours a role in the regulation of cell cohesion[11,23,102,108,111] through control of, for example, cell contact persistence[118] but this issue does need to be finally addressed. Substantial data exist to suggest that specific functions of 7TM-Cadherins require homophilic interaction.[49,84,108,111] Further studies are now required to examine in detail the molecular and cellular strategies through which homophilic interaction couples to changes in cell activity and cell signalling and to ask if similar mechanisms are utilised in different cell contexts. Is homophilic interaction the sole mechanism whereby 7TM-cadherins engage in cell signalling? What part do other regions of the 7TM-Cadherin extracellular domain play? For example, a number of studies[107,112,115] have implicated the HRM domain as a potential ligand-binding domain and data from chicken embryos suggests that a major tissue-specific form of c-Fmi1 protein exists in the membrane which does not contain atypical cadherin repeats (Formstone and Mason, unpublished).

Recent studies on Flamingo function during PCP establishment in the *Drosophila* wing are fuelling debate on whether Flamingo is a passive or active participant in the propagation of polarity information. In vertebrates however, a series of elegant studies on the role of *Celsr2* and *Celsr3* in neurite morphogenesis[108] have provided the strongest evidence to-date that mammalian Celsr proteins elicit active cell signalling through their 7TM domains. The authors implicate an arginine residue in the first intracellular loop of the TM region in functional differences between Celsr2 and Celsr3 in a dendritic growth assay. In this assay system, dendritic outgrowth was linked to the release of calcium from internal stores via phospholipase C and phosphoinositol signalling pathways. Calcium release was found to be reliant on the homophilic interaction of Celsr2 or Celsr3. Celsr2 homophilic activity had a greater effect on calcium release than that of Celsr3 and this difference was proposed to activate distinct second messenger pathways. Celsr2 activity stimulated CamKII signalling whereas Celsr3 activated a calcineurin pathway. Further studies should determine whether Celsr2 and Celsr3 directly interact with G protein effectors to activate these distinct enzyme pathways.

If the 7TM domain is an active signalling domain, what then is the role of the 7TM-Cadherin intracellular tail (IC)? This is an interesting question because in comparison to the extracellular domains, the IC domain exhibits little sequence conservation between invertebrate and vertebrate proteins. Moreover, between the different vertebrate family members, the IC domain is the least conserved protein domain. Since current data from Celsr1[96,106] and *Celsr3*[97,99] mouse mutants suggest distinct roles in PCP and nonPCP processes respectively, does the IC domain hold the key to their individual functions?

## CONCLUSION

The 7TM-cadherins are a unique group of adhesion-GPCRs which mediate important functional roles during embryonic development including the control of planar cell polarity (PCP), a process predicted to influence the formation and elaboration of multiple vertebrate organ systems. In the future these proteins will excite and fascinate as we advance our understanding of their roles during mammalian embryonic and postnatal development and gain insight into their cellular and molecular functions during PCP and neural development.

## ACKNOWLEDGEMENTS

I would like to thank Dr Hannah Mitchison for critical reading of the manuscript and Dr Anand Chandrasekhar for permission to use the zebrafish facial neuron migration data.

# REFERENCES

1. Hulpiau P, van Roy F. Molecular evolution of the cadherin superfamily. Int J Biochem Cell Biol 2009; 41:349-369.
2. Yagi T, Takeichi M. Cadherin superfamily genes: functions, genomic organization and neurologic diversity. Genes Dev 2000; 14:1169-1180.
3. Davis CG. The many faces of epidermal growth factor repeats. New Biol 1990; 2:410-419.
4. Timpl R, Tisi D, Talts JF et al. Structure and function of laminin LG modules Matrix Biol 2000; 19:309-317.
5. Oda H, Tsukita S. Nonchordate classic cadherins have a structurally and functionally unique domain that is absent from chordate classic cadherins. Dev Biol 1999; 216:406-422.
6. Hotta K, Takahashi H, Ueno N et al. A genome-wide survey of the genes for planar polarity signalling or convergent-extension-related genes in Ciona intestinalis and phylogenetic comparisons of evolutionary conserved signalling components. Gene 2003; 317:165-185.
7. Hadjantonakis A-K, Formstone CJ, Little PFR. mCelsr1 is an evolutionarily conserved seven-pass transmembrane receptor and is expressed during mouse embryonic development. Mech Dev 1998; 78:91-95.
8. Hill E, Broadbent ID, Chothia C et al. Cadherin Superfamily Proteins in Caenorhabditis elegans and Drosophila melanogaster. J Mol Biol 2001; 305:1011-1024.
9. Pettitt J. The cadherin superfamily (December 29, 2005). In: Moerman DG, Kramer JM, eds. The C. elegans Research Community, WormBook. Online review.
10. Hadjantonakis A-K, Sheward WJ, Harmar AJ et al. Celsr1, a neural-specific gene encoding an unusual seven-pass transmembrane receptor, maps to mouse chromosome 15 and human chromosome 22qter. Genomics 1997; 45:97-104.
11. Usui T, Shima Y, Shimada Y et al. Flamingo, a seven-pass transmembrane cadherin, regulates Planar Cell Polarity under the control of Frizzled. Cell 1999; 98:585-595.
12. Chae J, Kim MJ, Goo JH et al. The Drosophila tissue polarity gene starry night encodes a member of the protocadherin family. Development 1999; 126:5421-5429.
13. Nakayama N, Nakajima D, Nagase T et al. Identification of high-molecular- weight proteins with multiple EGF-like motifs by motif-trap screening. Genomics 1998; 51:27-34.
14. Formstone CJ, Little PFR. The flamingo-related Celsr family (Celsr1-3) genes exhibit distinct patterns of expression during embryonic development. Mech Dev 2001; 109:91-94.
15. Shima Y, Copeland NG, Gilbert DJ et al. Differential expression of the seven-pass transmembrane cadherin genes Celsr1-3 and distribution of the Celsr2 protein during mouse development. Dev Dyn 2002; 223:321-332.
16. Tissir F, De-Backer O, Goffinet AM et al. Developmental expression profiles of Celsr (flamingo) genes in the mouse. Mech Dev 2002; 112:157-160.
17. Formstone CJ, Mason I. Expression of the Celsr/flamingo homologue, c-fmi1, in the early avian embryo indicates a conserved role in neural tube closure and additional roles in asymmetry and somitogenesis. Dev Dyn 2005; 232:408-413.
18. Davies A, Formstone C, Mason I et al. Planar polarity of hair cells in the chick inner ear is correlated with polarized distribution of c-flamingo-1 protein. Dev Dyn 2005; 233:998-1005.
19. Formstone CJ, Mason I. Combinatorial activity of Flamingo proteins directs convergence and extension within the early zebrafish embryo via the planar cell polarity pathway. Dev Biol 2005; 282:320-335.
20. Wada H, Tanaka H, Nakayama S et al. Frizzled3a and Celsr2 function in the neuroepithelium to regulate migration of facial motor neurons in the developing zebrafish hindbrain. Development 2006; 133:4749-4759.
21. Morgan R, El-Kadi AM, Theokli C. Flamingo, a cadherin-type receptor involved in the Drosophila planar polarity pathway, can block signaling via the canonical wnt pathway in Xenopus laevis. Int J Dev Biol 2003; 47:245-252.
22. Crompton LA, Du Roure C, Rodriguez TA. Early embryonic expression patterns of the mouse flamingo and prickle homologues. Dev Dyn 2007; 236:3137-3143.
23. Beall SA, Boekelheide K, Johnson KA. Hybrid GPCR/Cadherin (Celsr) Proteins in Rat Testis Are Expressed With Cell Type Specificity and Exhibit Differential Sertoli Cell—Germ Cell Adhesion Activity. J Androl 2005; 26:529-538.
24. Gao F-B, Kohwi M, Brenman JE et al. Control of Dendritic field formation in Drosophila: The roles of Flamingo and competition between homologous neurons. Neuron 2000; 28:91-101.
25. Lee RC, Clandinin TR, Lee CH et al. The protocadherin Flamingo is required for axon target selection in the Drosophila visual system. Nat Neuroscience 2003; 6:57-563.
26. Senti K-A, Usui T, Boucke K et al. Flamingo Regulates R8 Axon-Axon and Axon-target Interactions in the Drosophila visual system. Curr Biol 2003; 13:828-832.

27. Lawrence PA, Shelton PMJ. The determination of polarity in the developing insect retina. J Embryol Exp Morphol 1975; 33:471-486.
28. Adler PN. Planar signalling and morphogenesis in Drosophila. Dev Cell 2002; 2:525-535.
29. Strutt H, Strutt D. Long-range coordination of planar polarity in Drosophila. Bioessays 2005; 27:1218-1227.
30. Strutt D. The planar polarity pathway. Curr Biol 2008; 18:R898-902.
31. Lawrence PA, Struhl G, Casal J. Planar Cell Polarity: one or two pathways? Nat Reviews Genet 2007; 8:555-562.
32. Ma D, Amonlirdviman K, Raffard RL et al. Cell Packing influences planar cell polarity signalling. PNAS 2008; 105:18800-18806.
33. Simons M, Mlodzik M. Planar Cell Polarity signalling: from fly development to human disease. Ann Rev Genet 2008; 42:25.1-25.24.
34. Wang Y, Guo N, Nathans J. The role of Frizzled3 and Frizzled6 in neural tube closure and in the planar polarity of inner-ear sensory hair cells. J Neurosci 2006; 26:2147-2156
35. Guo N, Hawkins C, Nathans J. Frizzled6 controls hair patterning in mice. Proc Natl Acad Sci USA 2006; 101:9277-9281.
36. Wallingford JB, Harland RM. Neural tube closure requires Dishevelled-dependent convergent extension of the midline. Development 2002; 129:5815-5825.
37. Hamblet NS, Lijam N, Ruiz-Lozano P et al. Dishevelled 2 is essential for cardiac outflow tract development, somite segmentation and neural tube closure. Development 2002; 129:5827-5838.
38. Wang J, Mark S, Zhang X et al. Regulation of polarized extension and planar cell polarity in the cochlea by the vertebrate PCP pathway. Nat Genet 2005; 37:980-985.
39. Wang J, Hamblet NS, Mark S et al. Dishevelled genes mediate a conserved mammalian PCP pathway to regulate convergent extension during neurulation. Development 2006; 133:1767-1778.
40. Etheridge SL, Ray S, Li S et al. Murine dishevelled 3 functions in redundant pathways with dishevelled 1 and 2 in normal cardiac outflow tract, cochlea and neural tube development. PLoS Genet 2008; 4:e1000259.
41. Park TJ, Mitchell BJ, Abitua PB et al. Dishevelled controls apical docking and planar polarization of basal bodies in ciliated epithelial cells. Nat Genet 2008; 40:871-879.
42. Cirone P, Lin S, Griesbach HL et al. A role for planar cell polarity signaling in angiogenesis. Angiogenesis 2008; 11:347-360.
43. Darken RS, Scola AM, Rakeman AS et al. The planar polarity gene strabismus regulates convergent extension movements in Xenopus. EMBO J 2002; 21:976-985.
44. Jessen JR, Topczewski J, Bingham S et al. Zebrafish trilobite identifies new roles for Strabismus in gastrulation and neuronal movements. Nat Cell Biol 2002; 4:610-615.
45. Voiculescu O, Bertocchini F, Wolpert L et al. The amniote primitive streak is defined by epithelial cell interaction before gastrulation. Nature 2007; 449:1049-1052.
46. Murdoch JN, Doudney K, Paternotte C et al. Severe neural tube defects in the loop-tail mouse result from mutation of Lpp1, a novel gene involved in floor plate specification. Hum Mol Gen 2001; 10:2593-2601.
47. Montcouquiol M, Rachel RA, Lanford PJ et al. Identification of Vangl2 and Scrb1 as planar polarity genes in mammals. Nature 2003; 423:173-177.
48. López-Schier H, Hudspeth AJ. A two-step mechanism underlies the planar polarization of regenerating sensory hair cells. Proc Natl Acad Sci USA 2006; 103:18615-18620.
49. Devenport D, Fuchs E. Planar polarisation in embryonic epidermis orchestrates global asymmetric morphogenesis of hair follicles. Nat Cell Biol 2008; 10:1257-1289.
50. Phillips HM, Murdoch JN, Chaudhry B et al. Vangl2 acts via RhoA signaling to regulate polarized cell movements during development of the proximal outflow tract. Circ Res 2005; 96:292-299.
51. Bingham SM, Higashijima S, Okamoto H et al. The zebrafish trilobite gene is essential for tangential migration of branchiomotor neurons. Dev Biol 2002; 242:149-160.
52. Ross AJ, May-Simera H, Eichers ER et al. Disruption of Bardet-Biedl syndrome ciliary proteins perturbs planar cell polarity in vertebrates. Nat Genet 2005; 37:1135-1140.
53. Lake BB, Sokol SY. Strabismus regulates asymmetric cell divisions and cell fate determination in the mouse brain. J Cell Biol 2009; 185:59-66.
54. Veeman MT, Slusarski DC, Kaykas A et al. Zebrafish prickle, a modulator of noncanonical Wnt/Fz signaling, regulates gastrulation movements. Curr Biol 2003; 13:680-685.
55. Takeuchi M, Nakabayashi J, Sakaguchi T et al. The prickle-related gene in vertebrates is essential for gastrulation cell movements. Curr Biol 2003; 13:674-679.
56. Carreira-Barbosa F, Concha M, Takeuchi M et al. Prickle 1 regulates cell movements during gastrulation and neuronal migration in zebrafish. Development 2003; 130:4037-4046.
57. Wallingford JB, Vogeli KM, Harland RM. Regulation of convergent extension in Xenopus by Wnt5a and Frizzled-8 is independent of the canonical Wnt pathway. Int J Dev Biol 2001; 45:225-227.

58. Kilian B, Mansukoski H, Barbosa FC et al. The role of Ppt/Wnt5 in regulating cell shape and movement during zebrafish gastrulation. Mech Dev 2003; 120:467-476.
59. Qian D, Jones C, Rzadzinska A et al. Wnt5a functions in planar cell polarity regulation in mice. Dev Biol 2007; 306:121-133.
60. Cervantes S, Yamaguchi TP, Hebrok M. Wnt5a is essential for intestinal elongation in mice. Dev Biol 2009; 326:285-294.
61. Vivancos V, Chen P, Spassky N et al. Wnt activity guides facial branchiomotor neuron migration and involves the PCP pathway and JNK and ROCK kinases. Neural Dev 2009; Epub ahead of print.
62. Tada M, Smith JC. Xwnt11 is a target of Xenopus brachyury: regulation of gastrulation movements via Dishevelled, but not through the canonical Wnt pathway. Development 2000; 127:2227-2238.
63. Ulrich F, Concha ML, Heid PJ et al. Slb/Wnt11 controls hypoblast cell migration and morphogenesis at the onset of zebrafish gastrulation. Development 2003; 130:5375-5384.
64. Hardy KM, Garriock RJ, Yatskievych TA et al. Noncanonical Wnt signaling through Wnt5a/b and a novel Wnt11 gene, Wnt11b, regulates cell migration during avian gastrulation. Dev Biol 2008; 320:391-401.
65. De Calisto J, Araya C, Marchant L et al. Essential role of noncanonical Wnt signalling in neural crest migration. Development 2005; 132:2587-2597.
66. Gros J, Serralbo O, Marcelle C. WNT11 acts as a directional cue to organize the elongation of early muscle fibres. Nature 2009; 457:589-593.
67. Murdoch JN, Henderson DJ, Doudney K et al. Disruption of Scribble (Scrb1) causes severe neural tube defects in the circletail mouse. Hum Mol Genet 2003; 12:87-98.
68. Phillips HM, Rhee HJ, Murdoch JN et al. Disruption of planar cell polarity signalling results in congenital heart defects and cardiomyopathy attributable to early cardiomyocyte disorganization. Circ Res 2007; 101:137-145.
69. Wada H, Iwasaki M, Sato T et al. Dual roles of zygotic and maternal Scribble1 in neural migration and convergent extension movements in zebrafish embryos. Development 2005; 132:2273-2285.
70. Lu X, Borchers AD, Jolicoeur C et al. PTK-7/CCK-4 is a novel regulator of planar cell polarity in vertebrates. Nature 2004; 430:93-98.
71. Shnitsar I, Borchers A. PTK7 recruits dsh to regulate neural crest migration. Development 2008; 135:4015-4024.
72. Montcouquiol M, Sans N, Huss D et al. Asymmetric localization of Vangl2 and Fz3 indicate novel mechanisms for planar cell polarity in mammals. J Neurosci 2006; 26:5265-5275.
73. Thorpe CJ, Schlesinger A, Bowerman B. Wnt signalling in C.elegans: regulating repressors and polarising the cytoskeleton. Trends Cell Biol 2000; 10:10-17.
74. Yamamoto S, Nishimura O, Misaki K et al. Cthrc1 selectively activates the planar cell polarity pathway of Wnt signalling by stabilising the Wnt-receptor complex. Dev Cell 2008; 15:23-26.
75. Lawrence PA, Casal J, Struhl G. Towards a model of the organisation of planar polarity and pattern in the Drosophila abdomen. Development 2002; 129:2749-2760.
76. Bilder D, Perrimon N. Localization of apical epithelial determinants by the basolateral PDZ protein Scribble. Nature 2000; 403:676-680.
77. Gubb D, Garcia-Bellido A. A genetic analysis of the determination of cuticular polarity during development in Drosophila melanogaster. J Embryol Exp Morphol 1982; 68:37-57.
78. Adler PN, Krasnow RE, Liu J. Tissue polarity points from cells that have higher Frizzled levels towards cells that have lower Frizzled levels. Curr Biol 1997; 7:940-949.
79. Widelitz R. Wnt signaling through canonical and noncanonical pathways: recent progress. Growth Factors 2005; 23:111-116.
80. Grigoryan T, Wend P, Klaus A et al. Deciphering the function of canonical Wnt signals in development and disease: conditional loss- and gain-of-function mutations of beta-catenin in mice. Genes Dev 2008; 22:2308-2341.
81. Gordon MD, Nusse R. Wnt signaling: multiple pathways, multiple receptors and multiple transcription factors. J Biol Chem 2006; 281:22429-22433.
82. Tree DR, Shulman JM, Rousset R et al. Prickle mediates feedback amplification to generate asymmetric planar cell polarity signalling. Cell 2002; 109:371-381.
83. Amonlirdviman K, Khare NA, Tree DR et al. Mathematical modeling of planar cell polarity to understand domineering nonautonomy. Science 2005; 307:423-426.
84. Lawrence PA, Casal J, Struhl G. Cell interactions and planar polarity in the abdominal epidermis of Drosophila. Development 2004; 131:4651-4664.
85. Das G, Reynolds-Kenneally J, Mlodzik M. The Atypical Cadherin Flamingo links Frizzled and Notch Signalling in Planar Polarity establishment in the Drosophila eye. Dev Cell 2002; 2:655-666.
86. Lu B, Usui T, Uemura T et al. Flamingo controls planar polarity of sensory bristles and asymmetric division of sensory organ precursors in Drosophila. Curr Biol 1999; 9:1247-1250.
87. Vinson CR, Adler PN. Directional noncell autonomy and the transmission of polarity information by the frizzled gene of Drosophila. Nature 1987; 329:549-551.

88. Adler PN, Taylor J, Charlton J. The domineering non-autonomy of frizzled and van Gogh clones in the Drosophila wing is a consequence of a disruption in local signalling. Mech Dev 2000; 96:197-207.
89. Shimada Y, Usui T, Yanagawa S et al. Asymmetric colocalisation of Flamingo, a seven-pass transmembrane cadherin and Dishevelled in planar cell polarisation. Curr Biol 2001; 11:859-863.
90. Strutt D. Asymmetric localisation of Frizzled and the establishment of cell polarity in the Drosophila wing. Mol Cell 2001; 7:367-375.
91. Chen WS, Antic D, Matis M et al. Asymmetric homotypic interactions of the atypical cadherin Flamingo mediate intercellular polarity signalling. Cell 2008; 133:1093-1105.
92. Strutt H, Strutt D. Differential stability of Flamingo protein complexes underlies the establishment of planar polarity in Drosophila. Curr Biol 2008; 18:1555-1564.
93. Wu J, Mlodzik M. The Frizzled extracellular domain is a ligand for Van Gogh/Stbm during non-autonomous planar cell polarity signalling. Dev Cell 2008; 15:462-469.
94. Lawrence PA, Struhl G, Casal J. Planar Cell polarity: a bridge too far? Curr. Biol 2008; 18:R959-960.
95. Mlodzik M. Planar polarity in the Drosophila eye: a multifaceted view of signaling specificity and cross-talk. EMBO J 1999; 18:6873-6879.
96. Curtin JA, Quint E, Tsipouri V et al. Mutation in Celsr1 disrupts Planar Polarity of Inner Ear Hair Cells and Causes Severe Neural Tube Defects in the Mouse. Curr Biol 2003; 13:1129-1133.
97. Tissir F, Bar I, Jossin Y et al. Protocadherin Celsr3 is crucial in axonal tract development. Nat Neuro 2005; 8:451-457.
98. Zhou L, Bar I, Tissir F et al. Early Forebrain Wiring: Genetic dissection using conditional mutant mice. Science 2008; 320:946-950.
99. Ying G, Wu S, Hou R et al. Protocadherin Celsr3 is required for interneuron migration in the mouse forebrain. Mol Cell Biol 2009; Epub ahead of print.
100. Warga RM, Kimmel CB. Cell movements during epiboly and gastrulation in zebrafish. Development 1990; 108:569-580.
101. Solnica-Krezel L. Conserved patterns of cell movements during vertebrate gastrulation. Curr Biol 2005; 15:R213-228.
102. Carreira-Barbosa F, Kajita M, Morel V et al. Flamingo regulates epiboly and convergence/extension movements through cell cohesive and signalling functions during zebrafish gastrulation. Development 2009; 136:383-392.
103. Ybot-Gonzalez P, Savery D, Gerrelli D et al. Convergent extension, planar-cell-polarity signalling and initiation of mouse neural tube closure. Development 2007; 134:789-799.
104. Ciruna B, Jenny A, Lee D et al. Planar cell polarity signalling couples cell division and morphogenesis during neurulation. Nature 2006; 439:220-224.
105. Deans MR, Antic D, Suyama K et al. Asymmetric distribution of prickle-like 2 reveals an early underlying polarization of vestibular sensory epithelia in the inner ear. J Neurosci 2007; 27:3139-3147.
106. Ravni A, Yibo Q, Goffinet A et al. Planar Cell Polarity Cadherin Celsr1 regulates skin hair patterning in the mouse. J Invest Derm 2009; Epub ahead of print.
107. Shima Y, Kengaku M, Hirano T et al. Regulation of dendritic maintenance and growth by a mammalian 7-pass transmembrane cadherin. Dev Cell 2004; 7:205-216.
108. Shima Y, Kawaguchi S-y, Kosaka K et al. Opposing roles in neurite growth control by two seven-pass transmembrane cadherins. Nat Neuroscience 2007; 10:963-969.
109. Tissir F, Goffinet AM. Expression of planar cell polarity genes during development of the mouse CNS. Eur J Neuro 2006; 23:5976-5607.
110. Chandrasekhar A. Turning heads: development of vertebrate branchiomotor neurons. Dev Dyn 2004; 229:143-161.
111. Chen P-L, Clandinin TR. The cadherin Flamingo mediates level-dependent interactions that guide photoreceptor target choice in Drosophila. Neuron 2008; 58:26-33.
112. Steinel MC, Whitington PM. The atypical cadherin Flamingo is required for sensory axon advance beyond intermediate target cells. Dev Biol 2009;327:447-457.
113. Zhou L, Qu Y, Tissir F et al. Role of the Atypical Cadherin Celsr3 during Development of the Internal Capsule. Cereb Cortex 2009; Epub ahead of print.
114. Sweeney NT, Li W, Gao F-B. Genetic manipulation of single neurons in vivo reveals specific roles of Flamingo in neuronal morphogenesis. Dev Biol 2002; 247:76-88.
115. Kimura H, Usui T, Tsubouchi A et al. Potential dual molecular interaction of the Drosophila 7-pass transmembrane cadherin Flamingo in dendritic morphogenesis. J Cell Sci 2006; 119:1118-1129.
116. Volynski KE, Silva JP, Lelianova VG et al. Latrophilin fragments behave as independent proteins that associate and signal on binding LTX (N4C). EMBO J 2004; 23:4423-4433.
117. Silva JP, Lelianova V, Hopkins C et al. Functional cross-interaction of the fragments produced by the cleavage of distinct adhesion G-protein-coupled receptors. J Biol Chem 2009; 284:6495-6506.

118. Witzel S, Zimyanin V, Carreira-Barbosa F et al. Wnt11 controls cell contact persistence by local accumulation of Frizzled7 at the plasma membrane. J Cell Biol 2006; 175:791-802.
119. Wasserscheid I, Thomas U, Knust E. Isoform-specific interaction of Flamingo/Starry Night with excess Bazooka affects Planar Cell Polarity in the Drosophila wing. Dev Dyn 2007; 236:1064-1071.
120. Ye B, Jan YN. The cadherin superfamily and dendrite development. Trends Cell Biol 2005; 15:65-67.
121. Reuter JE, Nardine TM, Penton A et al. A mosaic genetic screen for genes necessary for Drosophila mushroom body neuronal morphogenesis. Development 2003; 130:1203-1213.
122. Sakai T, Oshima A, Nozaki Y et al. Changes in density of calcium-binding-protein-immunoreactive GABAergic neurons in prefrontal cortex in schizophrenia and bipolar disorder. Neuropathology 2008; 28:143-150.

# CHAPTER 3

# LATROPHILIN SIGNALLING IN TISSUE POLARITY AND MORPHOGENESIS

Tobias Langenhan* and Andreas P. Russ*

**Abstract:** Understanding the mechanisms that coordinate the polarity of cells and tissues during embryogenesis and morphogenesis is a fundamental problem in developmental biology. We have recently demonstrated that the putative neurotoxin receptor *lat-1* defines a mechanism required for the alignment of cell division planes in the early embryo of the nematode *C. elegans*. Our analysis suggests that *lat-1* is required for the propagation rather than the initial establishment of polarity signals. Similar to the role of the flamingo/CELSR protein family in the control of planar cell polarity, these results implicate an evolutionary conserved subfamily of adhesion-GPCRs in the control of tissue polarity and morphogenesis.

## INTRODUCTION

A fundamental requirement in all multicellular organisms is a robust program to achieve the correct spatial arrangement of cells. Cell fate decisions, the orientation of mitotic divisions, the migration of individual cells and morphogenetic movements of cell groups have to be tightly coordinated. While our understanding of the molecular mechanisms controlling asymmetric cell fate decisions and mitotic spindle orientation in certain types of cell-cell interaction is advanced (reviewed in refs. 1,2), it is less well-understood how signals are propagated in larger groups of cells to align cell polarity and division plane orientation and how tissue polarity is coordinated with morphogenetic movements. The analysis of planar cell polarity (PCP) in epithelial sheets and the study of convergence and extension (C and E) movements during gastrulation in vertebrates have implicated signalling by the Wnt/PCP, Fat/Dachsous/Four-jointed

*Corresponding Authors: Tobias Langenhan—Institute of Physiology, University of Würzburg, 97070 Würzburg, Germany. Email: andreas.russ@bioch.ox.ac.uk;
Andreas Russ—Department of Biochemistry, University of Oxford, OX1 3QU, Oxford, UK. Email: tobias.langenhan@uni-wuerzburg.de

*Adhesion-GPCRs: Structure to Function*, edited by Simon Yona and Martin Stacey.
©2010 Landes Bioscience and Springer Science+Business Media.

(Fat/Ds/Fj) and anterior-posterior (a-p) tissue polarity pathways in the coordination of cell division plane orientation (reviewed by ref. 3, see Formstone, this volume).

Cells also have to find and maintain their correct positions in relation to surrounding cells. Since the pioneering experiments of Townes and Holtfreter[4] the molecular basis for the directed movement and selective adhesion of embryonic cells has been an area of intense interest. While substantial progress has been made in elucidating the formation and maintenance of boundaries between compartments and tissues, the segregation and sorting of mixed cell populations is much less well-understood.[5] Widely accepted hypotheses are the thermodynamic model, mainly based on differential adhesion mediated by cadherin-based mechanisms[6,7] and the activity of cell guidance systems transmitting attractive or repulsive cues to migrating cells.[8] However, the currently known mechanisms do not yet fully explain the developmental processes shaping embryos and organs.

## THE ROLE OF ADHESION-GPCRs IN DEVELOPMENT

An interesting class of candidate molecules for the control of cell-cell interactions are the adhesion-GPCRs,[9,10] which combine extracellular domain features of adhesion molecules with transmembrane regions characteristic for G protein-coupled receptors. Vertebrate genomes encode 30 or more adhesion-GPCRs with at least 8 different extracellular domain architectures[11] (see Schioth et al, this volume), making it the second largest group of seven-pass transmembrane (7TM) receptors. Adhesion-GPCRs are implicated in immune functions[12,13] and in rare inherited developmental disorders[14] but there is little information about the physiological function of most members of the protein family. A key role in development has been defined for the cadherin-like flamingo/starry night (FMI) and its vertebrate homologs (CELSR), which have essential and conserved functions in the PCP pathway and in neuronal development[15-23] (see Formstone, this volume).

Comparative genomics of the highly divergent adhesion-GPCR family shows that next to FMI only the domain architecture of latrophilins (LPHN; synonyms CL/CIRL/Lph/Lectomedin; see Ushkaryov, in this volume) is strictly conserved across phyla (see below).[11] The lectin-like latrophilins were originally described as cellular receptors for latrotoxin ($\alpha$-LTX), the main neurotoxin of the Black Widow spider *Latrodectus mactans*.[24,25] They have been implicated as modulators of neurotransmitter release[26-28] (Silva et al, this volume) and are thought to act as components of the fusion machinery that regulates discharge of the pool of biogenic amine vesicles (i.e., norepinephrin, GABA, glutamate) in several neuron types and vesicles carrying insulin in pancreatic $\beta$-cells. However, the physiological function of this highly conserved receptor is not well-defined and its endogenous ligands are unknown. Recent work from our laboratory has identified an unexpected role for latrophilins as essential regulators of tissue polarity in embryonic development.[29]

### Adhesion-GPCRs in *C. elegans*

A major challenge in the genetic analysis of orphan adhesion-GPCRs is the complexity of the gene family. The large number of different domain architectures raises issues about general conservation of function versus species-specific diversification. In addition, the

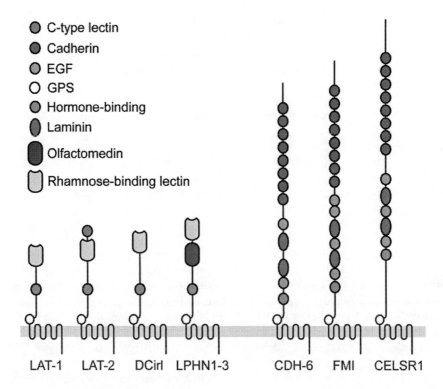

**Figure 1.** Adhesion-GPCR classes conserved between vertebrates and invertebrates. The domain architecture of adhesion-GPCRs is conserved from nematodes to mammals and characterized by an extracellular GPS motif in close proximity to the outer face of the 7TM region. The RBL domain is the hallmark for receptors of the latrophilin subfamily (LPHN), whereas the Flamingo (FMI) group is determined by the presence of cadherin, EGF and laminin domain repeats. Reprinted from Langenhan et al, Dev Cell 2009; 17(4):494-504,[29] ©2009 with permisson from Elsevier.

presence of up to 3-5 paralogs for some receptor subfamilies in vertebrates indicates possible functional overlap and compensation between paralogs. To investigate the physiological function of adhesion-GPCRs in a less complex system we turned to the nematode C. elegans.

The C. elegans genome contains two LPHN genes, *lat-1* and *lat-2*[28,30] and a single FMI homolog (*cdh-6*)[31,32] (Fig. 3). Similarly, FMI and LPHN (dCIRL) are the only conserved adhesion-GPCR architectures in Drosophila (Fig. 1). Other C. elegans or Drosophila genes showing similarity to adhesion-GPCRs are highly divergent with little sequence homology to adhesion-GPCRs in vertebrates,[11,33] while FMI and LPHN are conserved in other nematode and insect species. This suggests that FMI and LPHN represent the core functions of adhesion-GPCRs that are highly conserved in the evolution of bilateral animals.

The small number of adhesion-GPCRs implies a low level of functional redundancy in the worm and offers the possibility to separate and dissect the role of individual genes and to assign the physiological function to each member of the receptor class. Based on loss-of-function mutants, molecular requirements of different receptor domains can be tested by transgenic complementation. Quantitative assays provide a means to distinguish the different signalling properties of receptor mutants under physiological conditions even without knowing the identity of the endogenous ligand(s). Further, the interaction of adhesion-GPCR signalling with other molecular pathways can be tested by epistasis experiments.

## Introduction into *C. elegans* Embryonic Anatomy and Development

*C. elegans* has an essentially invariant embryonic cell lineage,[34] which unfolds by a sequence of asymmetric cell divisions and intercellular induction events.[35,36] Starting from the zygote (P0), the three body axes of the embryo are established within the first three cleavage divisions (Fig. 1). The first cleavage event generates the anterior AB and posterior P1 blastomers thereby assigning the primordial antero-posterior (a-p) axis to the early embryo. In the next round of cell divisions, AB is divided into an anterior (ABa) and posterior daughter (ABp), whereas P1 gives rise to the ventral EMS blastomere (Endoderm/MeSoderm) and the posterior P2 cell, thus defining the dorso-ventral (d-v) body axis. During the following third cleavage, ABa/p divide perpendicular to the a-p and d-v axes into ABal and ABpl on the left side of the embryo and ABar and ABpr on the right hand side. This establishes the left-right (l-r) axis and the slightly more anterior position of ABal/pl compared to ABar/pr defines a handed bilateral asymmetry (Fig. 2).

In subsequent asymmetric blastomere divisions, the P1-derived blastomere EMS divides into E and MS. P2 gives rise to C and P3 and the latter divides into D and P4. At the end of these first divisions all three body axes are laid down and 6 founder blastomeres have been generated, which eventually give rise to clonally expanding tissues that form the embryo: AB, MS, E, C, D, P4. Germ-line potential is always retained in the posterior blastomere Px. With the exception of E, which gives rise to all gut cells, i.e., endoderm, the founder blastomeres only loosely correspond to the classical germ layers. The AB, MS and C lineages can give rise to cell types with ectodermal and mesodermal characteristics (Fig. 3).

Contrary to a common misconception the invariant embryonic cell lineage of *C. elegans* is not a form of "mosaic" development determined exclusively by the segregation of preformed cell-autonomous determinants. Rather, it is established by a sequence of controlled asymmetric cell divisions and intercellular induction events very similar to the ones seen in the embryonic development of "higher" animals. Due to the small number of cells and their precisely reproduced locations and interactions in the nematode embryo, cell fates and cell division planes are coordinated so tightly that the lineage and fate of each cell appears to be invariant. The regulative features of *C. elegans* development have been identified by the analysis of mutations in signalling pathways and by the ablation of blastomeres with laser microbeams.[34,36-39]

The Wnt/β-catenin asymmetry pathway has been shown to be essential for cell fate decisions (reviewed by ref. 40) while a noncanonical Wnt/Frizzled (Wnt/Fz) pathway is required for the orientation of mitotic spindles (reviewed by ref. 41). The mechanisms controlling cell polarity in the first, second and third round of embryonic cell divisions are understood in considerable detail.[41,42] A posterior polarising centre is located in the descendants of the founder blastomere P1[43] and can orient the division planes of immediately adjacent cells.[44] The polarisation of EMS by P2 at the four-cell stage is thought to require an instructive Wnt/Fz signal and a permissive activity of *scr-1*/SRC

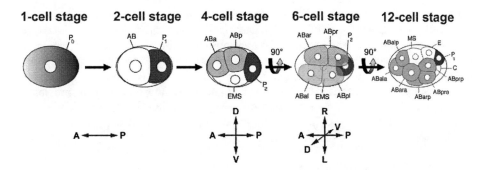

**Figure 2.** Establishment of principal body axes and founder blastomers through asymmetric divisions in the early *C. elegans* embryo. Within the first division rounds the three body axes are generated and after the fifth round (not depicted) all six founder lineages have been established. Germline precursors are labelled dark grey, AB lineage light grey, all other (P-derived) blastomeres white.

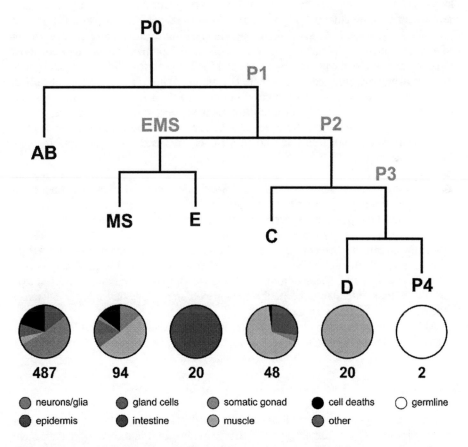

**Figure 3.** Embryonic lineage of *C. elegans*. Six founder blastomers give rise to all tissue types of the developed animal. Numbers below pie charts indicate total number of cells generated within this lineage during embryogenesis.[34] Transient blastomere names in grey.

oncogene and the receptor tyrosine kinase *mes-1*.[45-50] While the P2-EMS interaction at the 4-cell stage has served as an excellent paradigm to study the molecular mechanisms of a polarising induction, it is not well-understood how the polarising information is propagated and coordinated as the complexity of the embryo increases rapidly from the second (4 cells) to the 10th division cycle (~1000 cells). A *wnt*-dependent relay mechanism has been proposed,[51] but it is a matter of debate how this mechanism relates to existing models for PCP or a-p tissue polarity signalling.[52,53] A clear functional equivalent of PCP in *Drosophila* has not yet been described in *C. elegans*.

## LATROPHILINS AND TISSUE POLARITY

### Maternal and Zygotic *lat-1* Expression is Required for *C. elegans* Development

Homozygous offspring of nematodes heterozyous for a mutant *lat-1* allele develop with normal morphology to the first larval stage and display disturbed pharyngeal motor behaviour reminiscent of synaptic dysfunction.[28,30] A more detailed examination revealed that the offspring of *lat-1* homozygotes show additional severe defects in embryonic and larval development, leading to drastically reduced adult brood sizes for *lat-1* mutants.[29] The early defects in homozygous mutant embryos can be suppressed by the presence of maternal LAT-1 protein, while the phenotype is observed in heterozygous embryos created by mating of homozygous hermaphrodites with normal males, which lack maternal but not zygotic LAT-1. This indicates that maternal gene product is required and sufficient to support normal early development. The dependency on maternal *lat-1* gene product coincides with high levels of *lat-1* mRNA in the maternal germline and in all blastomeres during the first cleavage rounds of the zygote.[29]

The examination of embryos lacking maternal and zygotic gene product revealed a defect in the division plane orientation of 8-cell *lat-1* embryos that is distinct from polarity or patterning mutations described in the literature. In normal development, the division plane of ABal, the most anterior blastomere, is oriented in the anterior-posterior direction typical for most embryonic cell divisions. The mitotic spindle is skewed towards the putative a-p axis of the embryo, allowing only the posterior daughter ABalp to contact the posterior neighbour MS, while ABala assumes the most anterior position within the egg shell and does not touch MS (Fig. 4a,f). In *lat-1* mutants, the ABal axis is positioned perpendicular to the embryonic a-p axis, suggesting that anterior-posterior tissue polarity is defective (Fig. 4b,g).[29]

Although cell fate changes can be detected in several embryonic sublineages, the division of ABal appears to remain asymmetric, suggesting that *lat-1* is required for a-p cell polarity, but not for cell fate determination in ABal.

### *lat-1* is Required for Tissue Polarity in Anterior Blastomeres

A posterior signalling centre formed by descendants of the P1 blastomere polarizes the a-p axis of the *C. elegans* embryo.[43,44] Blastomere recombination experiments have suggested that Wnt-dependent signalling activity of P2 can provide an instructive cue to orient the EMS spindle[45] and that polarizing signals can be propagated to align cleavage planes in a larger group of cells by a Wnt-dependent relay mechanism.[51]

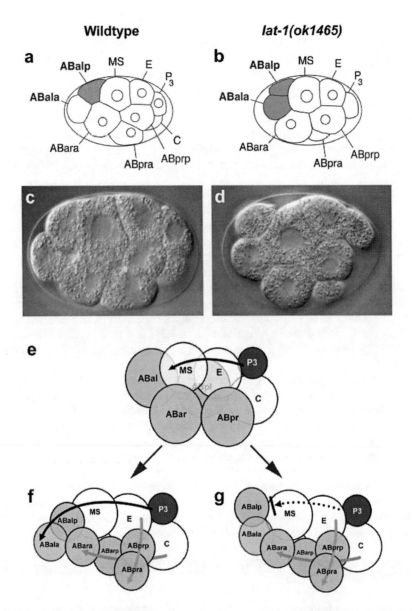

**Figure 4.** Division plane defect of *lat-1* mutants. a,c, during the transition from 8→12-cell stage the ABal blastomere divides in a plane (c) allowing only the posterior daughter ABalp to contact MS (grey cell), whereas ABala is separated from MS. b,d, in embryos deficient of *lat-1* the ABal division plane is skewed such that both daughters contact MS (grey cells). e, at the 4-8-cell stage a polarizing signal originating from P2/3 aligns the embryo along the the a-p axis (black arrow) f, when the embryo transits to the 12-cell stage ABal is furthest away from P3 and requires a polarizing signal to align the ABal daughters in an a-p direction. This putative signal is propagated via E and MS (black arrow). ABar, ABpr and ABpl are oriented by signals from E and C (grey arrows). g, in *lat-1* mutants, ABal daughters divide perpendicularly to the a-p axis indicating loss of the P3-polarizing signal. All other AB-derived blastomeres are in direct contact with primary or secondary polarizing cells (P3, E, C, MS) and thus still align appropriately. Reprinted from Langenhan et al, Dev Cell 2009; 17(4):494-504,[29] ©2009 with permission from Elsevier.

The current literature suggests that in normal development the division plane orientations of the blastomeres ABpl and ABpr are determined by Wnt/Fz-dependent signalling from E, while a different Wnt/Fz signal emanating from C orients the ABar spindle into its characteristic orientation perpendicular to ABpl/r.[41,49] However, the anterior ABal blastomere is only in contact with MS and AB descendants rather than E or C. It has been shown that the E, C and MS blastomeres acquire the capacity to transduce polarizing signals of different strength and quality, but that only E and C derived signals are equivalent to P2 signals.[53] Until recently no specific molecular requirements for the division plane orientation of ABal had been described, but a mechanism propagating the polarising signal from the primary source P2/3 via E to MS has been assumed.[51]

It could now be shown that *lat-1* is essential to align the mitotic spindles and division planes of the E-MS-ABal cell group to a common a-p axis (Fig. 4).[29] The blastomere ABal occupies the most anterior position in the 8-cell stage and fails to align in embryos lacking maternal and zygotic *lat-1* protein. Consistent with the model that a putative polarizing signal emanating from P2/3 would have to be transmitted through E and MS to reach ABal, the alignment of the MS spindle is also affected in *lat-1* mutants. The timing of spindle rotations suggests that successful alignment of E can "rescue" the alignment defect of MS, but not ABal, which has already undergone mitosis at this time. In contrast, *lat-1* is not required for the division plane orientation of EMS, E, or C, which are in direct contact to P2/3 and thus receive a polarizing signal directly. In turn, E and C retain most or all of their ability to orient ABar and ABpl/pr.[49,52]

While lack of *lat-1* function has little or no effect on the division planes of blastomeres that are in direct contact with the primary or secondary signalling cells P2/3, E and C, the spindle alignment of the next generation of ABalx or ABarp descendants is frequently delayed or failing. These results can be explained by a simple model in which *lat-1* is required to efficiently propagate spindle alignment cues from a posterior source towards the anterior through the growing cellular array.[29] In this model, ABal is a weak spot as its orientation relies on MS which is a "tertiary cell" not in direct contact to the primary source P2[53] and which itself shows delays and errors of a-p orientation in *lat-1* mutants. Later ABa descendants have more diverse cell contacts that could provide compensating signals and underlie stronger spatial constraints, leading to a lower penetrance of the overt spindle alignment phenotype.

## Interaction of Latrophilin Signalling and the Wnt/Frizzled Spindle Orientation Pathway

The genetic analysis of wnt/frizzled-dependent signalling in the early *C. elegans* embryo suggests that multiple parallel Wnt signals transmit the polarizing information.[46,48] *lat-1* might be specifically required to propagate one of these parallel signals, or an as-yet unknown Wnt-independent signal. Alternatively, *lat-1* function might be required for the efficient propagation of all parallel signals, e.g., for an essential response of the cells in the path of the signal(s). Alternative models for *lat-1* function are also plausible, but more complex. *lat-1* might be required for an anterior-to-posterior alignment activity overlapping and opposing the posterior-to-anterior signal, similar to the model recently presented for vulval precursor cell organisation.[54] The predictions made by the alternative models have not been tested in detail yet.

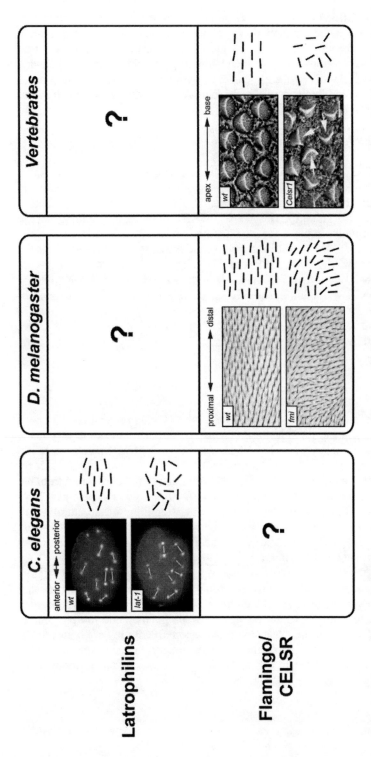

**Figure 5.** Adhesion-GPCRs and the control of tissue polarity. The FMI class of adhesion-GPCRs has been associated with defects in planar cell polarity: mutations in the fmi gene of Drosophila disturb the organwide polarity of wing bristles (middle panel), dendritic trees and ommatidial rotation (reproduced from Usui T, Cell 1999; 98(5):585-595,[15] ©1999 with permission from Elsevier); the vertebrate FMI homolog CELSR1 is involved in the establishment of inner ear sensory epithelium polarity (right panel: reproduced from Curtin et al, Curr Biol 2003; 13(13):1129-1133,[21] ©2003 with permission from Elsevier). Recent evidence shows that LPHN receptors are required for the correct establishment of tissue polarity in the developing C. elegans embryo (left panel: reproduced from Langenhan et al, Dev Cell 2009; 17(4):494-504,[29] ©2009 with permission from Elsevier). It is still unclear whether LPHNs act in similar phenomena in higher organisms and FMI has a polarizing role in C. elegans.

The analysis of differentiation markers and embryonic cell lineages shows that *lat-1* is not required for endoderm induction and does not appear to have a strong direct effect on cell fate in asymmetric cell divisions.[29] This indicates that *lat-1* is not an essential component of the transcription-dependent Wnt/β-catenin asymmetry pathway. In *lat-1* mutants, the ABal division still generates asymmetric cell fates in most cases and the normal ABala cell fate is surprisingly robust against altered cell position and ectopic cell contact to MS.

### The Molecular Mechanism of Latrophilin Signalling

Adhesion-GPCRs are heterodimers composed of an extracellular "adhesion" subunit and a GPCR-like domain with seven transmembrane helices. The heterodimers are derived from monomeric precursor proteins by cleavage at the GPS domain[24,55] (see chapter by Lin, this volume). The lectin-like RBL domain, the defining feature of LPHNs[56] is absolutely required for all functions of *lat-1*.[29] In contrast to results recently described for FMI,[57] constructs lacking the RBL domain but retaining the hormone-binding domain (HRM), GPS and 7TM domains have not shown partial activity. This is consistent with an essential role of the RBL domain in ligand binding and implies that the 7TM domain transduces an "outside-in" signal that is dependent on an extracellular interaction. Recent biochemical data argue strongly against a carbohydrate ligand for the lectin-like RBL domain and do not support homodimer formation mediated by the RBL domain.[29,56]

## CONCLUSION

The control of mitotic spindle orientation in the *C. elegans* embryo has been investigated intensively and the roles of PAR proteins and heterotrimeric G proteins in establishing zygotic polarity[35] and of Wnt/Fz and SRC-1/MES-1 pathways in P2/EMS signalling at the four-cell stage[41] have been identified. However, it is still poorly understood how spindle orientation and cell fate asymmetry are coordinated from the 8-cell stage onwards and clear equivalents of PCP or a-p tissue polarity pathways have not yet been defined in *C. elegans* embryogenesis.[49,52] Latrophilins are structurally very similar to FMI proteins, a related subfamily of highly conserved adhesion-GPCRs that are essential for PCP signalling in *Drosophila* and a-p tissue polarity in vertebrates[3,18] (see chapter by Formstone, this volume).

Unexpectedly, the study of *C. elegans* embryogenesis has revealed that the putative neurotoxin receptor *lat-1* defines a mechanism required for the alignment of cell division planes.[29] Similar to the role of FMI in PCP, this implicates an evolutionary conserved subfamily of adhesion-GPCRs in the control of tissue polarity and morphogenesis (Fig. 5). It also suggests that the expansion of adhesion-GPCRs in vertebrates might contribute to the larger variety of organ and tissue architectures in these species.[14,21,23,58] Further studies will be required to define the up- and downstream components of adhesion-GPCR signalling.

# REFERENCES

1. Gönczy P. Mechanisms of asymmetric cell division: flies and worms pave the way. Nat Rev Mol Cell Biol 2008; 9(5):355-366.
2. Siller KH, Doe CQ. Spindle orientation during asymmetric cell division. Nat Cell Biol 2009; 11(4):365-374.
3. Zallen JA. Planar polarity and tissue morphogenesis. Cell 2007; 129(6):1051-1063.
4. Townes, Holtfreter J. Directed movements and selective adhesion of embryonic amphibian cells J Exp Zool 1955; 128(1):53-120.
5. Tepass U, Godt D, Winklbauer R. Cell sorting in animal development: signalling and adhesive mechanisms in the formation of tissue boundaries. Curr Opin Genet Dev 2002; 12(5):572-582.
6. Steinberg MS. Differential adhesion in morphogenesis: a modern view. Curr Opin Genet Dev 2007; 17(4):281-286.
7. Halbleib JM, Nelson WJ. Cadherins in development: cell adhesion, sorting and tissue morphogenesis. Genes Dev 2006; 20(23):3199-3214.
8. Hinck L. The versatile roles of "axon guidance" cues in tissue morphogenesis. Dev Cell 2004; 7(6):783-793.
9. Harmar AJ. Family-B G-protein-coupled receptors. Genome Biol 2001; 2(12):REVIEWS3013.
10. Bjarnadóttir, Fredriksson R, Schiöth. The adhesion-GPCRs: A unique family of G protein-coupled receptors with important roles in both central and peripheral tissues. Cell Mol Life Sci 2007.
11. Nordström K, Lagerström M, Wallér L et al. The Secretin GPCRs descended from the family of adhesion-GPCRs. Molecular Biology and Evolution 2008.
12. Veninga H, Becker S, Hoek RM et al. Analysis of CD97 expression and manipulation: antibody treatment but not gene targeting curtails granulocyte migration. J Immunol 2008; 181(9):6574-6583.
13. Kwakkenbos MJ, Kop EN, Matmati M et al. The EGF-TM7 family: a postgenomic view. Immunogenetics 2004; 55(10):655-666.
14. Piao X, Hill RS, Bodell A et al. G protein-coupled receptor-dependent development of human frontal cortex. Science 2004; 303(5666):2033-2036.
15. Usui T, Shima Y, Shimada Y et al. Flamingo, a seven-pass transmembrane cadherin, regulates planar cell polarity under the control of Frizzled. Cell 1999; 98(5):585-595.
16. Chae J, Kim MJ, Goo JH et al. The Drosophila tissue polarity gene starry night encodes a member of the protocadherin family. Development 1999; 126(23):5421-5429.
17. Formstone CJ, Mason I. Combinatorial activity of Flamingo proteins directs convergence and extension within the early zebrafish embryo via the planar cell polarity pathway. Dev Biol 2005; 282(2):320-335.
18. Lawrence PA, Struhl G, Casal J. Planar cell polarity: one or two pathways? Nat Rev Genet 2007; 8(7):555-563.
19. Strutt D. The planar polarity pathway. Curr Biol 2008; 18(19):R898-902.
20. Shima Y, Kengaku M, Hirano T et al. Regulation of dendritic maintenance and growth by a mammalian 7-pass transmembrane cadherin. Dev Cell 2004; 7(2):205-216.
21. Curtin J, Quint E, Tsipouri V et al. Mutation of Celsr1 disrupts planar polarity of inner ear hair cells and causes severe neural tube defects in the mouse. Curr Biol 2003; 13(13):1129-1133.
22. Shima Y, Kawaguchi Sy, Kosaka K et al. Opposing roles in neurite growth control by two seven-pass transmembrane cadherins. Nat Neurosci 2007; 10(8):963-969.
23. Tissir F, Bar I, Jossin Y, De Backer O et al. Protocadherin Celsr3 is crucial in axonal tract development. Nat. Neurosci 2005; 8(4):451-457.
24. Krasnoperov VG, Bittner MA, Beavis R et al. Alpha-Latrotoxin stimulates exocytosis by the interaction with a neuronal G-protein-coupled receptor. Neuron 1997; 18(6):925-937.
25. Krasnoperov VG, Beavis R, Chepurny OG et al. The calcium-independent receptor of alpha-latrotoxin is not a neurexin. Biochem Biophys Res Commun 1996; 227(3):868-875.
26. Südhof TC. Alpha-Latrotoxin and its receptors: neurexins and CIRL/latrophilins. Annu Rev Neurosci 2001; 24:933-962.
27. Capogna M, Volynski KE, Emptage NJ et al. The alpha-latrotoxin mutant LTXN4C enhances spontaneous and evoked transmitter release in CA3 pyramidal neurons. J Neurosci 2003; 23(10):4044-4053.
28. Willson J, Amliwala K, Davis A et al. Latrotoxin receptor signaling engages the UNC-13-dependent vesicle-priming pathway in C. elegans. Curr Biol 2004; 14(15):1374-1379.
29. Langenhan T, Prömel S, Mestek L et al. Latrophilin signalling links anterior-posterior tissue polarity and oriented cell divisions in the C. elegans embryo. Dev Cell 2009; 17:494-504.
30. Mee CJ, Tomlinson SR, Perestenko PV et al. Latrophilin is required for toxicity of black widow spider venom in Caenorhabditis elegans. Biochem J 2004; 378(Pt 1):185-191.
31. Hutter H, Vogel BE, Plenefisch JD et al. Conservation and novelty in the evolution of cell adhesion and extracellular matrix genes. Science 2000; 287(5455):989-994.
32. Pettitt J. The cadherin superfamily. WormBook : the online review of C elegans biology 2005:1-9.

33. Lin YJ, Seroude L, Benzer S. Extended life-span and stress resistance in the Drosophila mutant methuselah. Science 1998; 282(5390):943-946.
34. Sulston JE, Schierenberg E, White JG et al. The embryonic cell lineage of the nematode Caenorhabditis elegans. Dev Biol 1983; 100(1):64-119.
35. Gönczy P, Rose LS. Asymmetric cell division and axis formation in the embryo. WormBook : the online review of C elegans biology 2005:1-20.
36. Priess JR. Notch signaling in the C. elegans embryo. WormBook : the online review of C elegans biology 2005:1-16.
37. Schnabel R. Why does a nematode have an invariant cell lineage? Seminars in Cell and Developmental Biology 1997; 8(4):341-349.
38. Kaletta T, Schnabel H, Schnabel R. Binary specification of the embryonic lineage in Caenorhabditis elegans. Nature 1997; 390(6657):294-298.
39. Lin R, Hill RJ, Priess JR. POP-1 and anterior-posterior fate decisions in C. elegans embryos. Cell 1998; 92(2):229-239.
40. Mizumoto K, Sawa H. Two betas or not two betas: regulation of asymmetric division by beta-catenin. Trends Cell Biol 2007; 17(10):465-473.
41. Walston TD, Hardin J. Wnt-dependent spindle polarization in the early C. elegans embryo. Seminars in Cell and Developmental Biology 2006; 17(2):204-213.
42. Cowan CR, Hyman AA. Asymmetric cell division in C. elegans: cortical polarity and spindle positioning. Annu Rev Cell Dev Biol 2004; 20:427-453.
43. Hutter H, Schnabel R. Specification of anterior-posterior differences within the AB lineage in the C. elegans embryo: a polarising induction. Development 1995; 121(5):1559-1568.
44. Goldstein B. Cell contacts orient some cell division axes in the Caenorhabditis elegans embryo. J Cell Biol 1995; 129(4):1071-1080.
45. Goldstein B, Takeshita H, Mizumoto K et al. Wnt signals can function as positional cues in establishing cell polarity. Dev Cell 2006; 10(3):391-396.
46. Rocheleau CE, Downs WD, Lin R et al. Wnt signaling and an APC-related gene specify endoderm in early C. elegans embryos. Cell 1997; 90(4):707-716.
47. Schlesinger A, Shelton CA, Maloof JN et al. Wnt pathway components orient a mitotic spindle in the early Caenorhabditis elegans embryo without requiring gene transcription in the responding cell. Genes Dev 1999; 13(15):2028-2038.
48. Thorpe CJ, Schlesinger A, Carter JC et al. Wnt signaling polarizes an early C. elegans blastomere to distinguish endoderm from mesoderm. Cell 1997; 90(4):695-705.
49. Walston T, Tuskey C, Edgar L et al. Multiple Wnt signaling pathways converge to orient the mitotic spindle in early C. elegans embryos. Dev Cell 2004; 7(6):831-841.
50. Bei Y, Hogan J, Berkowitz LA et al. SRC-1 and Wnt signaling act together to specify endoderm and to control cleavage orientation in early C. elegans embryos. Dev Cell 2002; 3(1):113-125.
51. Bischoff M, Schnabel R. A posterior centre establishes and maintains polarity of the Caenorhabditis elegans embryo by a Wnt-dependent relay mechanism. PLoS Biol 2006; 4(12):e396.
52. Park FD, Tenlen JR, Priess JR. C. elegans MOM-5/frizzled functions in MOM-2/Wnt-independent cell polarity and is localized asymmetrically prior to cell division. Curr Biol 2004; 14(24):2252-2258.
53. Park FD, Priess JR. Establishment of POP-1 asymmetry in early C. elegans embryos. Development 2003; 130(15):3547-3556.
54. Green JL, Inoue T, Sternberg PW. Opposing Wnt pathways orient cell polarity during organogenesis. Cell 2008; 134(4):646-656.
55. Lin H-H, Chang G-W, Davies JQ et al. Autocatalytic cleavage of the EMR2 receptor occurs at a conserved G protein-coupled receptor proteolytic site motif. J Biol Chem 2004; 279(30):31823-31832.
56. Vakonakis I, Langenhan T, Prömel S et al. Solution structure and sugar-binding mechanism of mouse latrophilin-1 RBL: a 7TM receptor-attached lectin-like domain. Structure 2008; 16(6):944-953.
57. Steinel MC, Whitington PM. The atypical cadherin Flamingo is required for sensory axon advance beyond intermediate target cells. Dev Biol 2009; 327(2):447-457.
58. Davies B, Baumann C, Kirchhoff C et al. Targeted deletion of the epididymal receptor HE6 results in fluid dysregulation and male infertility. Mol Cell Biol 2004; 24(19):8642-8648.

# CHAPTER 4

# GPS PROTEOLYTIC CLEAVAGE OF ADHESION-GPCRs

Hsi-Hsien Lin,* Martin Stacey, Simon Yona and Gin-Wen Chang

**Abstract:** The stability and functional diversity of proteins can be greatly modulated by posttranslational modification. Proteolytic cleavage at the GPCR proteolysis site (GPS) has been identified as an intrinsic protein modification process of many adhesion-GPCRs. In recent years, the conserved cleavage site, molecular mechanism and the potential functional implication of the GPS proteolysis have been gradually unveiled. However, many aspects of this unique cleavage reaction including its regulation, the relationship between the cleaved fragments and the functional pathways mediated by the cleaved receptor subunits, remain unanswered. Further investigation of the GPS proteolytic modification shall shed light on the biology of the adhesion-GPCRs.

## INTRODUCTION

Site-specific proteolytic cleavage is an important posttranslational modification for the maturation, trafficking, activation and function of many soluble and cell surface proteins. One of the most characteristic protein modifications in adhesion-GPCRs is proteolytic cleavage at the GPCR proteolysis site (GPS).[1-3] GPS proteolysis occurs within a highly conserved Cys-rich motif located at the membrane-proximal region, dissecting the receptor to produce an extracellular α-subunit and a 7TM β-subunit.[1,2] Intriguingly, both subunits somehow stay closely on the cell surface, likely by noncovalent association or other unspecified means. As almost all adhesion-GPCRs possess the consensus GPS sequence, it is thought the majority of adhesion-GPCRs are expressed on the membrane as a two-subunit complex.[4,5]

*Corresponding Author: Hsi-Hsien Lin—Department of Microbiology and Immunology, College of Medicine, Chang Gung University, 259 Wen-Hwa 1st Road, Kwei-San, Tao-Yuan, Taiwan. Email: hhlin@mail.cgu.edu.tw

*Adhesion-GPCRs: Structure to Function*, edited by Simon Yona and Martin Stacey.
©2010 Landes Bioscience and Springer Science+Business Media.

Unlike the majority of protease-mediated cleavage events, GPS proteolysis occurs through an intramolecular self-catalytic reaction.[6] Similar to other auto-proteolytic molecules such as Ntn hydrolases and hedgehog (Hh) proteins, hydrolysis of the peptide bond at the consensus GPS cleavage site is likely mediated by an N→O or N→S acyl shift brought upon by nucleophilic attack on critical residues at or around the cleavage site.[7-9] This unique proteolytic modification takes place in the ER during early protein biosynthesis and is probably an essential step in the production of mature receptor proteins.[6] Indeed, some reports have shown the GPS proteolysis might be a prerequisite for efficient receptor trafficking.[10] Furthermore, point mutations that disrupted GPS proteolysis of receptors have been linked to human genetic disorders.[11] Functional studies in recent years have further confirmed the significant role of GPS proteolysis in receptor function[12,13]

Although closely associated with adhesion-GPCRs, the GPS domain is also found in other non-GPCR proteins, such as the sea urchin sperm receptor for egg jelly-1 (suREJ1), suREJ3 and polycystin-1.[12,14,15] This suggests that the GPS domain and its associated proteolytic modification are widely used by receptor molecules. In this chapter, we will summarize the current view on the GPS proteolytic cleavage and its role in receptor function, and human diseases.

## THE IDENTIFICATION OF GPS PROTEOLYTIC MODIFICATION IN ADHESION-GPCRs

The earliest data of a unique proteolytic modification in an adhesion-GPCR was reported by Kelly and her colleagues in 1996.[16] They demonstrated that CD97, a leukocyte activation marker, was a novel two-subunit GPCR-like molecule composed of an extracellular protein fragment and a 7TM moiety. The two noncovalently associated subunits were derived from a proprotein precursor that was processed intracellularly, most probably in the ER or early Golgi. Soon after, many more adhesion-GPCRs were found to be similarly processed into the two-subunit structure.

In 1999, Petrenko et al coined the term GPS to describe the proteolytic processing of CIRL/latrophilin, of which the exact cleavage site was determined by the same authors earlier.[3,17] The consensus GPS cleavage site was found to locate within a Cys-rich region immediately N-terminal to the first TM segment. This Cys-rich region was later called the GPS domain. Almost all adhesion-GPCRs, except GPR123, contain the conserved GPS domain suggesting that GPS cleavage is prevalent, if not ubiquitous, among adhesion-GPCRs.[5] Indeed, most but not all adhesion-GPCRs examined to date were proteolytically processed[6,14,16,18-28](Fig. 1).

The GPS domain is an extracellular segment of ~50-60 amino acids located at a distance of ~15-30 residues from the first TM domain (Fig. 1). It is defined by a consensus tripeptide cleavage motif, 4 constrained Cys and 2 invariable Try residues.[2] In addition, the 6-8 amino acids immediately C-terminal to the cleavage site are usually small and hydrophobic. The cleavage tripeptide almost always starts with His (the $P^{-2}$ residue), followed by Leu (or Ile) (the $P^{-1}$ residue) and Ser/Thr (the $P^{+1}$ residue). Thus, the GPS cleavage will produce an extracelluar subunit with a new C-terminus (Leu or Ile) and a 7TM subunit with Ser/Thr at its N-terminal end. The 4 Cys residues in the GPS domain are believed to form two disulfide linkages, while other conserved

# GPS PROTEOLYSIS

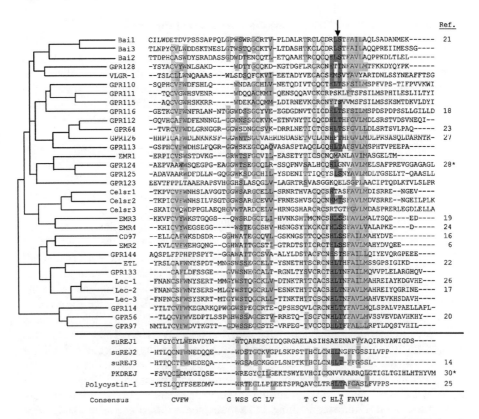

**Figure 1.** The alignment of the GPS domain sequences. The amino acid sequences of the GPS domain of all human adhesion-GPCRs and the GPS domain-containing proteins are aligned using the Vector NTI program (version 10.3, InVitrogen). The sequence of GPR123 was derived from a fragment at the similar location as of other GPS domains. The conserved cysteines are indicated in light yellow background. The conserved cleavage regions are showed in red background. The conserved and similar residues are showed in blue and green background, respectively. The arrow indicates the cleavage site. References that reported the GPS proteolysis of receptors are listed, while the one marked by asterisk reported the lack of GPS cleavage. A color version of this image is avaliable at www.landesbioscience.com/curie.

residues are likely involved in intramolecular bonding. Taken together, the GPS domain probably adopts a very similar conformation that is important for the proceeding of the GPS proteolytic reaction.

Interestingly, the GPS domain was found not only in adhesion-GPCRs but also in other receptor molecules. Sequence comparison has identified the presence of GPS domain in several cell surface proteins with either one- or eleven-pass TM topology. These include suREJ1,[15] suREJ3[14] and hPKDREJ[15,29] and polycystin-1,[12] the mammalian homologs of suREJ3 (Fig. 1). However, it was noted that the GPS domain of these receptors is less conserved than that of adhesion-GPCRs, with the absence of some conserved residues. Furthermore, some of these non adhesion-GPCR molecules are not cleaved.[28,30] What they all have in common is that the GPS domain is always embedded at the C-terminal end of a long extracellular stalk rich in Ser, Thr and Pro residues.

## THE MOLECULAR MECHANISM OF GPS PROTEOLYSIS

Proteolytic modification of cell surface receptor molecules is a well-known phenomenon and is usually mediated by proteases that target and cleave specific amino acid sequences. At first, the highly conserved GPS sequence seems to suggest the involvement of a unique but uncharacterized endopeptidase. However, our own study on the processing of EMR2 has yielded a very surprising conclusion, which indicated the GPS cleavage is the result of receptor auto-proteolysis.[6]

By changing the $P^{+1}$ residue (Ser) of EMR2 to 19 different amino acids, we first confirmed only three residues including Ser, Thr and Cys can be recognized as the active cleavage site. The highly specific cleavage residue requirement and conserved His at the $P^{-2}$ site is reminiscent of a group of auto-proteolytic proteins such as the *hedgehog* morphogens[31,32] and Ntn-hydrolases.[33-36] These molecules shared a unique auto-proteolytic reaction that cleaves an internal peptide bond via the deprotonation of a nucleophilic residue at the cleavage site. Earlier studies on Ntn hydrolases have shown that point mutations at the $P^{-2}$ residue can greatly reduced the auto-proteolytic reaction.[37] Indeed, when similar $P^{-2}$ site point mutants were made, EMR2 was found mostly as an uncleaved single-chain polypeptide.[6] This has allowed detailed examination of the cleavage reaction using purified protein in a defined condition. The EMR2 His mutants were found to undergo very slow proteolysis, but the cleavage is greatly enhanced upon the addition of a strong nucleophile such as hydroxylamine (HA). The HA-promoted cleavage reaction occurs even at 0°C and can not be inhibited by any protease inhibitor tested. Following more biochemical analyses, it was finally concluded that the GPS cleavage is most likely mediated by an auto-proteolytic reaction, analogous to that of Ntn-hydrolases.[6] Importantly, a similar finding has recently also been reported for polycystin-1, a GPS motif-containing 11TM receptor responsible for the autosomal dominant polycystic kidney disease (ADPKD).[25]

Based upon the strong similarity in the proteolytic reaction of these self-cleaving molecules, it was believed that the GPS cleavage of EMR2 (and probably other adhesion-GPCRs) is initiated by the deprotonation of the hydroxyl group of the $P^{+1}$ residue (Ser) by the $P^{-2}$ His residue. A subsequent *cis*-nucleophilic attack on the α-carbonyl carbon of the $P^{-1}$ Leu residue then leads to the formation of a tetrahedral intermediate. This is followed immediately by the generation of an ester intermediate via an N→O acyl shift. A final attack by water molecules hydrolyzes the ester bond, leading to the production of two separate protein subunits (Fig. 2).[6]

The molecular mechanism of GPS cleavage described above suggested the reaction occurs during or soon after protein synthesis. Therefore, it is of no surprise the reaction was found to take place in the lumen of ER, where the folding of cell surface receptors is initiated. Pulse chase experiments on several different adhesion-GPCRs indicated that the cleavage occurs within 10-15 minutes of chase.[10,16,18] Deglycosylation experiments further showed that the cleavage proceeds in parallel with early glycosylation in the ER, ahead of the addition of complex carbohydrate chains in the Golgi.[10,16] More evidence was provided by the use of inhibitor such as BFA that prevents the ER-to-Golgi vesicular trafficking and recombinant receptors with an ER-retention signal.[6,18,25] Overall, the current consensus is that GPS auto-proteolysis occurs soon after the receptor molecule is translocated into the ER lumen, probably when the protein is still being properly folded.

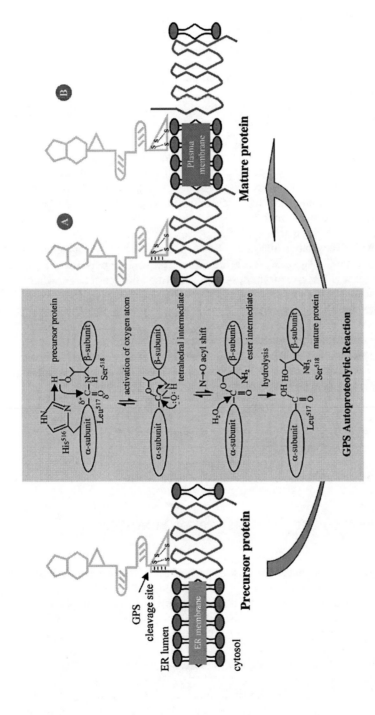

**Figure 2.** The autoproteolytic cleavage of adhesion-GPCRs at the GPS domain. The newly synthesized adhesion-GPCRs is translocated into the ER lumen where a correctly folded conformation is achieved allowing the autoproteolytic reaction to proceed and produce a mature product. The extracellular protein modules are represented by pentagonal, hexagonal and triangular shapes, the stalk region by a pair of half-ovals, the GPS motif by a triangle with two disulfide bonds and the TM regions by zig-zag lines. The hypothesized 'cleavage' conformation introduces a tight bend at the GPS cleavage site, where autoproteolysis occurs. The autoproteolytic reaction is shown inside a box using EMR2 as an example. For simplicity, only the cleavage tripeptide is shown. The arrow represents electron transfer and nucleophilic attack. His$_{516}$ is the proton donor/acceptor for the generation of the tetrahedral intermediate. The formation of ester intermediate is derived from the cleavage of C-N bond through the protonation of the amino group of Ser$_{518}$ (N→O acyl shift). The final hydrolysis of the ester bond produces the extracellular α-subunit and the TM7 β-subunit that lead to either the formation of a noncovalently associated cell surface heterodimer (A) or two independent membrane protein entities (B).

## THE REGULATION OF GPS PROTEOLYSIS

Although the GPS cleavage is self-catalytic, it is thought that the reaction might not be 100% efficient. If this is the case, it would be reasonable to expect the coexistence of both cleaved and uncleaved molecules of the same receptor. Indeed, recent reports have shown the presence of processed as well as unprocessed full-length receptors including GPR56 and polycystin-1 in vivo.[12,13,38] These data imply strongly that the GPS cleavage reaction is somehow regulated.

Little is currently known about the regulation of GPS cleavage; however earlier results have suggested that GPS cleavage is probably highly dependent on protein conformation. The proteolysis of EMR2 was impaired if any part of the mucin-like stalk N-terminal to the GPS domain is deleted.[39] Likewise, it was reported that the receptor for egg jelly (REJ) domain adjacent to the GPS sequence is required for the cleavage of polycystin-1.[12] It was thought that the GPS domain is necessary but not sufficient for the proteolytic reaction. The mucin-like domain (and REJ domain) next to the GPS region probably provides an important structural configuration that promotes the necessary nucleophilic attack to initiate the auto-proteolytic reaction.

By using various N-glycosylation inhibitors, Wei et al recently showed that the initial N-glycan attachment in the ER plays a key role in the efficient cleavage of polycystin-1.[25] It seems that the first step of N-glycosylation but not the subsequent modification of carbohydrate chains is critically essential for the GPS cleavage of polycystin-1. Likewise, our latest results also indicated that site-specific N-glycosylation can regulate or determine the efficiency of GPS cleavage of CD97.[19] From these results, it was suggested that the receptor molecules might adapt two different folding pathways; one promotes the GPS cleavage and the other hinders the proteolytic reaction. It is likely that the initial attachment of N-glycans is critical for some receptors to proceed through the "GPS proteolysis" folding pathway. Whether there are other factors regulating GPS proteolysis remains to be investigated.

The differential expression levels and tissue distribution patterns of the unprocessed and GPS-cleaved receptors detected in vivo is likely the result of the regulation of GPS cleavage. The fact that both forms of receptor coexist suggests either the two receptor forms potentially have distinct functions or the normal receptor function can be modulated by the different expression levels of cleaved and uncleaved receptors. Some of the latest data have started to answer these questions and are discussed below.

## THE ROLE OF GPS PROTEOLYSIS IN RECEPTOR FUNCTION AND HUMAN DISEASE

As GPS cleavage takes place in the ER at a very early phase of receptor biosynthesis, it has long been thought to play a role in receptor maturation, trafficking and function. By the generation of uncleavable GPS point mutants, Krasnoperov et al have suggested that GPS cleavage is essential for the transport of CIRL (latrophilin) to the plasma membrane.[10] However, other studies using similar GPS point mutants of adhesion-GPCRs have indicated that GPS cleavage per se might not have direct effects on the cell surface expression of receptors.[6,12,39] The trafficking failure observed for certain GPS cleavage mutants is most likely due to protein misfolding.

For example, several ADPKD-associated point mutations located in the REJ domain disrupt the GPS cleavage, but not the cell surface expression of polycystin-1.[12] On the other hand, among a dozen or so GPR56 point mutations responsible for the development of bilateral frontoparietal polymicrogyria (BFPP), two (C346S and W349S) render the receptor uncleavable and seriously impair its exit from the ER.[20] Therefore, it seems that the ability of the receptor to undergo GPS cleavage does not necessarily reflect its capacity for normal trafficking to the cell surface.

The best evidence for a functional role of GPS cleavage in receptor activity can be found in human disorders, noticeably ADPKD. As mentioned above, some disease-associated point mutations were found to completely inhibit GPS cleavage but not surface expression of polycystin-1.[12] When tested in an in vitro tubulogenesis model, these mutant receptors showed impaired tubulogenic functions that are normally exhibited by the wild-type receptor. These results indicated that proper cleavage at the GPS domain is essential for the biological functions of the receptors. Indeed, the role of GPS cleavage in vivo was further confirmed recently in an elegant knockin mouse model where the wild type *Pkd1* gene was replaced by an allele expressing uncleavable polycystin-1.[13] The mutant knockin mice displayed different phenotypes form those of the *Pkd1* null knockout animals, which died before birth. It was concluded that both uncleaved full-length and GPS-cleaved polycystin-1 receptors possess distinct functions; the full-length polycystin-1 is critical for embryonic development, while the cleaved receptor is essential for the postnatal development of distal nephron segment of mouse kidney as well as the common bile duct and biliary tracts.[13]

In addition, our recent unpublished data also points to an essential role of GPS cleavage in the cell migration function mediated by EMR2 and CD97 receptors. Altogether, GPS auto-proteolysis not only is an inherent posttranslational modification of adhesion-GPCRs, but also plays an important role in receptor functions.

## THE FATE AND FUNCTIONAL INTERACTION OF THE EXTRACELLULAR AND 7TM SUBUNITS FOLLOWING GPS PROTEOLYSIS

It has been well-accepted that after GPS cleavage, the N-terminal α-subunit is membrane-bound due to the noncovalent association with the 7TM β-subunit. Several lines of evidence strongly supported this conclusion. Firstly, immunoprecipitation experiments using tissue samples or transfected cells labeled with radioisotopes or surface biotinylation showed that the α-subunit can be readily pulled down by antibodies specific to the β-subunit.[16-18] Secondly, soluble chimeric proteins containing the α-subunit fused to a tag such as the immunoglobulin Fc region was found to be cleaved normally and the entire chimeric protein could be affinity-purified by Protein A column chromatography.[6,39] It was therefore believed that the α-subunit is tightly but noncovalently bound to the cleaved β-subunit (Fig. 2).

Nevertheless, recent studies on the proteolytic processing of latrophilin by Volynski et al have challenged this belief.[40] A low but specific binding of isotope-labeled α-latrotoxin to the surface of cells transfected with a soluble latrophilin molecule was observed, suggesting that the cleaved N-terminal latrophilin subunit can somehow anchor to the plasma membrane by itself. Furthermore, the surface bound N-terminal subunit can be efficiently dissociated from the plasma membrane by 0.1-0.2% perfluorooctanoic acid (PFO), a nondenaturing surfactant, with no apparent effect on the C-terminal receptor subunit. Confocal immunofluorescence examination showed incomplete cell surface colocalization of the N- and C-terminal latrophilin subunits. Moreover, both subunits formed distinct Ab crosslinked-patches and displayed different lateral diffusion rates. The N- and C-terminal subunits can be internalized independently and the different subcellular distribution patterns of the two subunits seem to further suggest that they traveled separately to, or from, the cell surface. Finally, the binding of α-latrotoxin to the α-subunit induced its re-association with the 7TM subunit, which in turn transmitted intracellular signals. In conclusion, it was suggested that the cleaved latrophilin α-subunit is an independent molecular entity that anchors itself into the plasma membrane, probably via a short segment of hydrophobic residues (Fig. 2).[40]

A recent follow-up study from the same laboratory demonstrated that it is even possible to detect the cross-interaction of independent α- and β-subunit derived from different adhesion-GPCRs.[41] Recently, we have obtained evidence confirming that the α- and β-subunit of EMR2 are not always bound together, but instead are separately located on the membrane. These results have pointed to possible multiple outcomes with regards to the functional interaction of adhesion-GPCR subunits.

Several functional consequences can be proposed. The independent α-subunit might interact with its binding partner and signal or internalize on its own, while the cleaved but "unoccupied" 7TM subunit might behave as a classical GPCR and bind to its own specific ligand. In this model, the GPS cleavage dissects one receptor into two membrane molecules, which react with two different ligands and signal via distinct pathways. Alternatively, the α-subunit might re-associate with the 7TM receptor subunit following the binding of its specific ligand(s) and transmit intracellular signal(s) via the 7TM receptor subunit as in the case of latrotoxin and latrophilin.[40] Thus, a tethered ligand-receptor pair is produced by the GPS autoproteolytic modification of a single polypeptide chain. Finally, a much more complex scenario can emerge when the cross-interaction of the α- and β-subunit from different adhesion-GPCRs is considered (Fig. 3).

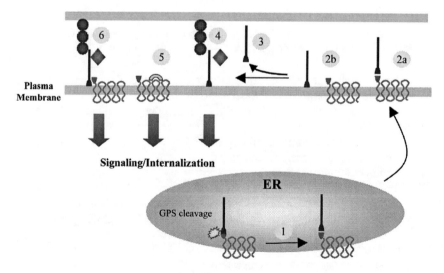

**Figure 3.** Potential functional outcomes of the cleaved subunits of adhesion-GPCRs. Several functional consequences can be proposed for the GPS proteolytic modification. After the GPS cleavage reaction in the ER (1), the final mature receptor might be a noncovalent heterodimer (2a) or two independent proteins (2b). The membrane-bound N-terminal extracellular domain subunit might be released from the cell surface under certain conditions (3) or interacts with its specific ligand(s) and transmits signals on its own (4). On the other hand, the cleaved but "unoccupied" C-terminal TM7 subunit might behave as a classical class B GPCR with a hormone-like ligand (5). Alternatively, the TM7 receptor subunit might re-associate with the N-terminal subunit following biding of specific ligand(s) to the N-terminal subunit and transmit intracellular signal(s) as in the case of latrotoxin and latrophilin (6).

## CONCLUSION

Ever since the proteolytic modification of CD97 was identified over a decade ago, many basic aspects of GPS cleavage have been delineated. However, many more characteristics of this novel cleavage reaction still remain unanswered. In the next decade, we hope to understand the detailed structural information of the GPS domain and the auto-proteolytic process, how the extracellular subunit stays on the membrane, the functional interaction between the α- and β-subunit and the signaling pathways resulting from their interaction. In this way, we should be able to unravel the mystery of the GPS proteolysis in adhesion-GPCRs.

## ACKNOWLEDGEMENTS

The authors would like to thank Cheng-Chih Hsiao for Figure 1 and the grant support from Chang Gung Memorial Hospital (CMRPD170012 and CMRPD160383) and National Science Council, Taiwan (NSC97-2628-B-182-030 and NSC98-2320-B-182-028).

## REFERENCES

1. Stacey M, Lin HH, Gordon S et al. LNB-TM7, a group of seven-transmembrane proteins related to family-B G-protein-coupled receptors. Trends Biochem Sci 2000; 25(6):284-289.
2. Yona S, Lin HH, Siu WO et al. Adhesion-GPCRs: emerging roles for novel receptors. Trends Biochem Sci 2008; 33(10):491-500.
3. Krasnoperov V, Bittner MA, Holz RW et al. Structural requirements for alpha-latrotoxin binding and alpha-latrotoxin-stimulated secretion. A study with calcium-independent receptor of alpha-latrotoxin (CIRL) deletion mutants. J Biol Chem 1999; 274(6):3590-3596.
4. Bjarnadottir TK, Fredriksson R, Hoglund PJ et al. The human and mouse repertoire of the adhesion family of G-protein-coupled receptors. Genomics 2004; 84(1):23-33.
5. Bjarnadottir TK, Fredriksson R, Schioth HB. The adhesion GPCRs: a unique family of G protein-coupled receptors with important roles in both central and peripheral tissues. Cell Mol Life Sci 2007; 64(16):2104-2119.
6. Lin HH, Chang GW, Davies JQ et al. Autocatalytic cleavage of the EMR2 receptor occurs at a conserved G protein-coupled receptor proteolytic site motif. J Biol Chem 2004; 279(30):31823-31832.
7. Brannigan JA, Dodson G, Duggleby HJ et al. A protein catalytic framework with an N-terminal nucleophile is capable of self-activation. Nature 1995; 378(6555):416-419.
8. Perler FB. Breaking up is easy with esters. Nat Struct Biol 1998; 5(4):249-252.
9. Perler FB, Xu MQ, Paulus H. Protein splicing and autoproteolysis mechanisms. Curr Opin Chem Biol 1997; 1(3):292-299.
10. Krasnoperov V, Lu Y, Buryanovsky L et al. Post-translational proteolytic processing of the calcium-independent receptor of alpha-latrotoxin (CIRL), a natural chimera of the cell adhesion protein and the G protein-coupled receptor. Role of the G protein-coupled receptor proteolysis site (GPS) motif. J Biol Chem 2002; 277(48):46518-46526.
11. Piao X, Hill RS, Bodell A et al. G protein-coupled receptor-dependent development of human frontal cortex. Science 2004; 303(5666):2033-2036.
12. Qian F, Boletta A, Bhunia AK et al. Cleavage of polycystin-1 requires the receptor for egg jelly domain and is disrupted by human autosomal-dominant polycystic kidney disease 1-associated mutations. Proc Natl Acad Sci USA 2002; 99(26):16981-16986.
13. Yu S, Hackmann K, Gao J et al. Essential role of cleavage of Polycystin-1 at G protein-coupled receptor proteolytic site for kidney tubular structure. Proc Natl Acad Sci USA 2007; 104(47):18688-18693.
14. Mengerink KJ, Moy GW, Vacquier VD. suREJ3, a polycystin-1 protein, is cleaved at the GPS domain and localizes to the acrosomal region of sea urchin sperm. J Biol Chem 2002; 277(2):943-948.

15. Moy GW, Mendoza LM, Schulz JR et al. The sea urchin sperm receptor for egg jelly is a modular protein with extensive homology to the human polycystic kidney disease protein, PKD1. J Cell Biol 1996; 133(4):809-817.
16. Gray JX, Haino M, Roth MJ et al. CD97 is a processed, seven-transmembrane, heterodimeric receptor associated with inflammation. J Immunol 1996; 157(12):5438-5447.
17. Krasnoperov VG, Bittner MA, Beavis R et al. Alpha-latrotoxin stimulates exocytosis by the interaction with a neuronal G-protein-coupled receptor. Neuron 1997; 18(6):925-937.
18. Abe J, Fukuzawa T, Hirose S. Cleavage of Ig-Hepta at a "SEA" module and at a conserved G protein-coupled receptor proteolytic site. J Biol Chem 2002; 277(26):23391-23398.
19. Hsiao CC, Cheng KF, Chen HY et al. Site-specific N-glycosylation regulates the GPS auto-proteolysis of CD97. FEBS Lett 2009; 583(19):3285-3290.
20. Jin Z, Tietjen I, Bu L et al. Disease-associated mutations affect GPR56 protein trafficking and cell surface expression. Hum Mol Genet 2007; 16(16):1972-1985.
21. Koh JT, Kook H, Kee HJ et al. Extracellular fragment of brain-specific angiogenesis inhibitor 1 suppresses endothelial cell proliferation by blocking alphavbeta5 integrin. Exp Cell Res 2004; 294(1):172-184.
22. Nechiporuk T, Urness LD, Keating MT. ETL, a novel seven-transmembrane receptor that is developmentally regulated in the heart. ETL is a member of the secretin family and belongs to the epidermal growth factor-seven-transmembrane subfamily. J Biol Chem 2001; 276(6):4150-4157.
23. Obermann H, Samalecos A, Osterhoff C et al. HE6, a two-subunit heptahelical receptor associated with apical membranes of efferent and epididymal duct epithelia. Mol Reprod Dev 2003; 64(1):13-26.
24. Stacey M, Chang GW, Sanos SL et al. EMR4, a novel epidermal growth factor (EGF)-TM7 molecule up-regulated in activated mouse macrophages, binds to a putative cellular ligand on B-lymphoma cell line A20. J Biol Chem 2002; 277(32):29283-29293.
25. Wei W, Hackmann K, Xu H et al. Characterization of cis-autoproteolysis of polycystin-1, the product of human polycystic kidney disease 1 gene. J Biol Chem 2007; 282(30):21729-21737.
26. Ichtchenko K, Bittner MA, Krasnoperov V et al. A novel ubiquitously expressed alpha-latrotoxin receptor is a member of the CIRL family of G-protein-coupled receptors. J Biol Chem 1999; 274(9):5491-5498.
27. Moriguchi T, Haraguchi K, Ueda N et al. DREG, a developmentally regulated G protein-coupled receptor containing two conserved proteolytic cleavage sites. Genes Cells 2004; 9(6):549-560.
28. Vallon M, Essler M. Proteolytically processed soluble tumor endothelial marker (TEM) 5 mediates endothelial cell survival during angiogenesis by linking integrin alpha(v)beta3 to glycosaminoglycans. J Biol Chem 2006; 281(45):34179-34188.
29. Hughes J, Ward CJ, Aspinwall R et al. Identification of a human homologue of the sea urchin receptor for egg jelly: a polycystic kidney disease-like protein. Hum Mol Genet 1999; 8(3):543-549.
30. Butscheid Y, Chubanov V, Steger K et al. Polycystic kidney disease and receptor for egg jelly is a plasma membrane protein of mouse sperm head. Mol Reprod Dev 2006; 73(3):350-360.
31. Lee JJ, Ekker SC, von Kessler DP et al. Autoproteolysis in hedgehog protein biogenesis. Science 1994; 266(5190):1528-1537.
32. Porter JA, von Kessler DP, Ekker SC et al. The product of hedgehog autoproteolytic cleavage active in local and long-range signalling. Nature 1995; 374(6520):363-366.
33. Guan C, Cui T, Rao V et al. Activation of glycosylasparaginase. Formation of active N-terminal threonine by intramolecular autoproteolysis. J Biol Chem 1996; 271(3):1732-1737.
34. Oinonen C, Tikkanen R, Rouvinen J et al. Three-dimensional structure of human lysosomal aspartylglucosaminidase. Nat Struct Biol 1995; 2(12):1102-1108.
35. Tikkanen R, Riikonen A, Oinonen C et al. Functional analyses of active site residues of human lysosomal aspartylglucosaminidase: implications for catalytic mechanism and autocatalytic activation. EMBO J 1996; 15(12):2954-2960.
36. Xu Q, Buckley D, Guan C et al. Structural insights into the mechanism of intramolecular proteolysis. Cell 1999; 98(5):651-661.
37. Guan C, Liu Y, Shao Y et al. Characterization and functional analysis of the cis-autoproteolysis active center of glycosylasparaginase. J Biol Chem 1998; 273(16):9695-9702.
38. Iguchi T, Sakata K, Yoshizaki K et al. Orphan G protein-coupled receptor GPR56 regulates neural progenitor cell migration via a G alpha 12/13 and Rho pathway. J Biol Chem 2008; 283(21):14469-14478.
39. Chang GW, Stacey M, Kwakkenbos MJ et al. Proteolytic cleavage of the EMR2 receptor requires both the extracellular stalk and the GPS motif. FEBS Lett 2003; 547(1-3):145-150.
40. Volynski KE, Silva JP, Lelianova VG et al. Latrophilin fragments behave as independent proteins that associate and signal on binding of LTX(N4C). EMBO J 2004; 23(22):4423-4433.
41. Silva JP, Lelianova V, Hopkins C et al. Functional cross-interaction of the fragments produced by the cleavage of distinct adhesion G-protein-coupled receptors. J Biol Chem 2009; 284(10):6495-6506.

# CHAPTER 5

# THE LATROPHILINS, "SPLIT-PERSONALITY" RECEPTORS

John-Paul Silva and Yuri A. Ushkaryov*

**Abstract:** Latrophilin, a neuronal "adhesion-G protein-coupled receptor", is the major brain receptor for α-latrotoxin, a black widow spider toxin which stimulates strong neuronal exocytosis in vertebrates. Latrophilin has an unusual structure consisting of two fragments that are produced by the proteolytic cleavage of the parental molecule and that behave independently in the plasma membrane. On binding an agonist, the fragments reassociate and send an intracellular signal. This signal, transduced by a heterotrimeric G protein, causes release of calcium from intracellular stores and massive release of neurotransmitters. Latrophilin represents a phylogenetically conserved family of receptors, with orthologues found in all animals and up to three homologues present in most chordate species. From mammalian homologues, latrophilins 1 and 3 are expressed in neurons, while latrophilin 2 is ubiquitous. Latrophilin 1 may control synapse maturation and exocytosis, whereas latrophilin 2 may be involved in breast cancer. Latrophilins may play different roles during development and in adult animals: thus, LAT-1 determines cell fate in early embryogenesis in *Caenorhabditis elegans* and controls neurotransmitter release in adult nematodes. This diversity suggests that the functions of latrophilins may be determined by their interactions with respective ligands. The finding of the ligand of latrophilin 1, the large postsynaptic protein lasso, is the first step in the quest for the physiological functions of latrophilins.

## INTRODUCTION

Latrophilin was isolated in 1996.[1,2] This was a result of extensive efforts of a number of laboratories trying to identify the functional receptor(s) of α-latrotoxin, a neurotoxin from black widow spider venom whose study had begun more than thirty years ago.

*Corresponding Author: Yuri A. Ushkaryov—Division of Cell and Molecular Biology, Imperial College London, Exhibition Road, London SW7 2AZ, UK. Email: y.ushkaryov@imperial.ac.uk

*Adhesion-GPCRs: Structure to Function*, edited by Simon Yona and Martin Stacey.
©2010 Landes Bioscience and Springer Science+Business Media.

Alpha-latrotoxin causes exhaustive release of neurotransmitters from nerve terminals of vertebrates even in the absence of extracellular $Ca^{2+}$.[3,4] Due to its stimulating effect on all types of synapses and endocrine cells, the toxin has been widely used to study the mechanisms of regulated exocytosis (for a review see refs. 5,6). When the toxin was found to require specific cell-surface receptors for its actions, several groups began looking for a high-affinity $Ca^{2+}$-independent α-latrotoxin receptor[7-9] which was envisaged to play a key role in exocytosis.

This protracted search eventually resulted in the isolation of latrophilin which not only became one of the first members of the adhesion G protein-coupled receptors (GPCRs) to be identified, but also the only member with a known exogenous agonist (α-latrotoxin). Latrophilin research has greatly contributed to the general understanding of this peculiar receptor family. Thus, the idea of posttranslational cleavage of adhesion-GPCRs, although first described for CD97,[10] was developed into a conceptualised theory based on the studies of latrophilin.[11] Latrophilin has also served as a model to propose the "split personality" hypothesis stating that the fragments of adhesion-GPCRs can behave as independent proteins capable of ligand-induced reassociation and concomitant signalling.[12,13] In addition, latrophilin is the first of the few adhesion-GPCRs for which physical and functional association with G proteins has been directly demonstrated,[12,14-16] justifying the name of the whole family.

## THE ISOLATION OF LATROPHILIN

In 1990, by means of α-latrotoxin affinity chromatography in the presence of $Ca^{2+}$, Petrenko et al isolated several proteins from solubilised bovine brain.[9] Two large components of this mixture were subsequently sequenced, cloned and termed neurexins.[17] Neurexins, a polymorphic family of neuronal cell-adhesion proteins, have later been shown to participate in synapse formation/stabilisation[18,19] and to contribute to predisposition to autism.[20] Although neurexin Iα was able to bind α-latrotoxin and mediate some of its toxic effects,[21,22] its binding was $Ca^{2+}$-dependent and its structural features did not suggest a signal transduction capability. This indicated the existence of another receptor that would bind α-latrotoxin in the absence of $Ca^{2+}$ and have an ability to send intracellular signals.

In fact, another major protein was always present in the α-latrotoxin column eluate, but it was initially dismissed because, having the same molecular mass as α-latrotoxin (120-130 kDa), it was mistakenly considered to be the toxin leaching off the column. This protein actually was the N-terminal fragment (NTF) of latrophilin, a hitherto unknown brain protein which was simultaneously isolated by two groups, using α-latrotoxin affinity chromatography in the absence of calcium.[1,2] This protein's characteristics (see below) were consistent with the predicted receptor functions and, based on its high affinity for α-latrotoxin (~0.5 nM), one group termed it latrophilin,[1,14] while the other called it CIRL (calcium-independent receptor of α-latrotoxin).[2,11] This protein is also sometimes referred to as CL (CIRL/latrophilin) in the literature.[23]

The cDNA of latrophilin was cloned on the basis of peptide sequences from the 120 kDa protein (p120) isolated from bovine and rat brain.[11,14] One long open reading frame was detected that encoded a protein consisting of 1466[14] or 1471[11] amino acid residues. Since the predicted size of the cloned protein (~162 kDa) was significantly larger than 120 kDa, the presence of the entire latrophilin sequence in the α-latrotoxin column eluate was investigated using antibodies.[11] While an antibody against p120 recognised this band only, an antibody against the predicted C-terminal peptide failed to stain p120 but instead labelled some aggregated material at the top of the gel.[11] This aggregate was resolved by supplementing SDS-polyacrylamide gels with 8 M urea and by not boiling the electrophoretic samples. The resulting C-terminal fragment (CTF) appeared as a fuzzy band of ~85 kDa and was termed p85.[11] It has been subsequently shown, however, that CTF can be successfully analysed in standard SDS-gels if the samples are heated to 50°C only;[12,13] under these conditions the protein consistently migrates as a concise group of bands with an average molecular mass of ~69 kDa, exactly as predicted for this fragment (see below).

These results suggested that the translated protein was cleaved. This hypothesis was directly confirmed by N-terminal sequencing of CTF (p85) which demonstrated that the N-terminus of p85 corresponded to Thr-838 of the full-size protein.[11] Thus, the cleavage at this position produced two fragments, or subunits. The predicted molecular mass of the α subunit was ~95 kDa (this corresponded to p120, which was known to be highly glycosylated).[1,11] The predicted size of the β subunit was ~69 kDa,[11] and its aberrant migration as p85 could be explained by the effect of urea.[12] To avoid confusion and due to the partial independence of the two fragments, in the rest of this chapter we will refer to p120 (95 kDa fragment, α subunit) as NTF and to p85 (69 kDa fragment, β subunit) as CTF.

## THE LATROPHILIN FAMILY

When the cDNA encoding rat and bovine latrophilin/CIRL was isolated and sequenced, three types of sequences were identified and found to represent a family of homologous mRNAs.[23-25] One of these, encoding the protein purified by α-latrotoxin chromatography,[1,2,11,14] was termed latrophilin 1 (CIRL1), while the other two sequences were assigned numbers (2 and 3) based on their homology to latrophilin 1. Consistent with its higher resemblance of latrophilin 1, latrophilin 2 also binds α-latrotoxin but with a much lower affinity,[26] which does not allow its purification on α-latrotoxin columns, while latrophilin 3 does not bind the toxin appreciably.[24] The genes encoding these mRNAs were termed *lphn1*, *lphn2* and *lphn3*, the most widely used nomenclature. These genes are located, respectively, on chromosomes 19, 1 and 4 in humans and 8, 3 and 5 in mice. In a separate later study,[27] three homologous genes encoding lectomedins (proteins containing lectin and olfactomedin domains) were identified by genome sequencing and termed *lec1*, *lec2* and *lec3*. Sequence comparisons demonstrate that *lec1* is identical to *lphn2* and *lec2* has the same sequence as *lphn1*, while *lec3* corresponds to *lphn3*. Also, human "latrophilin 1" (*lphh1*)[28] is in fact the human orthologue of rat and bovine latrophilin 2.

The latrophilin mRNAs have several sites of alternative splicing (two have been directly identified by cDNA sequencing in *lphn1*, five in *lphn2* and four in *lphn3*).[23] The exon boundaries in the three *lphn* genes are essentially the same, with a few exceptions, and many alternative splice sites coincide. The most notable are splice sites 5 and 7. The former alters the sequence of the third cytoplasmic loop between transmembrane regions (TMRs) 5 and 6 and is thus likely to affect G protein coupling (see 'Latrophilin as a GPCR'). The latter splice site leads to the expression of different cytosolic domains in latrophilin 3. Alternative splicing at splice site 2 in latrophilin 1 truncates the protein immediately downstream of the N-terminal lectin-like domain (see 'NTF' below), producing a short protein which is probably secreted.

Variably spliced latrophilin gene orthologues are present in all animals, from coelenterates, nematodes and insects to tunicates and vertebrates,[29,30] and have apparently evolved from a primordial adhesion-GPCR gene. The nematodes possess two homologous genes, while one gene is found in arthropods. Coelenterates have three genes only weakly homologous to latrophilin and lacking many domains. Three proper latrophilin homologues are present in most chordates: from sea squirt to fish, platypus and humans; however, only two genes have been identified in the chicken genome. The family of three genes probably resulted from two rounds of gene duplication in early *Chordata*, with latrophilin 1 being the latest evolutionary acquisition. Given the early divergence of the three vertebrate *lphn* genes and their location on different chromosomes, it is not surprising that their introns differ vastly in size and sequence. However, the intron positions are highly conserved and almost precisely coincide with the borderlines between the domains of respective proteins (Fig. 1).

Interestingly, the latrophilin orthologues from such distant vertebrates as fish and humans are much more homologous than the three latrophilins within one organism, indicating that the three latrophilins possess different functions which are strictly preserved in the evolution of chordate animals. This is further supported by the different expression patterns of the three latrophilin homologues in various tissues.

## EXPRESSION PATTERNS OF LATROPHILINS

Northern blot analyses of different tissues have shown that latrophilins 1 and 3 are strictly brain specific, while latrophilin 2 is expressed in many tissues.[14,23,24,26] Similar to neurexin I$\alpha$,[17,31] very small levels of latrophilin 1 mRNA can be detected outside brain, in particular in kidneys and pancreas. It is possible that in the samples of these latter tissues latrophilin 1 mRNA is actually present in neurons from autonomic ganglia or in endocrine cells. Indeed latrophilin 1 mRNA has been found in many endocrine cells, e.g., pancreatic $\beta$-cells[32] and chromaffin cells.[33]

Recently, a real-time PCR profiling of mRNA levels of all known adhesion-GPCRs, including latrophilins, was conducted.[34] Consistent with Northern blot studies carried out previously,[14,23,24,26] it was shown in that study that latrophilin 1 (Lec2) and latrophilin 3 (Lec3) mRNAs are strongly enriched in the mouse brain and are essentially absent from, for example, mouse lung and liver. Again in line with the previous publications, latrophilin 2 (Lec1) mRNA was found in most mouse tissues. On the other hand, in the rat, the levels of latrophilin 1 and 3 mRNA in the liver and lung appeared as high as in the brain.[34] This result is rather surprising because (1) it contradicts the multiple direct mRNA hybridisation experiments; (2) it is inconsistent with the close evolutionary relationship between mouse and rat and (3) latrophilin 1 cannot be detected in rat liver using $\alpha$-latrotoxin chromatography or anti-latrophilin antibodies.[1,24] It is possible that either the primers used in this work were able to amplify the latrophilin 2 message or that the samples of rat tissues used were fortuitously enriched with neuronal (e.g., autonomic ganglia) or endocrine cells.

It is also conceivable that even if some latrophilin 1 mRNA is present in nonneuronal tissues, it is not translated efficiently. From this point of view, tissue distribution of latrophilin protein should be studied directly. Indeed, our analysis of different rat tissues by Western blotting[1] and especially immunohistochemistry (which permits unequivocal typifying of positive cells; paper in preparation) show that latrophilin 1 is absent from any nonneuronal cells and is present in very small amounts in adrenals, but not in liver, lung

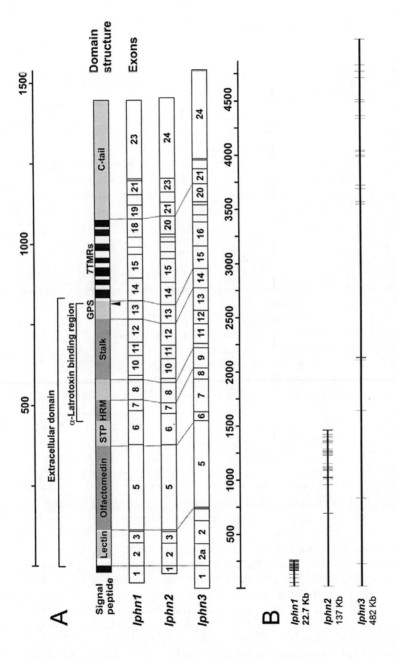

**Figure 1.** The structure of latrophilin proteins and genes. A) The distribution of protein domains in latrophilin 1 (top diagram) is shown in comparison with the distribution of exons in the mRNAs of latrophilins 1-3. Only the translated exons are shown (the numbering starts at the first protein-coding exon). Arrowhead, the site of proteolytic cleavage. The scales show the size of the mature protein (top) and the mRNA (bottom). Exon 2a (alternative) in *lphn3* is alternatively spliced. Note that many exons (or groups of exons) encode specific protein domains. B) Exon-intron structures and relative sizes of the three mouse latrophilin genes. Exons are depicted as vertical bars; introns, as horizontal lines. The size of each gene (including the translated exons and introns) is shown below each gene's name. The gene structures shown are from the 129/SvJ mouse (some intron sizes differ between the 129/SvJ and CL57BL/6 mouse strains).

or kidneys. This result has been confirmed by immunoelectron microscopy demonstrating latrophilin 1 presence in synaptic junctions only (data not shown).

## THE STRUCTURE OF LATROPHILIN

Latrophilin 1, due to its strong affinity for α-latrotoxin, is by far the most studied of the three latrophilins. However, the primary structures of these proteins are 48-63% identical and there is no reason to believe that the processing and behaviour of latrophilins 2 and 3 should be grossly different. Therefore, the general protein architecture and behaviour as described below apply to all latrophilins and may be relevant for all adhesion-GPCRs.

The primary structure of latrophilin comprises the following domains: an 851 residues-long extracellular domain; seven hydrophobic TMRs which, together with the intra- and extracellular loops, encompass 243 residues; and a cytoplasmic tail of 372 amino acids. Constitutive proteolysis within the extracellular domain (19 amino acids upstream of the first TMR) produces NTF (832 residues) and CTF (634 residues).

## NTF

The extracellular domain, which gives rise to NTF, begins with a hydrophobic signal peptide. Immediately downstream lies a 108 residues-long cysteine-rich region homologous to galactose-binding lectin (GBL).[11,14] GBL is present in most latrophilin orthologues found in animals from nematodes (*Caenorhabditis elegans*) to vertebrates (but not in coelenterates).[29] GBL is indeed able to bind D-galactose but shows much stronger preference for L-rhamnose[35,36] and ouabain.[35] The solution structure of this region, alternatively termed rhamnose-binding lectin, has been solved recently.[35] This study argues, however, that carbohydrates are unlikely to be the endogenous ligands of this domain[35] because (1) rhamnose is not normally found in animals; (2) the affinity of the monomeric lectin domain for rhamnose and especially other monosaccharides is insufficient for specific binding and (3) the residues critical for carbohydrate binding are not conserved in GBLs of latrophilin orthologues from different organisms. However, many GPCRs are dimeric and it would be interesting to determine the affinity of GBL for D-galactose-containing glycans when latrophilin dimerises in response to agonist (α-latrotoxin) binding.[12] In addition, the strong evolutionary conservation of GBL (amino acid sequence identity between latrophilins from nematodes, insects and mammals is 36-40%)[29] suggests that GBL plays a very important role in the function of this receptor.

The GBL domain is followed by the region (260 amino acids) of homology to olfactomedin, a glycoprotein of the extracellular matrix of the olfactory neuroepithelium (for a review see ref. 37). Olfactomedin domain is found in many different proteins all of which, apart from latrophilin, are secreted and most are expressed in the nervous system. Olfactomedin domain-containing proteins have been implicated in cell-cell interactions important for neurogenesis, neural crest formation, dorsoventral patterning and cell cycle regulation, while mutations in these proteins may be involved in various neurological diseases from glaucoma to psychiatric disorders. Interestingly, this domain is absent in all invertebrate orthologues of latrophilin and could be acquired during the early evolution of vertebrates.[29]

The 79 amino acid sequence downstream of the olfactomedin domain contains multiple serines, threonines and prolines (STP)[25] but shows no significant homology to any known proteins, except some proline-rich bacterial proteins with low sequence complexity and unknown functions. The STP domain is found in insect and vertebrate orthologues of latrophilin, but not in *C. elegans*.

The STP region is linked to a domain (~60 residues) characteristically found in Class II (or secretin family) peptide hormone GPCRs.[38] It is variably called "hormone-binding domain",[38] "signature domain"[24] or "hormone receptor motif" (HRM, adopted here).[29] HRM contains two conserved tryptophan residues and three to four conserved cysteines, which may form internal disulphide bridges. This region appears in many but not all adhesion-GPCRs; it is absent from the latrophilin orthologues from insects and in latrophilin 3 from the chicken, but is present in the *C. elegans* latrophilins.[29]

A unique 68 amino acid region downstream of HRM connects it to a glycosylated domain of 180 residues which is analogous (although only 20% identical) to the "Stalk" region of another adhesion-GPCR, EMR3[39] and its homologues. In EMR proteins, this region is essential for the cleavage of the ectodomain[39,40] and is thought to be an autoproteolytic enzyme.[41,42] Truncations of this domain in latrophilin render this receptor unable to bind its ligand, α-latrotoxin.[25]

The Stalk domain is attached to a short, highly conserved[29] sequence containing four cysteine residues and termed GPCR proteolysis site (or GPS).[11] To avoid confusion with the actual site of cleavage, we will be referring to this region as the "GPS motif". It is difficult to define the borderlines of this motif because of the variable conservation of the sequences surrounding this region in different adhesion-GPCRs. Therefore, we suggest to set the limits of the GPS motif according to the ends of the exon encoding this entire domain (see 'The latrophilin family' above). This would mean that the GPS motif in latrophilin is 57 amino acids-long (starting at Ala-788 and ending at Ile-844). Most importantly, this motif contains the site of posttranslational cleavage that divides latrophilin into the noncovalently bound NTF and CTF. The cleavage occurs between Leu-832 and Thr-833,[11] which is 8 amino acids upstream of the C-terminal end of the GPS motif or 19 amino acids upstream of the first TMR. As a result of this cleavage, the GPS motif itself becomes unequally split between NTF and CTF.

Both the Stalk domain and the GPS motif are present in latrophilin orthologues from all animal taxa,[29] as well as in all other adhesion-GPCRs. Sequence identity between latrophilins from vertebrates, insects and worms is 16-33% within the Stalk domain and 45-49% within the GPS motif.[29] These domains are even more conserved among the three latrophilin homologues found in any vertebrate animal, where sequence identity is 50-60% between the Stalks and 72-82% between the GPS motifs. It is possible that the Stalk and GPS motif form a single functional unit that is involved in the posttranslational cleavage of latrophilin.

## CTF

CTF begins at the site of cleavage within the GPS motif (Thr-833).[11,12] The most prominent feature of this fragment is the presence of seven TMRs that are highly homologous to those of the secretin family GPCRs (50-60% sequence similarity and 30% identity). Similar to peptide hormone GPCRs, extracellular loops 1 and 2 contain two cysteine residues which are believed to form an intramolecular disulphide bridge. In fact, CTF possesses many other features thought to be important for GPCRs, e.g., a negatively charged amino acid within the third TMR, proline residues in the fourth and fifth TMRs and potential sites of palmitoylation in the N-terminal part of the cytoplasmic tail.

The cytoplasmic tail is the least conserved domain among latrophilins. Thus, sequence identity in this region is 13-28% among latrophilin orthologues from worms, insects and vertebrates and 35-49% among the three latrophilin homologues found in mammals. For comparison, the average sequence identity of the whole protein molecules among the three homologues is 49-63%, while within the TMRs (including loop regions) it is 69-80%.

There are numerous potential phosphorylation sites for several types of protein kinases on the cytoplasmic C-terminal portion of CTF. In fact, CTF of latrophilin isolated from rat brain is phosphorylated on multiple positions and this explains the behaviour of this fragment in SDS-electrophoresis (see Section 'The isolation of latrophilin'). Phosphorylation does not normally occur in cultured cells expressing latrophilin, which suggests that this modification in the brain is a result of normal latrophilin function. The phosphorylation plays an important role in the interaction of latrophilin fragments: the phosphorylated forms of CTF bind NTF much stronger than the nonphosphorylated CTF (paper in preparation). This may have important implications for the behaviour of the two fragments after their reassociation and signalling.

## Cleavage and Unusual Behavior of Latrophilin Fragments

Latrophilin was the first adhesion-GPCR for which the site of intramolecular posttranslational cleavage was identified by direct sequencing of CTF.[11] Since then all adhesion-GPCRs studied in respect of cleavage have been proven to undergo proteolysis at a strictly conserved position within the GPS motif. This autoproteolytic cleavage and the two-subunit structure probably represent a common feature of all the members of the adhesion-GPCR family.[11,43,44] Moreover, the cleavage site between NTF and CTF coincides with the tentative borderline between the "adhesion" and the "GPCR" halves of these chimerical receptors.[43,44]

The cleavage occurs constitutively in the endoplasmic reticulum and is required for latrophilin trafficking to the cell surface[12,43] Full-size, noncleaved adhesion-GPCRs are apparently short-lived and not normally detectable in live tissues. Although full size GRP56 (another member of the adhesion-GPCR family) was reported to appear in large quantities in some mouse tissues,[45] this was later disproved.[46]

As described above, the cleavage site is localised in the GPS motif, 19 amino acids upstream of TMR1. Thus NTF contains no TMRs, but in most cells it is not released into the medium and remains noncovalently associated with the membrane. Given that NTF and CTF have an ability to interact strongly with each other when isolated by α-latrotoxin chromatography and immunoprecipitation,[11,12,43] NTF has been thought[11,43] to attach to the cell surface through its interaction with CTF, a transmembrane protein. This is also supported by the fact that proteolysis at a second site (located between the GPS and TMR1) releases some NTF into the medium.[47]

However, this does not seem to be true for a large proportion of NTF and CTF. Mutagenesis of CTF showed that only eight[43] or even seven[12] residues in the C-terminal part of the GPS motif (after cleavage forming the N-terminus of CTF) are both necessary and sufficient for the cleavage. These seven amino acids are also sufficient for the interaction between NTF and CTF. Despite such a short sequence holding NTF and CTF together, NTF cannot be removed from the membranes by most chaotropic conditions: pH 2.5, pH 12, 4 M $Mg^{2+}$ or 8 M urea (unpublished observations). Most detergents, while solubilising the membrane, do not affect any existing NTF-CTF complexes and the latter can be isolated from solution. However, upon solubilisation of cells expressing latrophilin, a large percentage of each fragment remains free from the other. Furthermore, a weak detergent, perfluorooctanoic acid, which does not break up the membrane bilayer, removes a large amount of NTF from the plasma membrane, while leaving all CTF behind.[13] Together, these data suggest that at least some proportion of NTF is anchored in the membrane independently of CTF, perhaps via hydrophobic amino acids or modifications.[12,13]

This "split personality" hypothesis has been tested in a comprehensive series of experiments,[12,13] which demonstrate that the two fragments do not always colocalise with each other on the cell surface and can even migrate in the membrane and internalise independently. When patched on live cells using antibodies, the two fragments behave as non-interacting free proteins.[12] This corroborates the idea that NTF could have a hydrophobic anchor of its own.

Under certain conditions, e.g., the binding of agonists (see 'Extracellular ligands' below) and also upon membrane solubilisation with detergents, the free latrophilin fragments can reassociate.[12] Treatment of latrophilin-expressing cells with α-latrotoxin or its mutant $LTX^{N4C}$ results in the formation of large ternary complexes (α-latrotoxin-NTF-CTF) on the plasma membrane. This leads to intracellular signal transduction to intracellular $Ca^{2+}$ stores (described in 'Latrophilin as a GPCR'). The mechanism of this reassociation apparently involves dimerisation of NTF domains, which increases the affinity between NTF and CTF.[12,13]

Most intriguingly, the ligand-induced reassociation of latrophilin fragments does not always occur within the same cleaved receptor molecule, as the ligand-bound NTF can interact with CTF from another latrophilin molecule.[13] Moreover, due to the high conservation of GPS motifs in all adhesion-GPCRs, NTF of latrophilin can even bind to CTF from another member of this receptor family. Such criss-cross association of NTFs and CTFs produced by the cleavage of different receptors creates functional complexes capable of intracellular signalling and has the potential of greatly diversifying the transduced signal.[13]

This "split personality" architecture of receptors, consisting of two independent modules that associate interchangeably on binding their ligands, is rather enigmatic but not entirely unprecedented. Several other signalling systems require coreceptors. For example, there are two receptors for the Wnt signalling proteins: Frizzled (a GPCR) and low-density-lipoprotein receptor-related protein.[48] Normally such coreceptors are phylogenetically unrelated proteins, both of which bind the same ligand molecule. However, in the case of latrophilin and possibly other adhesion-GPCRs, both "coreceptors" are the complementary fragments of the same (or structurally related) parental proteins. In addition, at least when α-latrotoxin is used, the ligand apparently only interacts with NTF, which then serves as an activated ligand of CTF.[12,13] Pleiotropy of downstream effects in this case is achieved not by one ligand activating two different receptors (as in the Wnt pathway), but by the ligand-bound NTF activating CTF from one or another adhesion-GPCR. This mechanism may have important implications for the biology of all adhesion-GPCRs.

## LATROPHILIN AS A GPCR

Latrophilin was classified as a GPCR[14] on the basis of its high sequence homology with the TMRs of the secretin family GPCRs and the features within CTF that are important for G protein coupling (see Section 'CTF' above). However, this cannot be considered a proof that this protein signals through G proteins. This aspect of receptor function was studied in some detail and, as a result, latrophilin has become the first receptor of the adhesion-GPCR group for which specific binding to G proteins, namely $G\alpha_o$ and $G\alpha_{q/11}$, was demonstrated.[14-16] This interaction is strong because it persists through two consecutive affinity chromatographies on different adsorbents that bind NTF only.[14,15] (It needs to be pointed out that G proteins can only interact with CTF and their isolation on NTF-binding columns is only possible due to NTF and CTF forming strong complexes). Moreover, the interaction between CTF and G proteins is dynamic and depends on the ability of G proteins to cleave GTP.[15] Thus, G proteins copurify abundantly with latrophilin only when GDP and EDTA are added to the solubilisation buffer, i.e., under the conditions when the GTPase activity of G proteins is inhibited and they normnally interact with respective GPCRs. The addition of GTP and $Mg^{2+}$ to purification buffer supports the GTPase activity and causes the dissociation of G proteins, resulting, as expected, in their loss from the column eluate.[15] Thus, excess of GTP is able to reverse the interaction of latrophilin with its requisite G protein(s), suggesting that this association is not only physical but also functional.

Intracellular signalling mechanisms of latrophilin have so far been studied using its exogenous agonist, α-latrotoxin. The signalling induced by α-latrotoxin is also consistent with the activation of G protein pathways. In particular, nonneuronal cells expressing latrophilin respond by stimulation of adenylate cyclase and phospholipase C (PLC) and release of intracellular calcium.[14] Similarly, the toxin triggers activation of PLC and increase in cytosolic $Ca^{2+}$ in PC12 cells[49] and synaptosomes.[50] However, studies of signal transduction from α-latrotoxin have been complicated by the fact that the toxin binds to at least three disparate receptors (neurexin, latrophilin and PTP6) and also forms $Ca^{2+}$-permeable transmembrane pores.[5,6] The effect of $Ca^{2+}$ influx through the toxin pore may obscure any physiological signalling. A breakthrough has been achieved with the creation in Tom Südhof's laboratory of a mutant α-latrotoxin, $LTX^{N4C}$.[51] This mutant lacks the ability to form pores[33] but still stimulates neurotransmitter secretion in hippocampal slices, neuronal cultures, neuromuscular junctions and synaptosomes,[6,52-55] indicating that its effect is based on stimulation of a receptor. The subsequent receptor transduction pathway involves a G protein coupled to activation of PLC, production of inositol-trisphosphate and release of $Ca^{2+}$ from intracellular stores.[53] To determine which receptor is involved in this signalling, neuroblastoma cells expressing latrophilin or neurexin were stimulated with $LTX^{N4C}$. It was demonstrated that only the cells expressing latrophilin (but not neurexin or latrophilin mutant with a single TMR) reacted by activating PLC and producing cytosolic $Ca^{2+}$ waves.[12] These results unequivocally indicate that latrophilin—via its CTF—can send intracellular signals to PLC and intracellular $Ca^{2+}$ stores.

Until recently, however, it has been difficult to prove that the same signalling cascade ($LTX^{N4C}$ – latrophilin – G protein – PLC – $Ca^{2+}$ stores)[12,13] also underlies the $LTX^{N4C}$-induced neurotransmitter release in nerve terminals.[33,53] This is because $LTX^{N4C}$ also binds two other neuronal receptors (neurexin Iα and PTPσ) which might contribute to or mediate its effect in neurons. In addition, $LTX^{N4C}$ action in nerve terminals requires extracellular $Ca^{2+}$,[33,53] possibly suggesting that the toxin might induce influx of $Ca_e^{2+}$ rather emodepside inhibited pharyngeal pumping of the nematodes in a concentration-dependent manner.[58] In *C. elegans* LAT-1 knockout mutants emodepside had a decreased paralysing effect on the pharyngeal muscle.[59] These studies suggested that cyclodepsipeptides were exogenous agonists of the latrophilin-like proteins in nematodes, leading to the release of an unidentified inhibitory transmitter which acted postsynaptically to relax both pharyngeal and somatic body wall muscle, causing flaccid paralysis of the nematode.

However, the expression of depsiphilin, a LAT-1 orthologue from the canine hookworm *Ancylostoma caninum*, did not correlate with emodepside sensitivity.[60] Also, in *C. elegans* LAT-1 knockout worms, only pharyngeal pumping was resistant to the inhibitory effect of emodepside, while locomotion was blocked by the drug even in the double mutant *lat-1, lat-2* worms.[61] Ultimately, emodepside has been found to target directly a $Ca^{2+}$-activated potassium channel, SLO-1, which is expressed in both neurons and muscles. One pathway involving neuronal SLO-1 and controlled by LAT-1, is responsible for pharynx pumping. The second pathway, based on both neuronal and muscle SLO-1 that is independent of LAT-1 or LAT-2, is responsible for locomotion.[61]

*Putative Small Endogenous Ligands*

The 54 kDa N-terminal fragment of latrophilin-like receptor HC110-R from *H. contortus* has been tested for its ability to bind different FMRFamide-like neuropeptides.[62] Three of these peptides (AF1, AF10 and PF2) exhibited low-affinity interaction with the receptor with dissociation constants of 11 μM, 52 μM and 583 μM, respectively. These data suggest that AF1, AF10 and PF2 might represent natural ligands of HC110-R and might be involved in controlling pharyngeal pumping in nematodes.

*Endogenous Adhesion Ligand*

The structure of latrophilin, with its large adhesion-like N-terminal domain and the finding of large protein ligands for other adhesion-GPCRs[63,64] suggest that latrophilin may be capable of interacting with ligands on adjacent cells or in the extracellular matrix. Therefore, in our attempts to isolate an endogenous ligand of latrophilin, we used NTF. Several variants of recombinant NTF were expressed, purified and used to synthesise an affinity adsorbent. Out of these constructs, only one resulted in an active column, which allowed us to isolate a protein from solubilised rat brain that specifically binds latrophilin (paper in preparation). This protein, termed lasso (latrophilin-associated synaptic surface organiser), is a large glycoprotein expressed on the postsynaptic membrane. The binding of lasso to NTF of latrophilin involves multiple sites in each molecule and causes specific association of cells expressing these proteins. Moreover, the interaction between latrophilin and lasso is required for synapse formation and maturation.

**Intracellular Partners**

Most importantly for its signalling function, latrophilin has been found to bind two types of α-subunits of heterotrimeric G proteins[14-16] (described in detail in 'Latrophilin as a GPCR').

Close to the C-terminus of CTF, there is a proline-rich region,[14] which could bind SH3 domains of proteins involved in signalling.

In addition, CTF may stably or transiently interact with structurally important proteins. In a yeast two-hybrid system, the C-terminal cytoplasmic tail of latrophilin was able to bind Shank, a proline-rich postsynaptic scaffolding protein.[65,66] The significance of this interaction is unclear: Shank contains a PDZ domain that binds proteins with a consensus sequence Ser/Thr-X-φ, where φ is a bulky hydrophobic amino acid. However, this sequence is present in all three latrophilin homologues and consistently both latrophilins 1 and 2 were isolated in this artificial system.[66] It is possible that the ubiquitous latrophilin 2 is the physiological target of Shank in the postsynaptic density, while latrophilin 1 is normally found in the presynaptic terminals (discussed in Section 'Expression patterns of latrophilins').

Finally, CTF of latrophilin has been shown to interact with TRIP8b, tetratricopeptide repeat-containing Rab8b-interacting protein,[67,68] a cytosolic protein that binds clathrin and the adaptor protein AP2. This indicates that latrophilin may play a role in receptor-mediated endocytosis or trafficking of neuronal channels.[69]

## LATROPHILIN GENE KNOCKOUTS

### Mouse

To study the physiological role of latrophilin 1, its gene has been knocked out in mice by deleting exon 2[70] or the distal part of exon 1 plus the proximal part of exon 2 (unpublished data). In our laboratory the first $lphn1^{-/-}$ mouse only appeared after about 40 rounds of mating heterozygous animals, suggesting that the $lphn1$ deletion is actually embryonically lethal. This is supported by our consistent finding of dead embryos in pregnant heterozygous female mice. However, both knockout approaches eventually resulted in live and fertile $lphn1^{-/-}$ mice, indicating that the lack of latrophilin 1 can be effectively compensated for by a mutation or upregulation of another gene that either occurs spontaneously or is introduced by C57BL/6 backcrossing. Both compensated mutant strains display lack of maternal behaviour (pup nursing, nest building, etc.). In our colony, these maternal nurturing defects reciprocally correlate with the dose of $lphn1$ gene. In addition, our knockout mice show increased aggression. It should be noted that a very similar phenotype has been reported for mice lacking Gαq[71] or PLC-β1,[72] the proteins known to be involved in the downstream signalling cascade of latrophilin. Despite the mild phenotypic manifestation of latrophilin 1 deletion, which is apparently due to compensatory changes in the genetic background, further behavioural studies are needed to throw more light on the functions of latrophilin 1 in those cells and brain regions where the genetic compensation is less pronounced. In addition, it would be especially revealing to determine the nature of the compensatory mutation/s.

At the biochemical level, knockout mice demonstrated a decreased binding of α-latrotoxin and a great decrease in toxin-evoked glutamate release from nerve terminals, both in the presence and absence of $Ca^{2+}$,[70] indicating that latrophilin is the major receptor for α-latrotoxin. However, this study employed the wild-type toxin, whose ability to form $Ca^{2+}$-permeable pores complicated the results and made it impossible to detect an inhibition of latrophilin signalling in knockout mice. An in-depth exploration of the role of latrophilin in nerve terminals must be conducted using the nonpore-forming mutant $LTX^{N4C}$ or other tools.

## C. elegans

The orthologues of mammalian latrophilins in the nematode *C. elegans* are encoded by two genes: *lat-1* and *lat-2*. The LAT proteins are 25-28% identical to all latrophilins, but not particularly related to any one latrophilin.

The results obtained from *lat-1* knockout in *C. elegans* strongly support the hypothesis that LAT-1 is presynaptic in adult nematodes and that its stimulation, similar to the mammalian latrophilin pathway,[12] signals via activation of $G\alpha_q$ protein and phospholipase C-$\beta$1, leading to the mobilisation of diacylglycerol (DAG). DAG then activates UNC-13, an important protein that regulates the tethering of presynaptic vesicles to the plasma membrane and synaptobrevin, a vesicular protein that binds vesicles to the plasma membrane. This is thought to result in transmitter release.[59,73]

In addition, loss-of-function mutations in *lat-1* (but not in *lat-2*) have indicated a different role for LAT-1 in *C. elegans* development.[74] The lack of this protein results in defects in anterior-posterior polarity, leading to arrest of larval development and suggesting that LAT-1, in parallel with the *wnt* pathway, controls the polarity of cell division and cell migration during nematode embryogenesis. Both the extracellular N-terminal region (including the GBL/RBL domain) and the C-terminal domain are required for this mechanism. This indicates that in the process of early worm development LAT-1 acts by transforming the interaction of NTF with adjacent cells into intracellular signals. These signals are probably different from those sent by the protein in terminally differentiated cells of the adult worm.

## LATROPHILINS IN DISEASE

To our knowledge, genetic links between the *lphn1* gene and an inheritable disease have not been established yet. This may suggest—in line with our knockout results above—that most mutations in latrophilin 1, as well as its deletion, are embryonically lethal. Indirect evidence suggests that latrophilin 1 may be associated with such mental disorders as schizophrenia and bipolar disorder. Thus, chronic administration of risperidone, an antipsychotic drug often used to treat schizophrenia, led to an upregulation of *lphn1* in rats.[75] Also, the lack of latrophilin in mice, despite the compensatory changes, led to behaviours consistent with schizophrenia phenotypes.[76] Schizophrenia is a complex neuropsychiatric disease and multiple genes and environmental factors can contribute to its manifestation, making further research into latrophilin genes even more important.

On the other hand, mutations in the human gene *lphh1* encoding the ubiquitous latrophilin 2 have been associated with breast cancer.[28] Analyses of tumour cell lines showed that *lphh1* expression was variable and gene product variability was higher in the tumour than in normal breast tissue.

## CONCLUSION

Taken together these data suggest that the ancient physiological role of latrophilins in animals is to convert cell contacts into intracellular signals. However, the members of this family have distinct distributions and functions, from early patterning during embryogenesis to controlling release of neurotransmitters in neurons. The identification of specific ligands that bind different latrophilin homologues, or each latrophilin homologue during different stages of animal development, will bring about a new level of understanding of these unusual receptors.

## REFERENCES

1. Davletov BA, Shamotienko OG, Lelianova VG et al. Isolation and biochemical characterization of a $Ca^{2+}$-independent α-latrotoxin-binding protein. J Biol Chem 1996; 271:23239-23245.
2. Krasnoperov VG, Beavis R, Chepurny OG et al. The calcium-independent receptor of α-latrotoxin is not a neurexin. Biochem Biophys Res Commun 1996; 227:868-875.
3. Clark AW, Mauro A, Longenecker HE Jr et al. Effects of black widow spider venom on the frog neuromuscular junction. Effects on the fine structure of the frog neuromuscular junction. Nature 1970; 225:703-705.
4. Longenecker HE, Hurlbut WP, Mauro A et al. Effects of black widow spider venom on the frog neuromuscular junction. Effects on end-plate potential, miniature end-plate potential and nerve terminal spike. Nature 1970; 225:701-703.
5. Ushkaryov YA, Rohou A, Sugita S. α-Latrotoxin and its receptors. In: Sudhof TC, Starke K, Boehm S, eds. Pharmacology of Neurotransmitter Release, Handb Exp Pharmacol 184. Berlin: Springer-Verlag, 2008:171-206.
6. Silva JP, Suckling J, Ushkaryov Y. Penelope's web: using α-latrotoxin to untangle the mysteries of exocytosis. J Neurochem 2009; 111:275-290.
7. Tzeng MC, Siekevitz P. The binding interaction between α-latrotoxin from black widow spider venom and a dog cerebral cortex synaptosomal membrane preparation. J Neurochem 1979; 33:263-274.
8. Scheer H, Meldolesi J. Purification of the putative α-latrotoxin receptor from bovine synaptosomal membranes in an active binding form. EMBO J 1985; 4:323-327.
9. Petrenko AG, Kovalenko VA, Shamotienko OG et al. Isolation and properties of the α-latrotoxin receptor. EMBO J 1990; 9:2023-2027.
10. Gray JX, Haino M, Roth MJ et al. CD97 is a processed, seven-transmembrane, heterodimeric receptor associated with inflammation. J Immunol 1996; 157:5438-5447.
11. Krasnoperov VG, Bittner MA, Beavis R et al. α-Latrotoxin stimulates exocytosis by the interaction with a neuronal G-protein-coupled receptor. Neuron 1997; 18:925-937.
12. Volynski KE, Silva JP, Lelianova VG et al. Latrophilin fragments behave as independent proteins that associate and signal on binding of $LTX^{N4C}$. EMBO J 2004; 23:4423-4433.
13. Silva JP, Lelianova V, Hopkins C et al. Functional cross-interaction of the fragments produced by the cleavage of distinct adhesion G-protein-coupled receptors. J Biol Chem 2009; 284:6495-6506.
14. Lelianova VG, Davletov BA, Sterling A et al. α-Latrotoxin receptor, latrophilin, is a novel member of the secretin family of G protein-coupled receptors. J Biol Chem 1997; 272:21504-21508.
15. Rahman MA, Ashton AC, Meunier FA et al. Norepinephrine exocytosis stimulated by α-latrotoxin requires both external and stored $Ca^{2+}$ and is mediated by latrophilin, G proteins and phospholipase C. Phil Trans R Soc Lond B 1999; 354:379-386.
16. Serova OV, Popova NV, Deev IE et al. Identification of proteins in complexes with α-latrotoxin receptors. Bioorg Khim 2008; 34:747-753.
17. Ushkaryov YA, Petrenko AG, Geppert M et al. Neurexins: synaptic cell surface proteins related to the α-latrotoxin receptor and laminin. Science 1992; 257:50-56.
18. Scheiffele P, Fan J, Choih J et al. Neuroligin expressed in nonneuronal cells triggers presynaptic development in contacting axons. Cell 2000; 101:657-669.
19. Varoqueaux F, Aramuni G, Rawson RL et al. Neuroligins determine synapse maturation and function. Neuron 2006; 51:741-754.
20. Szatmari P, Paterson AD, Zwaigenbaum L et al. Mapping autism risk loci using genetic linkage and chromosomal rearrangements. Nat Genet 2007; 39:319-328.
21. Geppert M, Khvotchev M, Krasnoperov V et al. Neurexin Iα is a major α-latrotoxin receptor that cooperates in α-latrotoxin action. J Biol Chem 1998; 273:1705-1710.
22. Sugita S, Khvochtev M, Südhof TC. Neurexins are functional α-latrotoxin receptors. Neuron 1999; 22:489-496.
23. Sugita S, Ichtchenko K, Khvotchev M et al. α-Latrotoxin receptor CIRL/latrophilin 1 (CL1) defines an unusual family of ubiquitous G-protein-linked receptors. G-protein coupling not required for triggering exocytosis. J Biol Chem 1998; 273:32715-32724.
24. Matsushita H, Lelianova VG, Ushkaryov YA. The latrophilin family: multiply spliced G protein-coupled receptors with differential tissue distribution. FEBS Lett 1999; 443:348-352.

25. Krasnoperov V, Bittner MA, Holz RW et al. Structural requirements for α-latrotoxin binding and α-latrotoxin-stimulated secretion. A study with calcium-independent receptor of α-latrotoxin (CIRL) deletion mutants. J Biol Chem 1999; 274:3590-3596.
26. Ichtchenko K, Bittner MA, Krasnoperov V et al. A novel ubiquitously expressed α-latrotoxin receptor is a member of the CIRL family of G-protein-coupled receptors. J Biol Chem 1999; 274:5491-5498.
27. Lander ES, Linton LM, Birren B et al. Initial sequencing and analysis of the human genome. Nature 2001; 409:860-921.
28. White GR, Varley JM, Heighway J. Isolation and characterization of a human homologue of the latrophilin gene from a region of 1p31.1 implicated in breast cancer. Oncogene 1998; 17:3513-3519.
29. Rohou A, Nield J, Ushkaryov YA. Insecticidal toxins from black widow spider venom. Toxicon 2007; 49:531-549.
30. Nordstrom KJ, Lagerstrom MC, Waller LM et al. The Secretin GPCRs descended from the family of Adhesion-GPCRs. Mol Biol Evol 2009; 26:71-84.
31. Occhi G, Rampazzo A, Beffagna G et al. Identification and characterization of heart-specific splicing of human neurexin 3 mRNA (NRXN3). Biochem Biophys Res Commun 2002; 298:151-155.
32. Lang J, Ushkaryov Y, Grasso A et al. $Ca^{2+}$-independent insulin exocytosis induced by α-latrotoxin requires latrophilin, a G protein-coupled receptor. EMBO J 1998; 17:648-657.
33. Volynski KE, Capogna M, Ashton AC et al. Mutant α-latrotoxin ($LTX^{N4C}$) does not form pores and causes secretion by receptor stimulation. This action does not require neurexins. J Biol Chem 2003; 278:31058-31066.
34. Haitina T, Olsson F, Stephansson O et al. Expression profile of the entire family of Adhesion G protein-coupled receptors in mouse and rat. BMC Neurosci 2008; 9:43.
35. Vakonakis I, Langenhan T, Promel S et al. Solution structure and sugar-binding mechanism of mouse latrophilin-1 RBL: a 7TM receptor-attached lectin-like domain. Structure 2008; 16:944-953.
36. Terada T, Watanabe Y, Tateno H et al. Structural characterization of a rhamnose-binding glycoprotein (lectin) from Spanish mackerel (Scomberomorous niphonius) eggs. Biochim Biophys Acta 2007; 1770:617-629.
37. Tomarev SI, Nakaya N. Olfactomedin domain-containing proteins: possible mechanisms of action and functions in normal development and pathology. Mol Neurobiol 2009; 40:122-138.
38. Harmar AJ. Family-B G-protein-coupled receptors. Genome Biol 2001; 2:3013.1-3013.10.
39. Stacey M, Lin HH, Hilyard KL et al. Human epidermal growth factor (EGF) module-containing mucin-like hormone receptor 3 is a new member of the EGF-TM7 family that recognizes a ligand on human macrophages and activated neutrophils. J Biol Chem 2001; 276:18863-18870.
40. Chang GW, Stacey M, Kwakkenbos MJ et al. Proteolytic cleavage of the EMR2 receptor requires both the extracellular stalk and the GPS motif. FEBS Lett 2003; 547:145-150.
41. Lin HH, Chang GW, Davies JQ et al. Autocatalytic cleavage of the EMR2 receptor occurs at a conserved G protein-coupled receptor proteolytic site motif. J Biol Chem 2004; 279:31823-31832.
42. Hsiao CC, Cheng KF, Chen HY et al. Site-specific N-glycosylation regulates the GPS autoproteolysis of CD97. FEBS Lett 2009.
43. Krasnoperov V, Lu Y, Buryanovsky L et al. Post-translational proteolytic processing of the calcium-independent receptor of α-latrotoxin (CIRL), a natural chimera of the cell adhesion protein and the G protein-coupled receptor. Role of the G protein-coupled receptor proteolysis site (GPS) motif. J Biol Chem 2002; 277:46518-46526.
44. Fredriksson R, Lagerstrom MC, Lundin LG et al. The G-protein-coupled receptors in the human genome form five main families. Phylogenetic analysis, paralogon groups and fingerprints. Mol Pharmacol 2003; 63:1256-1272.
45. Iguchi T, Sakata K, Yoshizaki K et al. Orphan G protein-coupled receptor GPR56 regulates neural progenitor cell migration via a G alpha 12/13 and Rho pathway. J Biol Chem 2008; 283:14469-14478.
46. Huang Y, Fan J, Yang J et al. Characterization of GPR56 protein and its suppressed expression in human pancreatic cancer cells. Mol Cell Biochem 2008; 308:133-139.
47. Krasnoperov V, Deyev I, Serova O et al. Dissociation of CIRL subunits as a result of two-step proteolysis. Biochemistry 2009.
48. Kikuchi A, Yamamoto H, Kishida S. Multiplicity of the interactions of Wnt proteins and their receptors. Cell Signal 2007; 19:659-671.
49. Vicentini LM, Meldolesi J. α Latrotoxin of black widow spider venom binds to a specific receptor coupled to phosphoinositide breakdown in PC12 cells. Biochem Biophys Res Commun 1984; 121:538-544.
50. Davletov BA, Meunier FA, Ashton AC et al. Vesicle exocytosis stimulated by α-latrotoxin is mediated by latrophilin and requires both external and stored $Ca^{2+}$. EMBO J 1998; 17:3909-3920.
51. Ichtchenko K, Khvotchev M, Kiyatkin N et al. α-Latrotoxin action probed with recombinant toxin: receptors recruit α-latrotoxin but do not transduce an exocytotic signal. EMBO J 1998; 17:6188-6199.

52. Ashton AC, Volynski KE, Lelianova VG et al. α-Latrotoxin, acting via two $Ca^{2+}$-dependent pathways, triggers exocytosis of two pools of synaptic vesicles. J Biol Chem 2001; 276:44695-44703.
53. Capogna M, Volynski KE, Emptage NJ et al. The α-latrotoxin mutant $LTX^{N4C}$ enhances spontaneous and evoked transmitter release in CA3 pyramidal neurons. J Neurosci 2003; 23:4044-4053.
54. Deák F, Liu X, Khvochtev M et al. α-Latrotoxin stimulates a novel pathway of $Ca^{2+}$-dependent synaptic exocytosis independent of the classical synaptic fusion machinery. J Neurosci 2009; 29:8639-8648.
55. Lelyanova VG, Thomson D, Ribchester RR et al. Activation of α-latrotoxin receptors in neuromuscular synapses leads to a prolonged splash acetylcholine release. Bull Exp Biol Med 2009; 147:701-703.
56. Mee CJ, Tomlinson SR, Perestenko PV et al. Latrophilin is required for toxicity of black widow spider venom in Caenorhabditis elegans. Biochem J 2004; 378:185-191.
57. Saeger B, Schmitt-Wrede HP, Dehnhardt M et al. Latrophilin-like receptor from the parasitic nematode Haemonchus contortus as target for the anthelmintic depsipeptide PF1022A. FASEB J 2001.
58. Harder A, Schmitt-Wrede HP, Krucken J et al. Cyclooctadepsipeptides-an anthelmintically active class of compounds exhibiting a novel mode of action. Int J Antimicrob Agents 2003; 22:318-331.
59. Willson J, Amliwala K, Davis A et al. Latrotoxin receptor signaling engages the UNC-13-dependent vesicle-priming pathway in C. elegans. Curr Biol 2004; 14:1374-1379.
60. Kruger N, Harder A, von Samson-Himmelstjerna G. The putative cyclooctadepsipeptide receptor depsiphilin of the canine hookworm Ancylostoma caninum. Parasitol Res 2009; 105 Suppl 1:S91-100.
61. Guest M, Bull K, Walker RJ et al. The calcium-activated potassium channel, SLO-1, is required for the action of the novel cyclo-octadepsipeptide anthelmintic, emodepside, in Caenorhabditis elegans. Int J Parasitol 2007; 37:1577-1588.
62. Muhlfeld S, Schmitt-Wrede HP, Harder A et al. FMRFamide-like neuropeptides as putative ligands of the latrophilin-like HC110-R from Haemonchus contortus. Mol Biochem Parasitol 2009; 164:162-164.
63. Mikesch JH, Buerger H, Simon R et al. Decay-accelerating factor (CD55): a versatile acting molecule in human malignancies. Biochim Biophys Acta 2006; 1766:42-52.
64. Yona S, Lin HH, Siu WO et al. Adhesion-GPCRs: emerging roles for novel receptors. Trends Biochem Sci 2008; 33:491-500.
65. Tobaben S, Sudhof TC, Stahl B. The G protein-coupled receptor CL1 interacts directly with proteins of the shank family. J Biol Chem 2000; 275:36204-36210.
66. Kreienkamp HJ, Zitzer H, Gundelfinger ED et al. The calcium-independent receptor for α-latrotoxin from human and rodent brains interacts with members of the ProSAP/SSTRIP/Shank family of multidomain proteins. J Biol Chem 2000; 275:32387-32390.
67. Popova NV, Plotnikov A, Deev IE et al. Interaction of calcium-independent latrotoxin receptor with intracellular adapter protein TRIP8b. Dokl Biochem Biophys 2007; 414:149-151.
68. Popova NV, Plotnikov AN, Ziganshin RK et al. Analysis of proteins interacting with TRIP8b adapter. Biochemistry (Mosc) 2008; 73:644-651.
69. Santoro B, Piskorowski RA, Pian P et al. TRIP8b splice variants form a family of auxiliary subunits that regulate gating and trafficking of HCN channels in the brain. Neuron 2009; 62:802-813.
70. Tobaben S, Sudhof TC, Stahl B. Genetic analysis of α-latrotoxin receptors reveals functional interdependence of CIRL/Latrophilin 1 and neurexin Iα. J Biol Chem 2002; 277:6359-6365.
71. Wettschureck N, Moers A, Hamalainen T et al. Heterotrimeric G proteins of the Gq/11 family are crucial for the induction of maternal behavior in mice. Mol Cell Biol 2004; 24:8048-8054.
72. Bohm D, Schwegler H, Kotthaus L et al. Disruption of PLC-beta 1-mediated signal transduction in mutant mice causes age-dependent hippocampal mossy fiber sprouting and neurodegeneration. Mol Cell Neurosci 2002; 21:584-601.
73. Harder A, Holden-Dye L, Walker R et al. Mechanisms of action of emodepside. Parasitol Res 2005; 97(Suppl 1):S1-10.
74. Langenhan T, Promel S, Mestek L et al. Latrophilin signaling links anterior-posterior tissue polarity and oriented cell divisions in the C. elegans embryo. Dev Cell 2009; 17:494-504.
75. Chen ML, Chen CH. Microarray analysis of differentially expressed genes in rat frontal cortex under chronic risperidone treatment. Neuropsychopharmacology 2005; 30:268-277.
76. Kellendonk C, Simpson EH, Kandel ER. Modeling cognitive endophenotypes of schizophrenia in mice. Trends Neurosci 2009; 32:347-358.

# CHAPTER 6

## STUDIES ON THE VERY LARGE G PROTEIN-COUPLED RECEPTOR:
### From Initial Discovery to Determining its Role in Sensorineural Deafness in Higher Animals

D. Randy McMillan* and Perrin C. White

**Abstract**  The very large G protein-coupled receptor 1 (VLGR1), also known as MASS1 or GPR98, is most notable among the family of adhesion-GPCR for its size. Encoded by an 18.9 kb open reading frame, the ~700 kDa primary translation product is by far the largest GPCR and additionally, the largest cell surface protein known to date. The large ectodomain of the protein contains several repeated motifs, including some 35 calcium binding, Calx-β repeats and seven copies of an epitempin repeat thought to be associated with the development of epilepsy. The extreme carboxy-terminus contains a consensus PDZ ligand sequence, suggesting interactions with other cytosolic or cytoskeletal proteins. At least two spontaneous and two targeted mutant mouse lines are currently known. The mutant mice present with sensitivity to audiogenic seizures but also have cochlear defects and significant, progressive hearing impairment. Although its ligand is currently unknown, VLGR1 is one of the few adhesion-GPCR family members in which mutations have been shown to be responsible for a human malady. Mutations in VLGR1 in humans result in one form (2C) of Usher syndrome, the most common genetic cause of combined blindness and deafness.

## INTRODUCTION

The VLGR1 story actually began in 1986, with the serendipitous isolation of a 3.5 kb cDNA fragment during an unrelated attempt to clone steroid 11β-hydroxylase (CYP11B1), a cytochrome P450 enzyme.[1] Sequence analysis of the cross-hybridizing clone, recovered from a human fetal adrenal cDNA library, indicated an incomplete open reading frame

*Corresponding Author: D. Randy McMillan—Department of Pediatrics, UT Southwestern Medical Center, Dallas, TX 75390-9063 USA. Email: daniel.mcmillan@utsouthwestern.edu

*Adhesion-GPCRs: Structure to Function*, edited by Simon Yona and Martin Stacey.
©2010 Landes Bioscience and Springer Science+Business Media.

of approximately 3.4 kb that extended to the 5'-end. Analysis of the predicted protein sequence of clone 5A1 revealed seven hydrophobic segments near the carboxyl-terminus. There was no strong homology to any sequence in the database at that time. In retrospect, that was because the first two GPCRs, bovine opsin[2] and the β-adrenergic receptor[3] were only cloned in 1983 and 1986, respectively. It was subsequently determined in 1991 that 5A1 encoded the carboxyl-terminal portion of a novel putative GPCR.

Following several unsuccessful attempts to isolate longer clones from multiple cDNA libraries we extended the 5' end of clone 5A1 with anchored PCRs, Through careful sequence comparison with human genomic clones isolated from both bacteriophage-λ and yeast artificial chromosome (Mega-YAC) libraries, we concluded that we had obtained a full-length cDNA, spanning 6,503 bp. With an open reading frame of 5,901 bp, the cDNA was predicted to encode a protein of 1,967 amino acid residues. As the protein was determined to be a very large GPCR, it was tentatively named VLGR1, pending functional studies.[4] During the ensuing attempt to isolate the homologous mouse VLGR1 clone it became apparent how fitting this initial designation was, as the original clone was ultimately found to be a species-specific isoform encoding only approximately one-third of the full-length protein.

Only upon cloning the full-length cDNA, was it apparent that work from two unrelated groups converged with ours. Both groups, via different methods, consequently cloned additional isoforms of VLGR1. While investigating the audiogenic seizure (AGS) susceptible *Frings* mouse, used as a model of reflex epilepsy, Skradski et.al. discovered a mutation in the responsible gene they termed monogeneic audiogenic seizure-susceptible (MASS1).[5] Additionally, while attempting to identify genes expressed in the rat ventricular zone during early corticogenesis, Yagi et al, used mRNA differential display to clone a gene they initially termed neuroepithelium-notable (Neurepin).[6]

## GENE STRUCTURE

The full-length, human VLGR1 gene locus spans 605 kb of chromosome 5 with the primary RNA transcript containing 90 exons, (Fig. 1).[7] The full-length, mature mRNA is 19.3 kb with an 18.9 kb open reading frame encoding 6307 amino acids. The primary translation product is predicted to be 693 kDa, but as there are some 90 sites appropriate for Asn-linked glycosylation the apparent molecular weight of the native protein could be significantly larger. The gene structure is highly conserved in the mouse, as the mouse VLGR1 gene consists of 90 exons and covers 538 kb of chromosome 13. The mouse open reading frame encodes 6299 amino acid residues.

### Splice Variants and Protein Isoforms

To date, there have been eight splice variants reported representing eight potential protein isoforms. Upon discovery of the full-length VLGR1 cDNA, the original human clone encoding 1967 amino acid residues was denoted VLGR1a and the full-length clone VLGR1b. The transcript encoding the VLGR1a isoform results from initiation within intron 64 and subsequent translation begins at an ATG, 22 codons upstream of exon 65. Expression of the VLGR1a isoform is apparently restricted to humans, as no ATG is found in the similar position in the mouse or zebrafish genomes. Subsequently, we discovered an additional isoform due to alternative splicing, VLGR1c, that results in

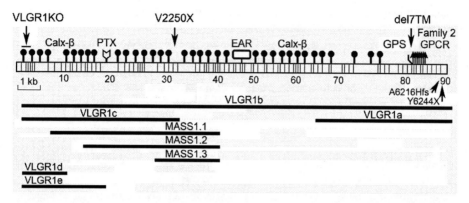

**Figure 1.** The full-length VLGR1 cDNA and protein isoforms. The long rectangle represents the cDNA; the vertical lines within the rectangle depict the location of the introns. Every tenth exon is numbered. The filled circles represent the Calx-β repeats; the chevron denotes the PTX domain; the rectangle denotes the EAR domain; the position of the GPS is indicated by a segmented line and the cluster of seven triangles represents the region encoding the seven putative membrane-spanning segments. The relative position and size of the protein isoforms are depicted as bars beneath the cDNA. The arrows above the cDNA depict the position of the mouse mutations; arrows below the cDNA depict two human mutations. (Adapted from ref. 7.)

a partial deletion of exon 31 in both human and mouse. The use of an alternative splice donor within the exon deletes the final 83 bp of exon 31, resulting in a reading frame shift and introduction of a stop codon at exon 32. Due to the loss of the transmembrane region and cytoplasmic tail, the VLGR1c isoform is predicted to be a soluble protein of 2296 amino acids with a primary translation product of ~251 kDa. Interestingly, the DNA sequence identity between the human and mouse genes at the final 113 bp of exon 31 is much higher than overall (100% versus 81%), perhaps suggesting an important function for the alternatively spliced product.[8]

Two additional soluble isoforms originating at the native amino-terminus of VLGR1b, but with carboxyl-terminal truncations have been identified.[6] VLGR1d, at 616 amino acids is the smallest isoform, terminating in intron 9 and VLGR1e, 1218 amino acids, terminates in intron 19.

Three internal isoforms have been reported, with corresponding transcripts that initiate short of the 5' boundary of the VLGR1 locus and terminate within intron 39. Although purportedly originating from the MASS1 locus,[5] MASS1.1, initiates within intron 5, MASS 1.2 initiates within intron 11 and MASS 1.3 initiates within intron 25 of the VLGR1 locus.

A thorough analysis of the tissue specificity and expression levels of all the isoforms has not been performed; however, some isoforms are known to predominate in particular tissues, e.g., VLGR1c is expressed almost exclusively instead of the a or b form in human embryonic testes.[7] Of note, the absolute expression of some isoforms is still questionable, as the level of MASS 1.1 was shown by real time PCR to be less than 1% the level of VLGR1b.[6] In addition, comparison of the MASS 1.3 sequence with the mouse genomic database indicates the purported ORF initiates at a putative polymorphic site.

While we have postulated that the expression of multiple extracellular isoforms may promote novel interactions with other cell surface proteins,[4] as is the case with many large gene products, the functional significance of the various isoforms is unknown.

## PROTEIN STRUCTURE

VLGR1 contains certain structural features consistent with other members of the adhesion-GPCR family, including the large extracellular domain, conserved amino acid homology in the seven transmembrane region to the secretin family (family B)[9] and the GPCR proteolysis site (GPS).[10,11] However, analyzing the domain structure of the VLGR1a isoform[4] and determining the initial features restricted to VLGR1 provided the first potential clues to its function.

The extracellular portion, or ectodomain, of VLGR1a contained seven conserved (and two less well-conserved) repeats with homology to the Calx-β motif, a calcium binding domain found in $Na^+/Ca^{2+}$ exchangers.[12] Subsequent calcium overlay experiments utilizing isolated groups of repeats expressed in bacteria, confirmed their calcium binding ability in this context and suggested the ectodomain of VLGR1 could bind calcium in its native form. Upon the subsequent cloning of the VLGR1b isoform it was determined the full-length protein contained 35 Calx-β motifs spaced along the length of its extracellular domain.

Interestingly, homology to the Calx-β motif is found in proteins that are known to be involved in either cell-matrix or cell-cell adhesion reactions, such as integrin β4[13] and the marine sponge aggregation factor MAFp3.[14] Although calcium is required for the extracellular interaction of integrins and their ligand, a functional analogy with integrin β4 is difficult to conceptualize, due to the intracellular localization of its calcium binding motif. However, the sponge aggregation factor forms the species-specific protein core of large extracellular proteoglycan complexes, called glyconectins,[15] that form homotypic interactions and regulate colony aggregation in a calcium-dependent manner.[16] This data, in conjuction with observations that CD97,[17] a related GPCR contained extracellular calcium binding motifs and bound a large protein ligand, prompted us to propose the hypothesis that the function of the calcium binding domains in the VLGR1 ectodomain was to mediate receptor-ligand interactions.[4] Although a specific ligand for VLGR1 has not been identified to date, recent data suggesting the calcium binding domains mediate homotypic interactions (see below) substantiate this hypothesis.

VLGR1b also contains a single Laminin G (LamG)/amino-terminal thrombospondin-like (TspN)/Pentraxin (PTX) domain[18] within the amino-terminal one-fourth of the ectodomain. Such domains have been implicated in interactions with a variety of cellular receptors and extracellular proteins and may be involved in a wide range of cellular functions. A functional role for this particular domain in VLGR1 is suspect, as ligand interactions often require at least tandem modules and conserved amino acids required for calcium binding are not present in the VLGR1 PTX domain.[19] Presently, no published studies have investigated the role of this domain in VLGR1.

Near the center of the ectodomain are seven copies of a short, ~50 amino acid repeat, variously termed epitempin (EPTP)[20] or epilepsy associated repeat (EAR).[21] This domain, first noted in the leucine-rich glioma inactivated (LGI) family of secreted proteins,[22] is thought to consist of β-sheets folded into a seven bladed β-propeller structure. While its function is still unclear, mutations in LGI1 have been shown to be responsible for the unusual human epilepsy, autosomal-dominant partial epilepsy with auditory features (ADPEAF).[23] In order to explain the epileptic phenotype resulting for mutations in either VLGR1 or LGI1, an interesting—but as yet unproven—hypothesis suggests that the EPTP domains of LGI1 and VLGR1 interact with functionally equivalent ligands during neurogenesis.[21]

The extreme carboxyl-terminus of VLGR1b (and VLGR1a in humans) contains a consensus sequence motif (Ser/Thr)-Xaa-(Val/Ile/Leu) that binds PDZ (Postsynaptic density protein 95/Drosophila Disks large/Zona occludens-1) domain containing proteins.[24] PDZ containing proteins are cytoplasmic scaffolding proteins[25] involved in diverse functions from receptor trafficking and clustering to linking receptors with their downstream signaling proteins. A similar PDZ ligand motif is additionally found in other adhesion-GPCRs, including CD97,[26] BAI1,[27] GPR123 and 124.[28]

## GENE EXPRESSION

VLGR1 expression can be detected by RT-PCR in most adult tissues. However, the highest expression is found in the developing central nervous system during embryogenesis, strongly suggesting a role for VLGR1 in neurogenesis.[6,7] Using in situ hybridization, we have shown that VLGR1 expression begins in the developing neuroepithelium between E6.75 and E8.0, coincident with development of the neural groove. Expression levels increase and spread through mid-gestation to include the length of the spinal cord, all layers of the eye, except the cornea and the ventricular zone of the neuroepithelium in the developing brain. By late gestation, coincident with the narrowing of the ventricular zone and subsequent slowing of neurogenesis, VLGR1 expression declines. In adulthood, VLGR1 expression in the brain is restricted to a subpopulation of neurons in the mammillary nuclei of the hypothalamus. Currently, no published studies have investigated the significance of this latter observation.

## MICE WITH MUTATIONS IN VLGR1

There are currently four known mouse lines containing neutralizing mutations in VLGR1. The two spontaneous mutant lines, *Frings*[29] and BUB/BnJ were discovered to have VLGR1 mutations during the positional cloning of the *Mass1* locus.[5] These mice share a common Swiss albino ancestry which likely explains an identical single base deletion in exon 31. The deletion results in a shift in reading frame, subsequently replacing the valine at position 2250 with a stop codon. While the new stop lies within the ORF of all *Mass1* transcripts, effectively truncating each by ~700 amino acids, it would have differential effects on other isoforms. A stop at position 2250 would truncate the full-length b form by approximately two-thirds, whereas only the carboxyl-terminal 63 amino acids would be lost from the c form. The amino-terminal d and e forms would not be affected.

Two, unrelated, engineered mutant lines have been developed: (1) Vlgr1del7TM, a knock-in mutation, created by adding a cassette containing both antigenic and purification tags together with a stop codon 19 bp 5′ of the GPS[30] and (2) a knock-out mutation created by deletion of exons 2-4.[6] The Vlgr1del7TM mutant would selectively effect only the full-length b form, deleting the transmembrane and cytoplasmic domains, thus resulting in the expression of a soluble ectodomain devoid of G protein-coupling activity. The knock-out mutant would effectively delete the b, c, d and e forms, but would leave the internal MASS1 isoforms untouched.

As VLGR1 is temporally expressed in the ventricular zone during development, one might predict a phenotype impacting neurogenesis and development of the CNS. However, all four lines are viable and fertile, with no apparent growth defects and no obvious histological defects detected in the brain. A variety of experiments targeting CNS development and function, including neuronal birth date analysis to analyze cortical lamination (McMillan and White unpublished results) in vivo cellular proliferation assays and in vitro neurosphere formation assays[6] and motor function analysis by Rota-rod testing[31] have detected no differences between wild-type and mutant mice.

The effects of the mutations are manifested in all mutant mice by the susceptibility to audiogenic seizure (AGS) development. AGS, considered to be a model for human reflex type epileptic seizures, are self-sustaining and characterized by three separate phases: wild running, followed by a clonic seizure and then a tonic extension that can be fatal without resuscitation.[32] It is well-known that AGS and hearing defects are related and in fact all four mutants have a profound hearing impairment that is both early in onset and progressive. Auditory brain-stem response (ABR) thresholds, a measure of auditory sensitivity, are elevated in all mutants, but to different degrees. BUB/BnJ are hearing impaired by 21 days of age[33] and are sufficiently impaired as to be unable to respond to high intensity auditory stimulus that initiates AGS after 25 days of age.[5] *Frings* are less impaired, with an early moderate hearing impairment that is relatively stable,[34] allowing them to remain susceptible to AGS into adulthood.[35] As would be expected, Vlgr1del7TM and the VLGR1 knock-out mutant are both profoundly hearing impaired by 21 days of age.[31,36] The apparent differences in progression between these two mutants, insensitivity of the former to AGS beyond 42 days of age,[30] while the latter maintains AGS sensitivity to 6 months of age,[6] is likely a result of the small number of animals tested at advanced ages.

Quite serendipitously, the positions of the mutations provide valuable information on the functional importance of the various VLGR1 isoforms. It is apparent that VLGR1b plays a significant role in the development and maintenance of normal hearing, as it is the only isoform affected by a mutation in all the four mutants. Interestingly, the close positioning of the VLGR1del7TM mutation to the GPS should allow for expression of a near full length ectodomain, indicating that association of the ectodomain with the authentic transmembrane portion is required for function. It is important to note that while the current data does not exclude a related function for the other isoforms, it is clear their expression cannot not substitute for VLGR1b in hearing function.

The differences in severity and progression of the hearing impairment between the *Frings* mouse and the other three mutants is due to the presence or absence of genetic modifiers of hearing fixed in the mouse strain background (for a review see ref. 37). In particular, a substitution (753G->A) in cadherin 23 (CDH23), the gene at the *ahl* locus, responsible for age related hearing loss[38] is found in many common strains of laboratory mice BUB/BnJ, C57Bl6/J and 129P1/ReJ,[39] whereas the *Frings* mouse is wild-type at the *ahl* locus.[40]

## EXPRESSION IN THE COCHLEA AND RETINA

VLGR1 is the mammalian ortholog of the avian ankle link antigen (ALA), a calcium-dependent epitope associated with the stereocilia of the inner ear and the ciliary calyx of photoreceptors of the eye.[41] Amino acid analysis of tryptic peptides recovered an anti-ALA immunoprecipitate from chick retina confirmed a partial sequence homologous to human VLGR1. The VLGR1 protein has been localized to the base of cochlear hair cells by immunofluoresence microscopy in perinatal mice, chicks and rats.[31,36,42] The calcium dependence of the immunofluorescence signal, was confirmed by treating mouse cochlea with 1,2-bis(2-aminophenoxy)ethane-N,N,N',N'-tetraacetic acid (BAPTA), a calcium chelator.[31,36,43] As the structure of the Calx-β motif has now been determined, it is apparent that the calcium dependence of the ALA epitope is a function of the conformational change of the Calx-β repeats upon binding calcium.[44]

Temporal expression of VLGR1, beginning prior to E17, with continued perinatal expression near the base of the stereocilia until 10 days of age,[31,36,43] approximately corresponds to development of the inner ear.[45,46] By electron microscopy, the stereociliar ankle links appear as single stranded filaments, near the base of the hair cell bundles that span the distance (~150 nm) between adjacent stereocilia.[47] In mice with VLGR1 mutations, the ankle links are absent.[31,36,43] The structural characteristics of VLGR1 (described above) suggest that ankle links are formed by calcium-dependent, homotypic interactions of VLGR1 molecules between adjacent stereocilia.[36] Ankle links are found in all vertebrates, but they are only transiently present in mammalian cochlea, e.g., in mice they are lost by 12 days of age.[48] However, the loss of VLGR1 results in morphological defects in stereociliar organization that help to explain the deafness phenotype. Slight differences in the ordered structure of the auditory hair cell bundles are apparent as early as 2 days of age, but there is progressive disorganization of both inner and outer hair cell bundles with time.[31,36,43] The structure of the vestibular hair cell bundles is less ordered and there are no obvious differences seen in the utricle of VLGR1 mutant mice. However, by 2 months of age there is a complete loss of hair cells, as well as both inner and outer pillar cells in the basal turn of the cochlea. Although the function of ankle links is still unclear, their unusual temporal expression pattern, together with the above data would suggest they play some role in stereociliar maturation.

VLGR1 expression in the eye has been detected by immunofluorescence in the outer plexiform layer and the connecting cilium on the retina.[42,49] More detailed analysis by immunoelectron microscopy makes it apparent that VLGR1 is a component of the precilliary complex, a fibrous structure between the connecting cilium and the inner segment of photoreceptor cells, that is thought to be homologous to the ankle links between stereocilia.[41,50] In the VLGR1del7TM mouse the fibrous links are lost, but this surprisingly results in no gross retinal abnormalities and only a minor age-related visual phenotype.[36,50]

**VLGR1 in Usher Syndrome**

The discovery of Usher syndrome patients carrying mutations in VLGR1 highlighted the importance of VLGR1 in human biology.[8] Usher syndrome encompasses a group of auditory and visual disorders and sometimes vestibular dysfunction, that represent the most commonly diagnosed cause of human deaf-blindness (for a review see refs. 51-52). The syndrome is divided into three types (Type 1, 2, 3) in descending order of severity. Type 2 is the most common, representing approximately one-half of Usher syndrome patients. There are 11 clinical subtypes corresponding to 11 defined loci with nine responsible genes cloned thus far. Mutations in VLGR1 result in Usher 2C with symptoms including moderate to severe hearing loss, normal vestibular function and later onset retinitis pigmentosa (RP). With the exception of RP, the clinical symptoms are phenocopied quite well by the existing mouse mutants. The lack of a visual phenotype in mice is likely due to modest structural differences in the murine retina, or perhaps their inherent life-span differences.[36]

The initial 13 USH2C patients identified with mutations in VLGR1 were all females, which potentially suggested a sexually dimorphic phenotype.[8] This has recently been disproved, as male patients have been found by three separate groups[53,54] and including large deletions, there have been 14 VLGR1 mutations reported. As in the mouse, comparing the location of the reported mutations with the putative VLGR1 isoforms (supposing all the reported murine isoforms exist in humans) confirms that only the full-length VLGR1b isoform is required for normal auditory function. Most importantly, two mutations (Y6244X and A6216Hfs) have been found within the carboxyl-terminal 100 amino acids, suggesting the complete cytoplasmic tail is required for normal VLGR1 function.[8,55] In vitro protein interaction studies have now explained the basis for this requirement.

A series of studies have now shown a complex association, a so-called interactome, of many of the Usher proteins, which likely explains the common phenotypes of Usher syndrome.[42,43,49,50,56] Additionally, these studies have shown direct physical interactions, via yeast two hybrid and in vitro protein association experiments, between the carboxyl-terminus of VLGR1 and cytoplasmic PDZ-containing scaffolding proteins, thus illustrating importance of the PDZ ligand domain to the function of the interactome. Protein interactions between the VLGR1 ectodomain and additional Usher proteins have been proposed, but no studies identifying direct interactions have been published. Likely these studies will be problematic due to the apparent tight association of the over-expressed VLGR1a ectodomain with the cell membrane, even in the absence of the transmembrane and cytoplasmic tail and the nonselective association of the ectodomain with a variety of unrelated, co-expressed cell surface proteins (McMillan and White, unpublished observations).

## CONCLUSION

Classic G protein-dependent signaling activity has currently been demonstrated in few of the adhesion-GPCR family members. However, in addition to an important, G protein independent, role in normal mammalian auditory and visual function through PDZ mediated protein interactions, VLGR1 may also have a classic signaling activity. A recent study[43] has shown that loss of VLGR1 results in a large increase in expression and a consequent redistribution of adenylate cyclase 6 (AC6) in the cochlea. While this is an exciting, albeit circumstantial observation, additional studies to show specific interactions are clearly warranted.

Our initial observations of high VLGR1 expression in the developing CNS[7] is certainly suggestive of a role in neurogenesis. VLGR1 does play some role in neurological function, as evident by the development of audiogenic seizures in mutant mice and the presentation of febrile and afebrile seizures in humans with VLGR1 mutations.[57] However, neurogenesis is clearly not dependant on VLGR1. Understandably, maintenance of functions that are critical to the existence of an organism usually involve redundant or adaptive mechanisms and these could potentially obscure the function of VLGR1 in neurogenesis.

The most obvious, albeit most puzzling, clues that suggest additional and fundamental functions of VLGR1 come from evolution. There must be strong selective pressure to maintain the strict conservation of VLGR1 from lower vertebrates like zebrafish[58] and even invertebrates like the sea urchin.[59] Elucidating such functions must await additional studies.

# REFERENCES

1. Chua SC, Szabo P, Vitek A et al. Cloning of cDNA encoding steroid 11 beta-hydroxylase (P450c11). Proc Natl Acad Sci USA 1987; 84:7193-7197.
2. Nathans J, Hogness DS. Isolation, sequence analysis and intron-exon arrangement of the gene encoding bovine rhodopsin. Cell 1983; 34:807-814.
3. Dixon RA, Kobilka BK, Strader DJ et al. Cloning of the gene and cDNA for mammalian beta-adrenergic receptor and homology with rhodopsin. Nature 1986; 321:75-79.
4. Nikkila H, McMillan DR, Nunez BS et al. Sequence similarities between a novel putative G protein-coupled receptor and $Na^+/Ca^{2+}$ exchangers define a cation binding domain. Mol Endocrinol 2000; 14:1351-1364.
5. Skradski SL, Clark AM, Jiang HM et al. A novel gene causing a mendelian audiogenic mouse epilepsy. Neuron 2001; 31:537-544.
6. Yagi H, Takamura Y, Yoneda T et al. Vlgr1 knockout mice show audiogenic seizure susceptibility. J Neurochem 2005; 92:191-202.
7. McMillan DR, Kayes-Wandover KM, Richardson JA et al. Very Large G Protein-coupled Receptor-1, the largest known cell surface protein, is highly expressed in the developing central nervous system. J Biol Chem 2002; 277:785-792.
8. Weston MD, Luijendijk MW, Humphrey KD et al. Mutations in the VLGR1 gene implicate G-protein signaling in the pathogenesis of Usher syndrome type II. Am J Hum Genet 2004; 74:357-366.
9. Attwood TK, Findlay JB. Fingerprinting G-protein-coupled receptors. Protein Eng 1994; 7:195-203.
10. Krasnoperov VG, Bittner MA, Beavis R et al. Alpha-latrotoxin stimulates exocytosis by the interaction with a neuronal G-protein-coupled receptor. Neuron 1997; 18:925-937.
11. Stacey M, Chang GW, Sanos SL et al. EMR4, a novel epidermal growth factor (EGF)-TM7 molecule up-regulated in activated mouse macrophages, binds to a putative cellular ligand on B-lymphoma cell line A20. J Biol Chem 2002; 277:29283-29293.
12. Schwarz EM, Benzer S. Calx, a Na-Ca exchanger gene of Drosophila melanogaster. Proc Natl Acad Sci USA 1997; 94:10249-10254.
13. Hogervorst F, Kuikman I, von dem B et al. Cloning and sequence analysis of beta-4 cDNA: an integrin subunit that contains a unique 118 kd cytoplasmic domain. EMBO J 1990; 9:765-770.
14. Fernandez-Busquets X, Burger MM. The main protein of the aggregation factor responsible for species-specific cell adhesion in the marine sponge Microciona prolifera is highly polymorphic. J Biol Chem 1997; 272:27839-27847.
15. Misevic GN. Molecular self-recognition and adhesion via proteoglycan to proteoglycan interactions as a pathway to multicellularity: atomic force microscopy and color coded bead measurements in sponges. Microsc Res Tech 1999; 44:304-309.
16. Jumblatt JE, Schlup V, Burger MM. Cell-cell recognition: specific binding of Microciona sponge aggregation factor to homotypic cells and the role of calcium ions. Biochemistry 1980; 19:1038-1042.
17. Hamann J, Vogel B, van Schijndel GMW et al. The seven-span transmembrane receptor CD97 has a cellular ligand (CD55,DAF). J Exp Med 1996; 184:1185-1189.
18. Beckmann G, Hanke J, Bork P et al. Merging extracellular domains: fold prediction for laminin G-like and amino-terminal thrombospondin-like modules based on homology to pentraxins. J Mol Biol 1998; 275:725-730.
19. Emsley J, White HE, O'Hara BP et al. Structure of pentameric human serum amyloid P component. Nature 1994; 367:338-345.
20. Staub E, Perez-Tur J, Siebert R et al. The novel EPTP repeat defines a superfamily of proteins implicated in epileptic disorders. Trends Biochem Sci 2002; 27:441-444.
21. Scheel H, Tomiuk S, Hofmann K. A common protein interaction domain links two recently identified epilepsy genes. Hum Mol Genet 2002; 11:1757-1762.
22. Gu W, Wevers A, Schroder H et al. The LGI1 gene involved in lateral temporal lobe epilepsy belongs to a new subfamily of leucine-rich repeat proteins. FEBS Lett 2002; 519:71-76.
23. Kalachikov S, Evgrafov O, Ross B et al. Mutations in LGI1 cause autosomal-dominant partial epilepsy with auditory features. Nat Genet 2002; 30:335-341.
24. Kennedy MB. Origin of PDZ (DHR, GLGF) domains. Trends Biochem Sci 1995; 20:350.
25. Tsunoda S, Sieralta J, Sun Y et al. A multivalent PDZ-domain protein assembles signalling complexes in a G-protein-coupled cascade. Nature 1997; 388:243-249.
26. Hamann J, Eichler W, Hamann D et al. Expression cloning and chromosomal mapping of the leukocyte activation antigen CD97, a new seven-span transmembrane molecular of the secretin receptor superfamily with an unusual extracellular domain. J Immunol 1995; 155:1942-1950.
27. Shiratsuchi T, Futamura M, Oda K et al. Cloning and characterization of BAI-associated protein 1: a PDZ domain-containing protein that interacts with BAI1. Biochem Biophys Res Commun 1998; 247:597-604.

28. Lagerstrom MC, Rabe N, Haitina T et al. The evolutionary history and tissue mapping of GPR123: specific CNS expression pattern predominantly in thalamic nuclei and regions containing large pyramidal cells. J Neurochem 2007; 100:1129-1142.
29. Frings H, Frings M. Development of strains of albino mice with predictable susceptibilities to audiogenic seizures. Science 1953; 117:283-284.
30. McMillan DR, White PC. Loss of the transmembrane and cytoplasmic domains of the very large G-protein-coupled receptor-1 (VLGR1 or Mass1) causes audiogenic seizures in mice. Mol Cell Neurosci 2004; 26:322-329.
31. Yagi H, Tokano H, Maeda M et al. Vlgr1 is required for proper stereocilia maturation of cochlear hair cells. Genes Cells 2007; 12:235 250.
32. Ross KC, Coleman JR. Developmental and genetic audiogenic seizure models: behavior and biological substrates. Neurosci Biobehav Rev 2000; 24:639-653.
33. Zheng QY, Johnson KR, Erway LC. Assessment of hearing in 80 inbred strains of mice by ABR threshold analyses. Hear Res 1999; 130:94-107.
34. Klein BD, Fu YH, Ptacek LJ et al. Auditory deficits associated with the frings mgr1 (mass1) mutation in mice. Dev Neurosci 2005; 27:321-332.
35. Skradski SL, White HS, Ptacek LJ. Genetic mapping of a locus (mass1) causing audiogenic seizures in mice. Genomics 1998; 49:188-192.
36. McGee J, Goodyear RJ, McMillan DR et al. The very large G-protein-coupled receptor VLGR1: a component of the ankle link complex required for the normal development of auditory hair bundles. J Neurosci 2006; 26:6543-6553.
37. Johnson KR, Zheng QY, Noben-Trauth K. Strain background effects and genetic modifiers of hearing in mice. Brain Res 2006; 1091:79-88.
38. Noben-Trauth K, Zheng QY, Johnson KR. Association of cadherin 23 with polygenic inheritance and genetic modification of sensorineural hearing loss. Nat Genet 2003; 35:21-23.
39. Johnson KR, Zheng QY, Erway LC. A major gene affecting age-related hearing loss is common to at least ten inbred strains of mice. Genomics 2000; 70:171-180.
40. Johnson KR, Zheng QY, Weston MD et al. The Mass1frings mutation underlies early onset hearing impairment in BUB/BnJ mice, a model for the auditory pathology of Usher syndrome IIC. Genomics 2005; 85:582-590.
41. Goodyear R, Richardson G. The ankle-link antigen: an epitope sensitive to calcium chelation associated with the hair-cell surface and the calycal processes of photoreceptors. J Neurosci 1999; 19:3761-3772.
42. Reiners J, van WE, Marker T et al. Scaffold protein harmonin (USH1C) provides molecular links between Usher syndrome type 1 and type 2. Hum Mol Genet 2005; 14:3933-3943.
43. Michalski N, Michel V, Bahloul A et al. Molecular characterization of the ankle-link complex in cochlear hair cells and its role in the hair bundle functioning. J Neurosci 2007; 27:6478-6488.
44. Hilge M, Aelen J, Vuister GW. $Ca^{2+}$ regulation in the $Na^+/Ca^{2+}$ exchanger involves two markedly different $Ca^{2+}$ sensors. Mol Cell 2006; 22:15-25.
45. Chen P, Johnson JE, Zoghbi HY et al. The role of Math1 in inner ear development: Uncoupling the establishment of the sensory primordium from hair cell fate determination. Development 2002; 129:2495-2505.
46. Ehret G. Postnatal development in the acoustic system of the house mouse in the light of developing masked thresholds. J Acoust Soc Am 1977; 62:143-148.
47. Tsuprun V, Goodyear RJ, Richardson GP. The structure of tip links and kinocilial links in avian sensory hair bundles. Biophys J 2004; 87:4106-4112.
48. Goodyear RJ, Marcotti W, Kros CJ et al. Development and properties of stereociliary link types in hair cells of the mouse cochlea. J Comp Neurol 2005; 485:75-85.
49. van Wijk E, van der Zwaag B, Peters T et al. The DFNB31 gene product whirlin connects to the Usher protein network in the cochlea and retina by direct association with USH2A and VLGR1. Hum Mol Genet 2006; 15:751-765.
50. Maerker T, van WE, Overlack N et al. A novel Usher protein network at the periciliary reloading point between molecular transport machineries in vertebrate photoreceptor cells. Hum Mol Genet 2008; 17:71-86.
51. Cohen M, Bitner-Glindzicz M, Luxon L. The changing face of Usher syndrome: clinical implications. Int J Audiol 2007; 46:82-93.
52. Petit C. Usher syndrome: from genetics to pathogenesis. Annu Rev Genomics Hum Genet 2001; 2:271-97.
53. Hmani-Aifa M, Benzina Z, Zulfiqar F et al. Identification of two new mutations in the GPR98 and the PDE6B genes segregating in a Tunisian family. Eur J Hum Genet 2009; 17:474-482.
54. Hilgert N, Kahrizi K, Dieltjens N et al. A large deletion in GPR98 causes type IIC Usher syndrome in male and female members of an Iranian family. J Med Genet 2009; 46:272-276.

55. Ebermann I, Wiesen MH, Zrenner E et al. GPR98 mutations cause Usher syndrome type 2 in males. J Med Genet 2009; 46:277-280.
56. Adato A, Michel V, Kikkawa Y et al. Interactions in the network of Usher syndrome type 1 proteins. Hum Mol Genet 2005; 14:347-356.
57. Nakayama J, Fu YH, Clark AM et al. A nonsense mutation of the MASS1 gene in a family with febrile and afebrile seizures. Ann Neurol 2002; 52:654-657.
58. Gibert Y, McMillan DR, Kayes-Wandover KM et al. Analysis of the very large G-protein coupled receptor gene (Vlgr1/Mass1/USH2C) in zebrafish. Gene 2005; 353:200-206.
59. Sodergren E, Weinstock GM, Davidson EH et al. The genome of the sea urchin Strongylocentrotus purpuratus. Science 2006; 314:941-952.

# CHAPTER 7

# ADHESION-GPCRs IN THE CNS

Natalie Strokes and Xianhua Piao*

**Abstract:** There are a total of 33 members of adhesion G protein-coupled receptors (GPCRs) in humans and 30 members in mice and rats. More than half of these receptors are expressed in the central nervous system (CNS), indicating their possible roles in the development and function of the CNS. Indeed, it has been shown that adhesion-GPCRs are involved in the regulation of neurulation, cortical development and neurite growth. Among the few adhesion-GPCRs being studied, GPR56 is so far the only member associated with a human brain malformation called bilateral frontoparietal polymicrogyria (BFPP). The histopathology of BFPP is a cobblestone-like brain malformation characterized by neuronal overmigration through a breached pial basement membrane (BM). Further studies in the *Gpr56* knockout mouse model revealed that GPR56 is expressed in radial glial cells and regulates the integrity of the pial BM by binding a putative ligand in the extracellular matrix of the developing brain.

## INTRODUCTION

Adhesion-GPCRs are a relatively new family of GPCRs with their large extracellular region linked to a seven-transmembrane-spanning domain via a GPCR proteolytic site (GPS)-motif. Their N-terminal segments often contain domains found in adhesion proteins, such as cadherin, lectin, immunoglobulin and thrombospondin domains. Thus, it is speculated that adhesion-GPCRs play dual roles in cellular adhesion and signaling. A recent extensive expression profiling of adhesion-GPCRs has discovered that 17 out of the 30 members of rodent adhesion-GPCRs are expressed in the CNS (Table 1). CELSR2/3, BAI1-3, GPR123 and latrophilin 3 are exclusively expressed in the CNS, while others are expressed more widely in different tissues. Despite its rather ubiquitous expression pattern, loss of GPR56 is only associated with brain malformation in both humans and mice.[1-3]

*Corresponding Author: Xianhua Piao— Division of Newborn Medicine, Children's Hospital Boston, Harvard Medical School, Boston, Massachusetts, USA. Email: xianhua.piao@childrens.harvard.edu

*Adhesion-GPCRs: Structure to Function*, edited by Simon Yona and Martin Stacey.
©2010 Landes Bioscience and Springer Science+Business Media.

**Table 1.** The CNS expression of adhesion-GPCRs

| Receptor | Expression | Method Used | References |
|---|---|---|---|
| *Bai1* | Cerebral Cortex (layers II-III), hippocampus, dentate gyrus, olfactory bulb, caudate putamen, medial septum, cerebellum | In Situ | 6 |
| *Bai2* | Cerebral Cortex (layers II-III), hippocampus, dentate gyrus, medial and lateral septum, medial and lateral habenular nucleus, cerebellum | In Situ | 5 |
| *Bai3* | Cerebral Cortex (layers II-III, IV), hippocampus, dentate gyrus, cerebellum | In Situ | 7 |
| *Vlgr1* | Cochlear hair bundles, fourth ventricle, telencephalic ventricle, hypothalamus, optic cup | IHC, In Situ | 15,16,20 |
| *Gpr126* | Ubiquitous CNS expression | RT-PCR | 60 |
| *Gpr64* | Hypothalamus, brain stem, substantia nigra, spinal cord | RT-PCR, In Situ | 60,61 |
| *Gpr56* | Cerebral Cortex, Ventricular Zone, Rostral cerebellum | In Situ, IHC | 1,3,41 |
| *Latrophilin1* | Ubiquitous CNS expression | WB, NB, RT-PCR | 8,9,60 |
| *Latrophilin2* | Ubiquitous CNS expression | WB, NB, RT-PCR | 8,9,60 |
| *Latrophilin3* | Ubiquitous CNS expression | NB, RT-PCR | 9,60 |
| *Etl* | Ubiquitous CNS expression | RT-PCR | 60 |
| *Gpr123* | Thalamic Nuclei, Cerebral Cortex (layers V and VI), amygdala, hypothalamus, spinal cord | RT-PCR, In Situ | 60,63 |
| *Gpr124* | Ubiquitous CNS Expression | RT-PCR | 60 |
| *Gpr125* | Choroid Plexus, Cerebral Cortex, piriform cortex area | RT-PCR, In Situ, IHC | 60,62 |
| *Celsr1* | Ventricular Zone, External Granular cell layer in Cerebellum | In Situ | 10 |
| *Celsr2* | Ventricular Zone, Cortical Plate, hippocampus, dentate gyrus, cerebellar cortex, pontine nuclei | In Situ | 10 |
| *Celsr3* | Cerebral and hippocampal plate, tectum, pontine nuclei, external granular cell layer in cerebellum | In Situ | 10 |
| *Gpr116* | Ubiquitous CNS Expression | RT-PCR | 60 |

Of those that have been studied, the most significant adhesion-GPCRs with functions within the CNS are BAI1-3, CELSR1-3, latrophilin1-3, VLGR1 and GPR56. Since there are individual chapters in this book devoted to each one of them, we will only briefly summarize their role in the CNS and focus our discussion mainly on the role of GPR56 in brain development.

## ADHESION-GPCRs IN BRAIN DEVELOPMENT

The brain angiogenesis inhibitors (BAI) 1-3 are expressed almost exclusively in the CNS. The transcript of human *BAI1* was found to be expressed in both the fetal and adult brain.[4] In mouse embryonic and adult brains, the expression of *Bai1* and *Bai2* was up-regulated following birth, with peak expression at postnatal day 10 (P10) and a high expression maintained until adulthood.[5,6] This allows for the possibility that these genes are involved in brain angiogenesis as well as other neuronal functions. Interestingly, *Bai3* was found to reach its highest level of expression at P1, thereafter its expression steadily decreased.[7] Latrophilin 1-3, on the other hand, have been linked to cell adhesion and cell signaling within the CNS, with latrophilin 1 and 2 predominantly expressed in the brain.[8,9]

*Celsr1-3* genes are expressed broadly in the neuroepithelium at early developmental stages, with distinct expression patterns in the developing CNS arising later.[10] At embryonic day 14 (E14), *Celsr1* was expressed only in the ventricular zone, an area rich in neural progenitor cells. Complementary to *Celsr1*, the expression of *Celsr3* is restricted to the cortical plate.[10] The expression of *Celsr2* is less specific and is detected throughout the developing cerebral cortex and spinal cord.[10]

CELSR2 and CELSR3 are activated by their homophilic interactions and exert opposing roles in the growth of dendrites and axons.[11] Through silencing of *Celsr3*, an overgrowth of dendrites and axons were found to occur.[12] Celsr3 thus plays a role in suppressing neurite growth while Celsr2 was found to enhance neurite growth.[12,13] Celsr1 is however implicated in neurulation as homozygous mutant *Celsr1* mice demonstrate severe neural tube defects.[14]

Very large G protein-coupled receptor-1 (VLGR1) is highly expressed in the CNS during embryogenesis.[15] The expression of VLGR1 is seen as early as E8, coinciding with neurulation.[15] By mid-gestation, strong VLGR1 expression is detected in the ventricular zone. *Vlgr1* mutant mice have been found to exhibit audiogenic seizures.[16,17] Morphological analysis of their brain, however, failed to reveal any obvious malformations.[16] Mutations in the *VLGR1* gene were linked to the pathogenesis of human Usher syndrome type II, an autosomal recessive disorder of congenital hearing loss and progressive retinitis pigmentosa.[18,19] Subsequent studies in the mouse model revealed that VLGR1 plays a crucial role in the normal development of auditory hair bundles.[20]

GPR56 is the only adhesion-GPCRs that has successfully been genetically linked and accountable for a specific human brain malformation. In addition to its role in brain development, GPR56 is also highly implicated in cancer progression, as discussed in the previous chapter.

# GPR56 IN BRAIN DEVELOPMENT AND MALFORMATION

## GPR56 and Bilateral Frontoparietal Polymicrogyria (BFPP)

Mutations in the human *GPR56* gene cause a specific human brain malformation called bilateral frontoparietal polymicrogyria (BFPP).[1,2] BFPP is a recessively inherited genetic disorder of human brain development.[1,2,21,22] Individuals with BFPP present clinically with mental retardation of a moderate to severe degree, motor developmental delay, seizure disorder, cerebellar signs consisting of truncal ataxia, finger dysmetria and rest tremor and dysconjugate gaze described as esotropia, nystagmus, exotropia and strabismus. Magnetic resonance imaging (MRI) of the BFPP brains demonstrates symmetric polymicrogyria with the anterior regions most severely affected, ventriculomegaly, bilateral white matter signal changes and small brainstem and cerebellar structures.[1,2,21,22] The incidence of BFPP remains unknown, largely due to the difficulties in making the diagnosis prior to the availability of high resolution MRI and molecular testing.

A total of thirteen distinct GPR56 mutations have been reported in BFPP patients, including one deletion, two splicing and ten missense mutations (Table 2).[1,2,23] The four missense mutations in the tip of GPR56$^N$ (R38Q, R38W, Y88C and C91S) produce proteins with reduced intracellular trafficking and poor cell surface expression. The two mutations in the GPS domain (C346S and W349S) produce proteins with dramatically impaired cleavage that fail to traffic beyond the endoplasmic reticulum. Chemical chaperons, thapsigargin and 4-phenylbutyrate, can partially rescue mutant GPR56 cell surface expression in cells expressing the mutant receptors, raising the possibility of potential therapeutic intervention for affected pregnancies.[24]

## BFPP is a Cobblestone-Like Brain Malformation

Genotype-phenotype analysis of individuals with BFPP and other similar polymicrogyria syndromes have demonstrated that *GPR56* sequence alterations define a characteristic clinical syndrome similar to cobblestone-like cortical malformation.[2]

**Table 2.** BFPP-associated mutations

| Nucleotide Change | Exon/Intron | Predicted Protein | References |
|---|---|---|---|
| 97C>G | 2 | R33P | 23 |
| 112C>T | 3 | R38W | 1,2 |
| 113G>A | 3 | R38Q | 2 |
| 235C>T | 3 | R79X | 23 |
| 263A>G | 3 | Y88C | 1 |
| 272G>C | 3 | C91S | 1 |
| 739_746delCAGGACC | 5 | Frame Shift | 1 |
| E5-1G>C | 5 | Splicing Mutation | 1 |
| 1036T>A | 8 | C346S | 1 |
| 1046G>C | 8 | W349S | 2 |
| IVS9+3G>C | Intron 9 | Splicing Mutation | 1 |
| 1693C>T | 13 | R565W | 1,2,23 |
| 1919T>G | 13 | L640R | 2 |

Studies in *Gpr56* knockout mice revealed a classic cobblestone-like cortical phenotype, confirming that the histopathology of BFPP is indeed a cobblestone-like cortical malformation (Fig. 1A).[3]

The cobblestone cortex results from aberrant neuronal migration through breaches in the basal lamina (Fig. 1B).[25] Cobblestone cortex is typically seen in three distinct human disorders: Muscle-eye-brain (MEB) disease, Fukuyama-type congenital muscular dystrophy (FCMD) and Walker-Warburg syndrome (WWS).[25] These three disorders are autosomal recessive diseases that encompass congenital muscular dystrophy, ocular malformations and cobblestone lissencephaly. MEB, FCMD and some WWS cases are caused by aberrant glycosylation of α-dystropglycan, a receptor for laminin.[26-28] Mutant mice with deletions in some members of integrin pathway molecules and the extracellular matrix constituents also exhibit cortical migration defects with deficiencies in basal lamina integrity and cortical ectopias, features that resemble the human cobblestone malformation.[29-34]

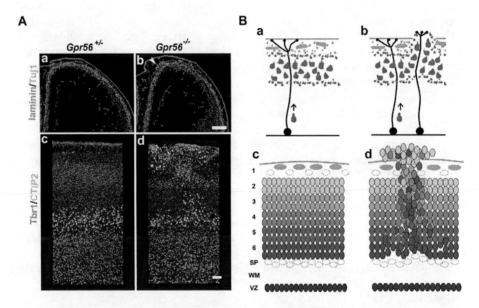

**Figure 1.** A) Cobblestone-like cortical malformation in *Gpr56* null mutant mice. a-b) Double-labeled immunohistochemistry (IHC) of Tuj1 (green) and laminin (red) on E13.5 coronal sections revealing intact pial basement membrane in *Gpr56* heterozygous brains (a) with Tuj1-positive neurons lining up neatly underneath the pial BM. Whereas, *Gpr56* homozygous brain (b) has a ruptured pial BM with Tuj1-positive neurons migrating through the fragmented pial BM (arrowhead). c-d) Immunostaining on P6 brain sections for cortical layer specific markers Tbr1 (Layers II-III, VI) and CTIP2 (Layer V) revealed the disorganized lamination in *Gpr56* null mutant brain (d). B) Development of normal cerebral cortex and cobblestone lissencephaly. a,c) Postmitotic neurons (blue) migrate along the radial glia process (black) to their respective layer. This formation occurs in an 'inside-out' manner, where progressively newer neurons migrate past the early-born neurons to occupy progressively more superficial layers of the cortex and eventually forms the six-layered cortex found in adult brains. All layers are formed directly beneath the marginal zone where Cajal-Retzius (C-R) cells (green) are found (c). b,d) The formation of cobblestone lissencephaly. Cobblestone lissencephaly has three essential features: first, the basal lamina is broken (red line); second, radial glial endfeet are abnormally positioned at the defective reagion; third, neurons migrate though the breach and form ectopic bumps on the surface of the brain. A color version of this image is available at www.landesbioscience.com/curie.

The suggested mechanism leading to cobblestone lissencephaly has been the defective pial basement membrane (BM).[25] However, recent literature has challenged this notion. For example, (1) mice lacking MEKK4 and the Ras family guanine nucleotide exchange factor, C3G, exhibited a neuronal migration defect and the development of a cobblestone-like cortical malformation;[35,36] (2) complete absence of Ena/VASP, an actin binding protein, resulted in the formation of cobblestone-like cortex.[37] Taken together, these observations raise the possibility that abnormal neuronal migration may account partially for the improper formation of cobblestone-like cortex.

## GPR56 is Expressed in Neuronal Progenitor Cells

*Gpr56* mRNA was detected in the ventricular zone of the mouse developing cortex.[1] Subsequent immunohistochemistry (IHC) revealed that GPR56 is highly expressed in radial glial cells.[3] Radial glia cells are special progenitor cells in the developing cortex. They have their somata in the ventricular zone and extend their long radial processes through the entire cortex, attaching via their endfeet to the pial BM. The proper anchorage of the radial glial endfeet is highly relevant to the integrity of the pial BM.[29-33,38-40] Double-labeled IHC showed a high expression of GPR56 in the radial glial endfeet.[3] The putative ligand of GPR56 is also present in the pial BM.[3] Thus, it is conceivable that GPR56 is involved in the dynamic regulation of the pial BM integrity by binding its ligand in the extracellular matrix of the developing brain.

## GPR56 Regulates Granule Cell Adhesion

BFPP patients demonstrate cerebellar signs consisting of truncal ataxia, finger dysmetria and rest tremor, as well as cerebellar hypoplasia on their brain MRIs. However, little is known about the characteristics of the cerebellar defects. Histological analysis of adult *Gpr56* knockout mice revealed a malformed rostral part of the cerebellum, encompassing lobules I-V in the form of fragmented pial BM and disruptions in folding of the cerebellar lobes.[41] Strikingly, the defects in the developing cerebellum are only seen in the regions where *Gpr56* is expressed between approximately E18.5 and P0.5.[41] Furthermore, despite the expansion of GPR56 expression to the caudal cerebellum at a later stage, the affected region remains restricted to lobules I-V at all ages, indicating the unique function of *Gpr56* in rostral cerebellar development during the perinatal period.[41]

Although GPR56 is expressed in other cell types during later developmental stage in the cerebellum, its expression is restricted to the external granule layer (EGL) between E18.5-P0.5.[41] Loss of GPR56 resulted in decreased granule cell adhesion to laminin and fibronectin.[41] Moreover, GPR56 does not mediate granule cell adhesion through direct binding, since neither addition of soluble N-terminal fragment of GPR56 to the medium in the granule cell adhesion assay nor overexpression of full-length mouse GPR56 in HEK 293T cell line altered cell adhesion to laminin and fibronectin. It is possible that GPR56 regulates cell adhesion together with other membrane proteins like integrins and tetraspanins (Fig. 2). Indeed, GPR56 has been shown to interact with tetraspanins in cultured cells and members of the tetraspanins family act as molecular scaffolds with intergrin.[42,43]

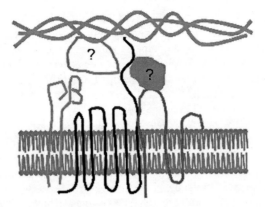

**Figure 2.** Possible interactions between GPR56 (black), integrins (orange), tetraspanin (red) and the unknown ligand(s) (blue). Extracellular membrane (ECM) is shown in gray with the cell membrane in green. A color version of this image is available at www.landesbioscience.com/curie.

## GPR56 Binds Tissue Transglutaminase (TG2)

GPR56 was shown to bind tissue transglutaminase (TG2).[44] However, the biological consequences of this binding are not clear. TG2 is an intriguing molecule that was identified more than 40 years ago as a liver enzyme that adds amines to proteins.[45] TG2 is expressed in both the cytosol and extracellular space. It 'moonlights' between several distinct biochemical roles at various cellular locations. In addition to crosslinking, TG2 can modify proteins by amine incorporation and deamidation and by acting as an isopeptidase in a $Ca^{2+}$-dependent manner.[46] Furthermore, TG2 mediates the interaction of integrins with fibronectin and crosslinks proteins of the extracellular matrix.[47,48] The intracellular TG2, a latent form due to the low cytosolic $Ca^{2+}$ concentration under physiological condition, functions as a G protein ($G_h$).[49] $G_h$/TG2 couples α1b- and α1d-adrenoreceptors, thromboxane and oxytocin receptors to activate phospholilpase C (PLCδ1).[50,51]

The expression of TG2 in the developing forebrain was demonstrated by RT-PCR and western blot analysis with a progressively decreased protein level from E12 to P7, yet it is not known whether it is expressed in the cytosol or the ECM.[52] Interestingly, the enzymatic activity of TG2 was low in the early embryonic mouse forebrain but increased fivefold to peak on the day of birth (P0), which corresponds with the beginning of the mouse brain growth spurt including neuronal differentiation, neurite outgrowth and synaptogenesis.[52] These observations raise the possibility that TG2 may play different roles at different stages of brain development, i.e., "moonlighting" between a signaling G protein and a transamidating enzyme.

## The Ligand of GPR56

Our unpublished data suggests that TG2 is not the only binding partner in the developing brain. Furthermore, we demonstrated that the putative ligand of GPR56 is expressed in the meninges and pial BM.[3] The meninges, differentiated from the immature meningeal fibroblasts of neural crest origin, consist of three distinct mesenchymal cell layers—the pia, arachnoid and dura. The pia produces the BM, a specialized extracellular matrix that overlies the surface of the developing brain. Meningeal fibroblasts contribute to the pial BM by secreting and organizing the majority of basal lamina constituents, including laminin, collagen IV, nidogen and the heparin sulfate proteoglycan.[53] The role of pial meningeal fibroblasts in cortical development was first observed after chemical ablation of these cells.[54] Recent studies in meningeal specific transcriptional factors demonstrated that meningeal fibroblasts regulate cortical development by actively maintaining the dynamic structure of the pial BM/marginal zone.[55-57]

### The Signaling of GPR56 in Cell Migration

Little is known about the signaling properties of GPR56. GPR56 has been shown to associate with the tetraspanin molecules CD9 and CD81, whose complex than associates with $G\alpha_{q/11}$ and $G\beta$ subunits.[42] More recently, it was suggested that GPR56 signals through the $G\alpha_{12/13}$ and RhoA pathway.[58] Identifying the ligand of GPR56 is the first step towards elucidating its signaling. It was described that the N-terminal fragment of GPR56 binds TG2.[44] However, it is unclear whether TG2 functions as the ligand of GPR56.

It is suggested that GPR56 plays a role in cell migration by the following evidences: (1) the presence of GPR56 at the leading edge of cell membrane filopodia;[59] (2) colocalized GPR56 and $\alpha$-actinin;[59] (3) activation of transcription factors, including NF-κB, by overexpression of GPR56;[59] and (4) GPR56 inhibit neural progenitor cell migration by coupling with $G\alpha_{12/13}$ to induce Rho-dependent transcription activation and actin fiber reorganization.[58]

## CONCLUSION

Many of the adhesion-GPCRs are expressed and most probably have functions within the CNS. Yet, only a few of them are studied for their functions in the CNS. GPR56 thus far is the only one implicated in a specific human brain malformation. Nevertheless, the unraveling of GPR56 signaling has just begun. With the emerging effort in studying the developmental insights in relation to the adhesion-GPCR family, more members will reveal their roles in the development and function of the CNS.

## ACKNOWLEDGEMENTS

Works in the author's laboratories are in part supported by the National Institute of Neurological Disorders and Stroke, K08NS045762 and R01NS057536.

## REFERENCES

1. Piao X, Hill RS, Bodell A et al. G protein-coupled receptor-dependent development of human frontal cortex. Science 2004; 303(5666):2033-2036.
2. Piao X, Chang BS, Bodell A et al. Genotype-phenotype analysis of human frontoparietal polymicrogyria syndromes. Ann Neurol 2005; 58(5):680-687.
3. Li S, Jin Z, Koirala S et al. GPR56 regulates pial basement membrane integrity and cortical lamination. J Neurosci 2008; 28(22):5817-5826.

4. Mori K, Kanemura Y, Fujikawa H et al. Brain-specific angiogenesis inhibitor 1 (BAI1) is expressed in human cerebral neuronal cells. Neurosci Res 2002; 43(1):69-74.
5. Kee HJ, Koh JT, Kim MY et al. Expression of brain-specific angiogenesis inhibitor 2 (BAI2) in normal and ischemic brain: involvement of BAI2 in the ischemia-induced brain angiogenesis. J Cereb Blood Flow Metab 2002; 22(9):1054-1067.
6. Koh JT, Lee ZH, Ahn KY et al. Characterization of mouse brain-specific angiogenesis inhibitor 1 (BAI1) and phytanoyl-CoA alpha-hydroxylase-associated protein 1, a novel BAI1-binding protein. Brain Res Mol Brain Res 2001; 87(2):223-237.
7. Kee HJ, Ahn KY, Choi KC et al. Expression of brain-specific angiogenesis inhibitor 3 (BAI3) in normal brain and implications for BAI3 in ischemia-induced brain angiogenesis and malignant glioma. FEBS Lett 2004; 569(1-3):307-316.
8. Ichtchenko K, Bittner MA, Krasnoperov V et al. A novel ubiquitously expressed alpha-latrotoxin receptor is a member of the CIRL family of G-protein-coupled receptors. J Biol Chem 1999; 274(9):5491-5498.
9. Matsushita H, Lelianova VG, Ushkaryov YA. The latrophilin family: multiply spliced G protein-coupled receptors with differential tissue distribution. FEBS Lett. 1999; 443(3):348-352.
10. Shima Y, Copeland NG, Gilbert DJ et al. Differential expression of the seven-pass transmembrane cadherin genes Celsr1-3 and distribution of the Celsr2 protein during mouse development. Dev Dyn 2002; 223(3):321-332.
11. Shima Y, Kengaku M, Hirano T et al. Regulation of dendritic maintenance and growth by a mammalian 7-pass transmembrane cadherin. Dev Cell 2004; 7(2):205-216.
12. Shima Y, Kawaguchi SY, Kosaka K et al. Opposing roles in neurite growth control by two seven-pass transmembrane cadherins. Nat Neurosci 2007; 10(8):963-969.
13. Tissir F, Bar I, Jossin Y et al. Protocadherin Celsr3 is crucial in axonal tract development. Nat Neurosci 2005; 8(4):451-457.
14. Curtin JA, Quint E, Tsipouri V et al. Mutation of Celsr1 disrupts planar polarity of inner ear hair cells and causes severe neural tube defects in the mouse. Current Biology 2003; 13(13):1129-1133.
15. McMillan DR, Kayes-Wandover KM, Richardson JA et al. Very large G protein-coupled receptor-1, the largest known cell surface protein, is highly expressed in the developing central nervous system. J Biol Chem 2002; 277(1):785-792.
16. McMillan DR, White PC. Loss of the transmembrane and cytoplasmic domains of the very large G protein-coupled receptor-1 (VLGR1 or Mass1) causes audiogenic seizures in mice. Mol Cell Neurosci 2004; 26(2):322-329.
17. Skradski SL, White HS, Ptacek LJ. Genetic mapping of a locus (mass1) causing audiogenic seizures in mice. Genomics 1998; 49(2):188-192.
18. Weston MD, Luijendijk MW, Humphrey KD et al. Mutations in the VLGR1 gene implicate G-protein signaling in the pathogenesis of Usher syndrome type II. Am J Hum Genet 2004; 74(2):357-366.
19. van Wijk E, van der Zwaag B, Peters T et al. The DFNB31 gene product whirlin connects to the Usher protein network in the cochlea and retina by direct association with USH2A and VLGR1. Hum Mol Genet 2006; 15(5):751-765.
20. McGee J, Goodyear RJ, McMillan DR et al. The very large G-protein-coupled receptor VLGR1: a component of the ankle link complex required for the normal development of auditory hair bundles. J Neurosci 2006; 26(24):6543-6553.
21. Piao X, Basel-Vanagaite L, Straussberg R et al. An autosomal recessive form of bilateral frontoparietal polymicrogyria maps to chromosome 16q12.2-21. Am J Hum Genet 2002; 70(4):1028-1033.
22. Chang BS, Piao X, Bodell A et al. Bilateral frontoparietal polymicrogyria: clinical and radiological features in 10 families with linkage to chromosome 16. Ann Neurol 2003; 53(5):596-606.
23. Parrini E, Ferrari AR, Dorn T et al. Bilateral frontoparietal polymicrogyria, Lennox-Gastaut syndrome and GPR56 gene mutations. Epilepsia 2008; 50(6):1344-1353.
24. Jin Z, Tietjen I, Bu L et al. Disease-associated mutations affect GPR56 protein trafficking and cell surface expression. Hum Mol Genet 2007; 16(16):1972-1985.
25. Olson EC, Walsh CA. Smooth, rough and upside-down neocortical development. Curr Opin Genet Dev 2002; 12(3):320-327.
26. Kobayashi K, Nakahori Y, Miyake M et al. An ancient retrotransposal insertion causes Fukuyama-type congenital muscular dystrophy. Nature 1998; 394(6691):388-392.
27. Yoshida A, Kobayashi K, Manya H et al. Muscular dystrophy and neuronal migration disorder caused by mutations in a glycosyltransferase, POMGnT1. Dev Cell 2001; 1(5):717-724.
28. Michele DE, Barresi R, Kanagawa M et al. Post-translational disruption of dystroglycan-ligand interactions in congenital muscular dystrophies. Nature 2002; 418(6896):417-422.
29. Georges-Labouesse E, Mark M, Messaddeq N et al. Essential role of alpha 6 integrins in cortical and retinal lamination. Curr Biol 1998; 8(17):983-986.

30. De Arcangelis A, Mark M, Kreidberg J et al. Synergistic activities of alpha3 and alpha6 integrins are required during apical ectodermal ridge formation and organogenesis in the mouse. Development 1999; 126(17):3957-3968.
31. Graus-Porta D, Blaess S, Senften M et al. Beta1-class integrins regulate the development of laminae and folia in the cerebral and cerebellar cortex. Neuron 2001; 31(3):367-379.
32. Beggs HE, Schahin-Reed D, Zang K et al. FAK deficiency in cells contributing to the basal lamina results in cortical abnormalities resembling congenital muscular dystrophies. Neuron 2003; 40(3):501-514.
33. Niewmierzycka A, Mills J, St-Arnaud R et al. Integrin-linked kinase deletion from mouse cortex results in cortical lamination defects resembling cobblestone lissencephaly. J Neurosci 2005; 25(30):7022-7031.
34. Costell M, Gustafsson E, Aszodi A et al. Perlecan maintains the integrity of cartilage and some basement membranes. J Cell Biol 1999; 147(5):1109-1122.
35. Sarkisian MR, Bartley CM, Chi H et al. MEKK4 signaling regulates filamin expression and neuronal migration. Neuron 2006; 52(5):789-801.
36. Voss AK, Britto JM, Dixon MP et al. C3G regulates cortical neuron migration, preplate splitting and radial glial cell attachment. Development 2008; 135(12):2139-2149.
37. Kwiatkowski AV, Rubinson DA, Dent EW et al. Ena/VASP is required for neuritogenesis in the developing cortex. Neuron 2007; 56(3):441-455.
38. Halfter W, Dong S, Yip YP et al. A critical function of the pial basement membrane in cortical histogenesis. J Neurosci 2002; 22(14):6029-6040.
39. Haubst N, Georges-Labouesse E, De Arcangelis A et al. Basement membrane attachment is dispensable for radial glial cell fate and for proliferation, but affects positioning of neuronal subtypes. Development 2006; 133(16):3245-3254.
40. Hu H, Yang Y, Eade A et al. Breaches of the pial basement membrane and disappearance of the glia limitans during development underlie the cortical lamination defect in the mouse model of muscle-eye-brain disease. J Comp Neurol 2007; 501(1):168-183.
41. Koirala S, Jin Z, Piao X et al. GPR56-regulated granule cell adhesion is essential for rostral cerebellar development. J Neurosci 2009; 29(23):7439-7449.
42. Little KD, Hemler ME, Stipp CS. Dynamic regulation of a GPCR-tetraspanin-G protein complex on intact cells: central role of CD81 in facilitating GPR56-G$\alpha_{q/11}$ association. Mol Biol Cell 2004; 15(5):2375-2387.
43. Levy S, Shoham T. The tetraspanin web modulates immune-signalling complexes. Nat Rev Immunol 2005; 5(2):136-148.
44. Xu L, Begum S, Hearn JD et al. GPR56, an atypical G protein-coupled receptor, binds tissue transglutaminase, TG2 and inhibits melanoma tumor growth and metastasis. Proc Natl Acad Sci USA 13 2006; 103(24):9023-9028.
45. Lorand L, Graham RM. Transglutaminases: crosslinking enzymes with pleiotropic functions. Nat Rev Mol Cell Biol 2003; 4(2):140-156.
46. Fesus L, Piacentini M. Transglutaminase 2: an enigmatic enzyme with diverse functions. Trends Biochem Sci 2002; 27(10):534-539.
47. Gaudry CA, Verderio E, Aeschlimann D et al. Cell surface localization of tissue transglutaminase is dependent on a fibronectin-binding site in its N-terminal beta-sandwich domain. J Biol Chem 1999; 274(43):30707-30714.
48. Telci D, Wang Z, Li X et al. Fibronectin-tissue transglutaminase matrix rescues RGD-impaired cell adhesion through syndecan-4 and beta1 integrin cosignaling. J Biol Chem 2008; 283(30):20937-20947.
49. Mhaouty-Kodja S. Ghalpha/tissue transglutaminase 2: an emerging G protein in signal transduction. Biol Cell 2004; 96(5):363-367.
50. Nakaoka H, Perez DM, Baek KJ et al. Gh: a GTP-binding protein with transglutaminase activity and receptor signaling function. Science 1994; 264(5165):1593-1596.
51. Kang SK, Yi KS, Kwon NS et al. Alpha1B-adrenoceptor signaling and cell motility: GTPase function of Gh/transglutaminase 2 inhibits cell migration through interaction with cytoplasmic tail of integrin alpha subunits. J Biol Chem 2004; 279(35):36593-36600.
52. Bailey CD, Johnson GV. Developmental regulation of tissue transglutaminase in the mouse forebrain. J Neurochem 2004; 91(6):1369-1379.
53. Sievers J, Pehlemann FW, Gude S et al. Meningeal cells organize the superficial glia limitans of the cerebellum and produce components of both the interstitial matrix and the basement membrane. J Neurocytol 1994; 23(2):135-149.
54. Super H, Martinez A, Soriano E. Degeneration of Cajal-Retzius cells in the developing cerebral cortex of the mouse after ablation of meningeal cells by 6-hydroxydopamine. Brain Res Dev Brain Res 1997; 98(1):15-20.
55. Zarbalis K, Siegenthaler JA, Choe Y et al. Cortical dysplasia and skull defects in mice with a Foxc1 allele reveal the role of meningeal differentiation in regulating cortical development. Proc Natl Acad Sci USA 2007; 104(35):14002-14007.

56. Paredes MF, Li G, Berger O et al. Stromal-derived factor-1 (CXCL12) regulates laminar position of Cajal-Retzius cells in normal and dysplastic brains. J Neurosci 2006; 26(37):9404-9412.
57. Inoue T, Ogawa M, Mikoshiba K et al. Zic deficiency in the cortical marginal zone and meninges results in cortical lamination defects resembling those in type II lissencephaly. J Neurosci 2008; 28(18):4712-4725.
58. Iguchi T, Sakata K, Yoshizaki K et al. Orphan G protein-coupled receptor GPR56 regulates neural progenitor cell migration via a $G\alpha_{12/13}$ and Rho pathway. J Biol Chem 2008; 283(21):14469-14478.
59. Shashidhar S, Lorente G, Nagavarapu U et al. GPR56 is a GPCR that is overexpressed in gliomas and functions in tumor cell adhesion. Oncogene 2005; 24(10):1673-1682.
60. Haitina T, Olsson F, Stephansson O et al. Expression profile of the entire family of Adhesion G protein-coupled receptors in mouse and rat. BMC Neurosci 2008; 9:43.
61. Lein ES, Hawrylycz MJ, Ao N et al. Genome-wide atlas of gene expression in the adult mouse brain. Nature 2007; 445(7124):168-176.
62. Pickering C, Hagglund M, Szmydynger-Chodobska J et al. The Adhesion-GPCR GPR125 is specifically expressed in the choroid plexus and is upregulated following brain injury. BMC Neurosci 2008; 9:97.
63. Lagerström MC, Rabe N, Haitina T et al. The evolutionary history and tissue mapping of GPR123: specific CNS expression pattern predominantly in thalamic nuclei and regions contraining large pyramidal cells. J Neurochem 2007; 100:1129-1142.

## CHAPTER 8

# GPR56 INTERACTS WITH EXTRACELLULAR MATRIX AND REGULATES CANCER PROGRESSION

Lei Xu*

**Abstract:** GPR56 is a relatively recent addition to the adhesion-GPCR family. Genetic and biochemical studies uncovered its roles in cancer and development and established its function as an adhesion receptor to mediate the interactions between cells and extracellular matrix. Despite of much progress on understanding its biological implications, the mechanism of its function remains elusive. It has not been firmly established whether GPR56 signals directly through G proteins and what its upstream stimuli and downstream effectors are to execute its various biological effects. This chapter will give an overview of the primary structures of the Gpr56 gene and its encoded protein and attempt to point out open questions in this research area, with an emphasis on its roles in cancer and signal transduction.

## INTRODUCTION

GPR56 is a relatively recent addition to the adhesion-GPCR family.[1,2] After its identification, significant progress has been made in understanding its biological roles.[3-5] GPR56 has been implicated in cancer progression and brain development. The expression levels of GPR56 were inversely correlated with the metastatic potential of human melanoma cell lines[2,3] and its expression suppressed melanoma metastasis and growth in a xenograft model.[3] Mutations in the *GPR56* gene were found in human patients with a specific brain malformation called BFPP (bilateral frontoperietal polymicrogyria).[5] The defects in BFPP patients were recapitulated in *Gpr56* knockout mice.[4,6] The effects of GPR56 on cancer and brain development converged on its predicted role as an adhesion receptor: it binds to extracellular matrix (ECM) and mediates the interactions between cells and ECM.[3,4] These functions of GPR56 and their implications on cancer will be discussed in details

*Lei Xu—Department of Biomedical Genetics, 601 Elmwood Ave., University of Rochester Medical Center, Rochester, New York 14642, USA. Email: lei_xu@urmc.rochester.edu

*Adhesion-GPCRs: Structure to Function*, edited by Simon Yona and Martin Stacey.
©2010 Landes Bioscience and Springer Science+Business Media.

in this chapter, following the introduction of the structures of the *GPR56* gene and its encoded protein. The functions of GPR56 in brain development will be elucidated in the chapter by N. Strokes et al.

## THE PRIMARY STRUCTURE OF GPR56 GENE AND ITS ENCODED PROTEIN

### The *GPR56* Gene

The *GPR56* gene was identified in 1999 in an effort to search for *EMR2*-like genes from a human cDNA library by degenerative PCR.[1] In the same year, a different group discovered an inverse relationship between a gene named TM7XN1 and the metastatic potential of human melanoma cell lines.[2] The TM7XN1 gene was later found to be identical to *GPR56*. Although the family of adhesion-GPCRs are conserved from sea squirt to mammals,[7] proteins homologous to GPR56 were only found in birds and mammals (www.ncbi.nlm.nih.gov). In both avian and mammalian genomes, the *Gpr56* gene is located between the *Gpr114* gene and the *Gpr97* gene on the same chromosome. GPR114 and GPR97 proteins share significant sequence similarity with GPR56 and were clustered with GPR56 in the same subclass (group VIII) of adhesion-GPCRs.[8] They might be products of ancestral gene duplications, but what roles they might play or whether they have redundant functions as GPR56 have not been investigated.

The human *GPR56* gene is localized on Chromosome 16q12.2-q21 and spans ~45 kb with 15 exons (GeneID 9289, Fig. 1). The *GPR56* mRNA is alternatively spliced and produces multiple mRNA variants that result in three main isoforms of GPR56 protein (Fig. 1). These GPR56 isoforms differ in their inclusion of an RVPLPC sequence in the first intracellular loop or an ASASS sequence in the signal peptide. Whether these variations result in any difference in protein functions has not been investigated. The longest isoform (isoform a) of GPR56 protein contains 693 amino acids.

*GPR56* mRNA is widely distributed in tissues and organs.[1] It is up-regulated in both neuronal and hematopoietic stem cells;[9,10] it is one of the few genes that are shared by both types of stem cells, suggesting its potential roles in stem cell maintenance. Consistent with this, *Gpr56* mRNA was shown to be expresssed in the subventricular zone of mouse cerebral cortex,[4,5] where the neuronal stem cells reside. It is not clear whether the distribution of GPR56 protein overlaps with that of its mRNA, but at least in cerebral cortex, both the GPR56 protein and its mRNA are expressed in neuronal progenitor cells.[4]

### The GPR56 Protein

Like other adhesion-GPCRs, GPR56 protein contains an extended N-terminus and a seven-pass transmembrane domain that shares significant homology with class B GPCRs.[2] The N-terminus of GPR56 contains seven N-glycosylation sites and many potential O-glycosylation sites.[11] Deglycosylation analyses in vitro showed that glycosylation occurs at all of the seven N-glycosylation sites and mutating all of them abolished cell surface localization of the receptor.[11] The N-terminus of GPR56 does not contain any protein domains that are known to be involved in cell adhesion. However, like many other adhesion-GPCRs, the intermediate segment within the GPR56 N-terminus is highly enriched in Serine, Threonine and Proline (STP region) (Fig. 2). STP region

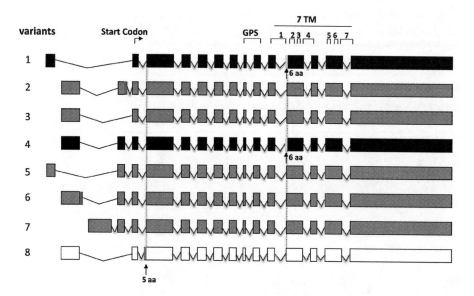

**Figure 1.** *GPR56* mRNA Variants. Human *GPR56* gene spans 15 exons and encodes a protein containing a seven-span transmembrane domain and a GPS motif. Sequences from human ESTs correspond to eight different variants of *GPR56* mRNA, presumably due to alternative splicing. These eight *GPR56* mRNA variants encode three isoforms of GPR56 proteins. In the diagram, boxes indicate exons and lines indicate introns. Exons in black constitute variant 1 and 4 mRNAs, encoding the isofrom a of GPR56 protein. Exons in grey constitute variant 2, 3, 5, 6, 7 mRNAs, all encoding the isoform b of GPR56 protein. Exons in white consistute variant 8 mRNA, encoding the isoform c of GPR56 protein. Isoform a of GPR56 protein differs from isoform b in that it includes an insertion of six amino acids in the loop between the first and second transmembrane domains (shown in light grey boxes). Isoform c of GPR56 protein differs from the isoform b protein in that it includes an insertion of five amino acids in the signal peptide (shown in a dark grey box). Variant 2 and 6 *GPR56* mRNAs include additional sequences at their 5' UTR (shown in green boxes), which do not cause any change in protein sequences. A color version of this image is available at www.landesbioscience.com/curie.

is typically found in mucins.[12,13] Its sequence is poorly defined but thought to act as docking sites for O-glycosylation and mediate protein-protein interactions. STP-rich regions have been found in many other adhesion-GPCR family members,[14] but their functions are not clear.

Immediately upstream of the seven-pass transmembrane domain of GPR56 is a cysteine-rich GPS (GPCR proteolytic site) motif. The GPS motif was initially discovered in latrophilin[15] and was later found in most adhesion-GPCRs as well as other multi-pass transmembrane proteins, such as the polycystin-1 family members.[7] GPS motif contains the highly conserved proteolytic site, His-Leu-Ser (Thr), with cleavage occurring between the Leu and Ser (Thr) residues. Proteolysis through the GPS motifs is thought to proceed via an auto-catalyzed mechanism.[15a] Autocatalytic cleavages occur in a diverse group of proteins reference including Hedgehog, glycosylasparaginase, nucleoporin and intein-containing proteins and are important for the normal functions of these proteins.[16] Similarly, the GPS motifs in adhesion-GPCRs also appear to be required for their functions. Mutations in the GPS motif of latrophilin resulted in the retention of the receptor in ER.[17] Presumably through its GPS motif,[3,11] the mature GPR56 protein is cleaved into two fragments. Similar to latrophilin, this cleavage is

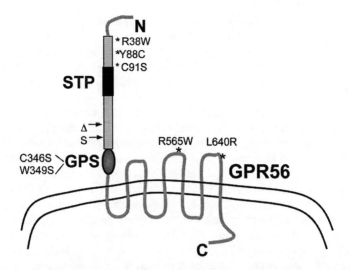

**Figure 2.** GPR56 Protein. GPR56 protein contains a seven-pass transmembrane domain and a GPS motif. The intermediate segment of its N-terminus presides a Serine Threonine Proline (STP)—rich region. Mutations found in BFPP patients are shown on the diagram. Most of the mutations are at the N-terminus of the protein, except the two that are in the extracellular loop and the beginning of the last transmembrane segment.

essential for the normal functions of GPR56. Two of the mutations in human BFPP patients were identified in the GPS motif of GPR56.[5] When these mutated forms of GPR56 were expressed in cell lines in vitro, they were found to remain uncleaved and fail to traffic to cell surface.[11]

Downstream from the cleavage site is the seven-span transmembrane domain followed by a short C-terminal tail. Two additional mutations in BFPP patients, R565W and L640R,[5] reside in the 2nd extracellular loop of the transmembrane domain and at the beginning of the last transmembrane segment, respectively. When expressed in cell lines in vitro, these mutations did not cause any defects in cleavage or secretion of the receptor, but appeared to differentially regulate the localization of the C-terminal fragment of GPR56 to cell surface.[11] The C-terminal tail and the 3rd intracellular loop in many conventional G protein-coupled receptors bind to Gα proteins and transmit the downstream signaling. However, these segments in GPR56 are unusually short, raising the question whether GPR56 indeed directly signals through Gα's as classical GPCRs.

## FUNCTIONS OF GPR56 IN CANCER

GPR56 was identified as a putative metastasis suppressor in an experimental metastasis model.[3,18] In this model, a pool of poorly metastatic human melanoma cells were injected directly into the blood stream of immunodeficient mice via their tail veins. The rare lung metastases were collected and amplified in vitro as cell lines. When these cell lines were injected into mice again, they formed many more lung metastases than the parental line. Using this method, several human melanoma cell lines with different metastatic potentials were derived and genes that were up- or down-regulated in the samples from highly metastatic cell lines compared with those from the poorly metastatic parental line were identified by gene expression analyses. *GPR56* mRNA was found to be down-regulated in tumor samples from highly metastatic melanoma cell lines, consistent with an earlier study.[2] Further investigations showed that GPR56 played suppressive roles in melanoma metastasis. Its overxpression led to reduced metastasis and its down-regulation led to enhanced metastasis.[3]

How does GPR56 suppress metastasis? Cancer cells must complete at least four steps to successfully metastasize: (1) their detachment from the primary tumor, (2) intravasation into circulation, (3) extravasation (exit) from the circulation, and (4) survival/growth in a distant organ.[18,19] The above experimental metastasis assay analyzes the last two steps of metastasis, since the cancer cells were injected directly into the circulation of animals. Both clinical and experimental data suggest that these last two steps are rate-limiting for metastasis. In human cancer patients, a large number of cancer cells are often detected in their circulation, but metastases are rarely detected, or detected only in selected organs.[18] Therefore understanding how metastatic cells exit from circulation and establish themselves as detectable metastases is critical for effective cancer treatment. The finding that GPR56 suppressed metastasis in the experimental metastasis model suggests that it might inhibit the last two steps during metastasis. Whether it affects the extravasation (step 3) of cancer cells has not been investigated. However, expression of GPR56 was shown to suppress the growth of tumor cells in vivo, indicating that it might block the metastatic growth (step 4) of cancer cells during metastasis.

The effects of GPR56 on inhibiting cancer metastasis may correlate with its potential roles in stem cells. During the establishment of metastases, metastatic cells need to proliferate as a tumor mass, a process similar to cancer-initiating process during primary tumor growth. The metastatic cells must therefore contain cancer-initiating cells, or cancer stem cells. *Gpr56* has been shown to be up-regulated in both neuronal and hematopoietic stem cells.[9] Whether it plays any causative roles in stem cells has not been reported. However, stem cells are slow cycling cells, therefore GPR56 might function to maintain the quiescent state of stem cells and, in the context of metastasis, to inhibit the proliferation of metastatic cells. In addition, GPR56 mediates the interaction between cells and ECM as an adhesion receptor. It might maintain the communications between stem cells and their microenvironment in normal tissues[20] and its down-regulation in cancer cells result in dysregulation of these interactions and favor the establishment of metastases.

Several reports also showed that *GPR56* was up-regulated in certain types of cancers compared with corresponding normal tissues.[21,22] In one additional report, reduction of *GPR56* was found to induce the transformation of fibroblast cells and inhibit apoptosis induced by anoikis (anchorge-independent growth).[23] These data suggest that GPR56 might promote cancer progression instead of inhibiting it. The discrepancies between these findings might suggest that GPR56 plays different roles in different cancer types or stages and need to be addressed by further investigation. However, up-regulation of GPR56 in cancer cells compared with normal tissues is not necessarily contradictory to its suppressive roles in metastasis and tumor growth. It is possible that in normal tissues GPR56 is expressed in a subset of cells (such as stem cells) that are amplified in a particular type of tumor, therefore the expression of GPR56 is increased in those tumors compared with normal tissues. Alternatively, up-regulation of GPR56 might act as a feedback mechanism to overcome the proliferation of cancer cells.

## SIGNALING PATHWAYS MEDIATED BY GPR56

GPR56 is predicted to signal through G proteins like conventional GPCRs. However, whether it does signal directly through G proteins and how it transmits its signals have not been established. Recently, protein factors that bind to GPR56 were identified and the pathways downstream GPR56 are beginning to be elucidated. These new findings and their implications will be discussed below.

### TG2, the Putative Ligand

Tissue transglutaminase (TG2) was identified as the first putative ligand for GPR56. In the report by Xu et al, the authors found that the suppressive roles of GPR56 in tumor growth and metastasis appeared to be mediated by a factor in tumor microenvironment.[3] GPR56 was an orphan G protein-coupled receptor and it was speculated that its unknown ligand might be present in the tumor stroma and might have mediated its suppressive effects on cancer progression. To identify such a ligand, a fusion protein between the N-terminus of GPR56 and the human Fc fragment (denoted as FcGPRN) was expressed and purified. It was used as an antibody to detect GPR56-binding partners by immunostaining and on western blots. Results from both immunostaining and western blots indicated that the ligand of GPR56 was part of ECM. Through a series of biochemical purifications and mass spectrometry, it was identified as tissue transglutaminase (TG2), a major crosslinking enzyme in ECM.

The interactions between GPR56 and TG2 established GPR56 as an adhesion receptor that mediates the signaling from ECM to intracellular pathways. Cell-ECM interactions are critical at multiple levels during metastasis and tumor growth.[24] Both the receptors, such as integrins[25] and their ECM ligands have been shown to profoundly affect cancer progression. The mechanisms, however, remain unclear. The link between GPR56 and TG2 provide additional avenues to dissect the functions of cell-ECM interactions in cancer progression.

Transglutaminases catalyze crosslinking of proteins by forming $\varepsilon(\gamma\text{-glutamyl})$lysine peptide bonds between two proteins in a $Ca^{2+}$-dependent manner.[26,27] TG2 is ubiquitously expressed and it is localized both intracellularly and extracellularly. The extracellular TG2 interacts with GPR56 and might mediate the suppressive roles of GPR56 in cancer progression.[28-30] Extracellular TG2 has been implicated in tumor suppression by many studies. One earlier example of such studies used a rat dorsal skin flap window chamber to analyze mammary tumor progression and found that atopic application of TG2 reduced the growth of mammary adenocarcinoma.[28] More recently, exogenous TG2 has been administered to MDA-MB-231 breast cancer cells and it inhibited their invasion through matrigel.[29] The contribution of TG2 in tumor suppression was perhaps most vigorously examined by Jones et al using $Tg2^{-/-}$ mice.[30] They found that the mouse melanoma cell line, B16-F1 cells, grew at a significantly enhanced rate in $Tg2^{-/-}$ mice than in wildtype mice, suggesting strongly that TG2 acts to suppress tumor growth.

How TG2 suppresses tumor progression is not clear. In the study from Jones et al, exogenous TG2 was shown to inhibit angiogenesis in vitro,[30] possibly due to accumulated deposition of collagen I surrounding the capillaries. Increased crosslinking by TG2 was shown to promote the resistance of ECM proteins to degradation.[27] Therefore elevated TG2 activity might stabilize and increase the deposition of ECM proteins, but whether and how these effects regulate tumor progression is not clear. Activated TG2 was

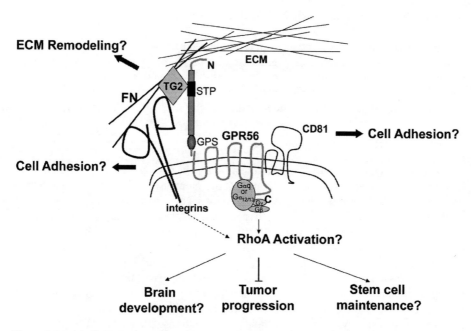

**Figure 3.** Model of GPR56 Function. GPR56 is involved in metastasis suppression, brain development and possibly stem cell maintenance. GPR56 protein is cleaved, presumably in the GPS domain, into two fragments that remain noncovalently bound. The N-terminal fragment contains a Serine-Threonine-Proline (STP)—rich region. The C-terminal fragment contains the seven transmembrane segments. GPR56 associates with CD81 and Gαq in a complex and may be involved in cell adhesion. The N-terminus of GPR56 interacts with TG2, a transglutaminase, which crosslinks the ECM proteins and interacts with fibronectin and integrins on cell surface. These functions of GPR56 and TG2 may lead to ECM remodeling and cell adhesion. TG2 might activate GPR56 and stimulate downstream signaling pathways that inhibit tumor progression and regulate brain development as well as stem cell maintenance. The intracellular signaling events that may mediate the functions of GPR56 are not known. Recently it was reported that GPR56 activates RhoA through Gα$_{12/13}$. Reproduced with permission from Xu L, Hynes RO. Cell Cycle 2007; 6(2):160-165.[18]

reported to enhance the incorporation of latent TGF-β binding protein-1 (LTBP-1) into ECM and thus promote the activity of TGF-β.[31,32] The activated TGF-β acts as a tumor suppressor in many cancer types and therefore it might inhibit cancer cell proliferation upon the activation of TG2. In addition to its crosslinking activity, extracellular TG2 also functions in a crosslinking-independent manner. TG2 binds to fibronectin[33] and integrins[34] and might affect cancer progression by regulating fibronectin-mediated cell adhesion.[35]

How might TG2 mediate the effects of GPR56 in tumor progression? TG2 might activate GPR56 as a ligand and induce the downstream signaling that inhibits cancer progression. Whether this is true needs to be tested. Nevertheless, $Tg2^{-/-}$ mice do not show defects in ECM assembly as observed in $Gpr56^{-/-}$ mice,[36,37] suggesting that either TG2 does not signal through GPR56 or TG2 is not the only ligand for GPR56. Other proteins might also signal through GPR56 and they are most likely ECM proteins, since the N-terminus of GPR56 was shown to bind to ECM in multiple tissues.[3,4] Alternatively, GPR56 might regulate the activity of TG2 via an inside-out mechanism and modulate ECM remodeling in the tumor microenvironment (Fig. 3). GPR56 on tumor cells might "recruit" extracellular TG2 from tumor stroma and thus increases the local concentration of TG2 and TG2 activity. This increased TG2 activity might alter the ECM property in tumors and inhibit tumor progression. Finally, GPR56 might form a complex with TG2 and its associated factors (e.g., fibronectin and integrins) to regulate cell adhesion during cancer progression (Fig. 3). Cell-cell adhesion and cell-ECM adhesion have been shown to play critical roles in cancer growth and metastasis. Therefore, GPR56 could interfere the cell adhesion during extravasation and metastatic growth and inhibit metastasis. All the above possibilities are not mutually exclusive and should be tested in future for further understanding of the roles of GPR56 in cancer.

## GPR56 and CD81

GPR56 was reported to form a complex with the tetraspanin protein, CD81 and Gαq in cell lysates.[38] Tetraspanin proteins are four-span transmembrane proteins that often form large protein complexes (called tetraspanin web) on cell surface and have been implicated in cell adhesion and cancer.[39] CD81 itself has not been directly linked to cancer, but many CD81-associated factors have. For example, CD81 tightly associates with another tetraspanin protein, CD9. CD9 was shown to impair invasion, metastasis and survival in many cancer types.[40-42] GPR56 might mediate suppression in cancer progression through CD9 after its binding with CD81. In addition, CD81 interacts with integrin α4 and strengthens the interaction between integrin α4 to VCAM on endothelial cells.[43] This function of CD81 could potentially facilitate the transmigration of cancer cells through blood vessels during metastasis, which might be inhibited after its binding with GPR56.

## Signaling Components Downstream of GPR56

Like other adhesion-GPCRs, the signaling components downstream of GPR56 have not been well-characterized. GPR56 was reported to associate with CD81 and Gαq in a complex,[38] suggesting that it signals through Gαq and stimulates $Ca^{2+}$ influx and $IP_3$ production. Recently, Iguchi et al proposed an alternative pathway downstream of GPR56. The authors reported that GPR56 signals through $Gα_{12/13}$ and RhoA in neural progenitor cells, but not through Gαq. They found that GPR56 inhibited the migration of neural progenitor cells and these effects were reversed by the addition of inhibitors to $Gα_{12/13}$ and RhoA.[44] Nevertheless, the direct interactions between GPR56 and $Gα_{12/13}$ were not shown. It is possible that GPR56 activates another GPCR that signals through $Gα_{12/13}$ and RhoA. In the same report, the authors also mentioned that ectopic TG2 did not affect the signaling of GPR56, therefore alternative ligands must mediate the signal through GPR56 in these cells. However, TG2 is ubiquitously expressed and therefore most likely was already present in the cells that were analyzed. Exogenous TG2 might not be effective in mediating signaling through GPR56 in those cells, since the receptor is already occupied by the endogenous TG2.

## CONCLUSION

The study on GPR56 is still at an early stage. Although much progress has been made in the last decade on its biological roles, many questions remain unanswered. GPR56 was identified as a suppressor to metastasis and tumor growth. The expression levels of *GPR56* were repeatedly reported to inversely correlate with the metastatic potential of human melanoma cell lines and the expression of GPR56 caused inhibition of melanoma growth and metastasis in a xenograft tumor model. The N-terminus of GPR56 was found to bind directly with tissue transglutaminase (TG2), a crosslinking enzyme in the extracellular matrix (ECM). This finding established GPR56 as an adhesion receptor that mediates the interactions between cells and ECM. It has a wide-range of implications on how GPR56 might affect cancer progression and should be investigated further. On the contrary to the suppressive roles of GPR56 in cancer progression, a few additional reports suggested that GPR56 might promote cancer progression. *GPR56* was found up-regulated in certain types of cancers compared with corresponding normal tissues and reduction of GPR56 was shown to induce transformation of fibroblast cells and inhibit apoptosis. The discrepancies between these findings need to be addressed by further investigations. In addition, GPR56 has been implicated in stem cell maintenance. The *Gpr56* mRNA was shown to be up-regulated in both neuronal stem cells and hematopoietic stem cells and it is localized in the subventricular zones of mouse brains, where neuronal progenitor cells preside. But whether and how GPR56 affects stem cells have not been explored and deserve attentions in future studies. Finally, the signaling pathway that is mediated by GPR56 remains a mystery. GPR56 was shown to signal through $G\alpha_{12/13}$ to activate RhoA in neuronal progenitor cells. Whether GPR56 directly binds to $G\alpha_{12/13}$ and whether TG2 or other ligands stimulates this signaling pathway await future investigations.

## ACKNOWLEDGEMENTS

I thank Dr. Liquan Yang for preparing Figure 1 and Dr. Liquan Yang and Dr. Xianhua Piao for reading the manuscript.

## REFERENCES

1. Liu M, Parker RM, Darby K et al. GPR56, a novel secretin-like human G-protein-coupled receptor gene. Genomics 1999; 55(3):296-305.
2. Zendman AJ, Cornelissen IM, Weidle UH et al. TM7XN1, a novel human EGF-TM7-like cDNA, detected with mRNA differential display using human melanoma cell lines with different metastatic potential. FEBS Lett 1999; 446(2-3):292-298.
3. Xu L, Begum S, Hearn JD et al. GPR56, an atypical G protein-coupled receptor, binds tissue transglutaminase, TG2 and inhibits melanoma tumor growth and metastasis. Proc Natl Acad Sci USA 2006; 103(24):9023-9028.
4. Li S, Jin Z, Koirala S et al. GPR56 regulates pial basement membrane integrity and cortical lamination. J Neurosci 2008; 28(22):5817-5826.
5. Piao X, Hill RS, Bodell A et al. G protein-coupled receptor-dependent development of human frontal cortex. Science 2004; 303(5666):2033-2036.
6. Koirala S, Jin Z, Piao X et al. GPR56-regulated granule cell adhesion is essential for rostral cerebellar development. J Neurosci 2009; 29(23):7439-7449.
7. Yona S, Lin HH, Siu WO et al. Adhesion-GPCRs: emerging roles for novel receptors. Trends Biochem Sci 2008; 33(10):491-500.
8. Lagerstrom MC, Schioth HB. Structural diversity of G protein-coupled receptors and significance for drug discovery. Nat Rev Drug Discov 2008; 7(4):339-357.
9. Terskikh AV, Easterday MC, Li L et al. From hematopoiesis to neuropoiesis: evidence of overlapping genetic programs. Proc Natl Acad Sci USA 2001; 98(14):7934-7939.
10. Terskikh AV, Miyamoto T, Chang C et al. Gene expression analysis of purified hematopoietic stem cells and committed progenitors. Blood 2003; 102(1):94-101.
11. Jin Z, Tietjen I, Bu L et al. Disease-associated mutations affect GPR56 protein trafficking and cell surface expression. Hum Mol Genet 2007; 16(16):1972-1985.

12. Cyster J, Somoza C, Killeen N et al. Protein sequence and gene structure for mouse leukosialin (CD43), a T-lymphocyte mucin without introns in the coding sequence. Eur J Immunol 1990; 20(4):875-881.
13. Corral L, Singer MS, Macher BA et al. Requirement for sialic acid on neutrophils in a GMP-140 (PADGEM) mediated adhesive interaction with activated platelets. Biochem Biophys Res Commun 1990; 172(3):1349-1356.
14. Stacey M, Lin HH, Gordon S et al. LNB-TM7, a group of seven-transmembrane proteins related to family-B G-protein-coupled receptors. Trends Biochem Sci 2000; 25(6):284-289.
15. Krasnoperov VG, Bittner MA, Beavis R et al. Alpha-latrotoxin stimulates exocytosis by the interaction with a neuronal G-protein-coupled receptor. Neuron 1997; 18(6):925-937.
15a. Lin H, Chang G, Davies J et al. Autocatalytic cleavage of the EMR2 receptor occurs at a conserved G protein-coupled receptor proteolytic site motif. JBC 2004; 279(30): 31823-31832.
16. Paulus H. Protein splicing and related forms of protein autoprocessing. Annu Rev Biochem 2000; 69:447-496.
17. Krasnoperov V, Bittner MA, Holz RW et al. Structural requirements for alpha-latrotoxin binding and alpha-latrotoxin-stimulated secretion. A study with calcium-independent receptor of alpha-latrotoxin (CIRL) deletion mutants. J Biol Chem 1999; 274(6):3590-3596.
18. Xu L, Hynes RO. GPR56 and TG2: possible roles in suppression of tumor growth by the microenvironment. Cell Cycle 2007; 6(2):160-165.
19. Fidler IJ. The pathogenesis of cancer metastasis: the 'seed and soil' hypothesis revisited. Nat Rev Cancer 2003; 3(6):453-458.
20. Morrison SJ, Spradling AC. Stem cells and niches: mechanisms that promote stem cell maintenance throughout life. Cell 2008; 132(4):598-611.
21. Shashidhar S, Lorente G, Nagavarapu U et al. GPR56 is a GPCR that is overexpressed in gliomas and functions in tumor cell adhesion. Oncogene 2005; 24(10):1673-1682.
22. Sud N, Sharma R, Ray R et al. Differential expression of G-protein coupled receptor 56 in human esophageal squamous cell carcinoma. Cancer Lett 2006; 233(2):265-270.
23. Ke N, Sundaram R, Liu G et al. Orphan G protein-coupled receptor GPR56 plays a role in cell transformation and tumorigenesis involving the cell adhesion pathway. Mol Cancer Ther 2007; 6(6):1840-1850.
24. Joyce JA, Pollard JW. Microenvironmental regulation of metastasis. Nat Rev Cancer 2009; 9(4):239-252.
25. Hynes RO. Cell-matrix adhesion in vascular development. J Thromb Haemost 2007; 5(Suppl 1):32-40.
26. Lorand L, Graham RM. Transglutaminases: crosslinking enzymes with pleiotropic functions. Nat Rev Mol Cell Biol 2003; 4(2):140-156.
27. Kotsakis P, Griffin M. Tissue transglutaminase in tumour progression: friend or foe? Amino Acids 2007; 33(2):373-384.
28. Haroon ZA, Lai TS, Hettasch JM et al. Tissue transglutaminase is expressed as a host response to tumor invasion and inhibits tumor growth. Lab Invest 1999; 79(12):1679-1686.
29. Mangala LS, Arun B, Sahin AA et al. Tissue transglutaminase-induced alterations in extracellular matrix inhibit tumor invasion. Mol Cancer 2005; 4:33.
30. Jones RA, Kotsakis P, Johnson TS et al. Matrix changes induced by transglutaminase 2 lead to inhibition of angiogenesis and tumor growth. Cell Death Differ 2006; 13(9):1442-1453.
31. Verderio E, Gaudry C, Gross S et al. Regulation of cell surface tissue transglutaminase: effects on matrix storage of latent transforming growth factor-beta binding protein-1. J Histochem Cytochem 1999; 47(11):1417-1432.
32. Kojima S, Nara K, Rifkin DB. Requirement for transglutaminase in the activation of latent transforming growth factor-beta in bovine endothelial cells. J Cell Biol 1993; 121(2):439-448.
33. Lorand L, Dailey JE, Turner PM. Fibronectin as a carrier for the transglutaminase from human erythrocytes. Proc Natl Acad Sci USA 1988; 85(4):1057-1059.
34. Akimov SS, Krylov D, Fleischman LF et al. Tissue transglutaminase is an integrin-binding adhesion coreceptor for fibronectin. J Cell Biol 2000; 148(4):825-838.
35. Akimov SS, Belkin AM. Cell surface tissue transglutaminase is involved in adhesion and migration of monocytic cells on fibronectin. Blood 2001; 98(5):1567-1576.
36. De Laurenzi V, Melino G. Gene disruption of tissue transglutaminase. Mol Cell Biol 2001; 21(1):148-155.
37. Nanda N, Iismaa SE, Owens WA et al. Targeted inactivation of Gh/tissue transglutaminase II. J Biol Chem 2001; 276(23):20673-20678.
38. Little KD, Hemler ME, Stipp CS. Dynamic regulation of a GPCR-tetraspanin-G protein complex on intact cells: central role of CD81 in facilitating GPR56-Galpha q/11 association. Mol Biol Cell 2004; 15(5):2375-2387.
39. Hemler ME. Targeting of tetraspanin proteins—potential benefits and strategies. Nat Rev Drug Discov 2008; 7(9):747-758.
40. Ikeyama S, Koyama M, Yamaoko MS et al. Suppression of cell motility and metastasis by transfection with human motility-related protein (MRP-1/CD9) DNA. J Exp Med 1993; 177(5):1231-1237.

41. Takeda T, Hattori N, Tokuhara T et al. Adenoviral transduction of MRP-1/CD9 and KAI1/CD82 inhibits lymph node metastasis in orthotopic lung cancer model. Cancer Res 2007; 67(4):1744-1749.
42. Saito Y, Tachibana I, Takeda Y et al. Absence of CD9 enhances adhesion-dependent morphologic differentiation, survival and matrix metalloproteinase-2 production in small cell lung cancer cells. Cancer Res 2006; 66(19):9557-9565.
43. Feigelson SW, Grabovsky V, Shamri R et al. The CD81 tetraspanin facilitates instantaneous leukocyte VLA-4 adhesion strengthening to vascular cell adhesion molecule 1 (VCAM-1) under shear flow. J Biol Chem 2003; 278(51):51203-51212.
44. Iguchi T, Sakata K, Yoshizaki K et al. Orphan G protein-coupled receptor GPR56 regulates neural progenitor cell migration via a G alpha 12/13 and Rho pathway. J Biol Chem 2008; 283(21):14469-14478.

# CHAPTER 9

# ADHESION-GPCRs IN TUMORIGENESIS

## Gabriela Aust*

**Abstract:**  Tumor growth is a highly complex, multistep process that involves tumor cell detachment, migration, invasion and metastasis accompanied by angiogenesis and extracellular matrix turn-over. Each of the steps is influenced by tumor cell interaction and interaction of the tumor cell with its microenvironment that consists of different cell types as tumor-associated fibroblasts, endothelial cells and leukocytes as well as the extracellular matrix produced by the tumor cells themselves or by the fibroblasts.
Cellular communication takes place by the regulated expression of adhesion receptors. Adhesion-GPCRs are characterized by very long extracellular N-termini that have multiple domains. When considering this complex structure it is only logical that adhesion-GPCRs are involved in tumor cell interactions. Moreover, these receptors function in cell guidance and/or trafficking, which, in addition to their structure, makes them interesting for tumorigenesis.
The aberrant expression of several adhesion-GPCRs on tumor cells and their involvement in tumor growth have been shown for some of the family members. This overview summarizes expression database data as well as data from original research articles of adhesion-GPCRs in tumors.

## DATABASES

Expression data on tumors are available for most of the adhesion-GPCRs (Table 1). Data on tumor cell lines will give further relevant information whether a particular adhesion-GPCR is expressed in a certain tumor entity. The following databases were utilized:

1. BioGPS (http://biogps.gnf; formerly the SymAtlas) is a database on gene function and structure funded by the Genomics Institute of the Novartis Research

---

*Gabriela Aust—University of Leipzig, Faculty of Medicine; Department of Surgery, Research Laboratories, Liebigstr. 20; D-04103, Leipzig, Germany. Email: gabriela.aust@medizin.uni-leipzig.de

*Adhesion-GPCRs: Structure to Function*, edited by Simon Yona and Martin Stacey.
©2010 Landes Bioscience and Springer Science+Business Media.

Foundation that especially contains mRNA expression analysis of tumor cell lines and primary tumors.[1]
2. HPA: The Swedish Human Proteome Resource program hosts the Human Protein Atlas (HPA, http://www.proteinatlas.org) portal with expression profiles of human proteins in normal and tumor tissues and cells. Data of one-third of the adhesion-GPCRs are available.[2]
3. AGCOH: The Atlas of Genetics and Cytogenetics in Oncology and Haematology (AGCOH; http://atlasgeneticsoncology.org) is a peer reviewed on-line journal and database with free access on the internet devoted to genes, cytogenetics and clinical entities in cancer and cancer-prone diseases. Latrophilin-2 and CD97 are the only adhesion-GPCRs included in AGCOH.[3]

## OVERVIEW ON THE ADHESION-GPCR FAMILY

The classification used in this chapter is according to Bjarnadottir et al.[4]

## GROUP I: LATROPHILIN-LIKE (LATROPHILIN 1-3, ETL)

Latrophilin-1 and -3 are present predominantly in normal brain whereas latrophilin-2 is expressed ubiquitously.[5,6] Latrophilins probably have a physiological function in normal synaptic cell adhesion.[7] Only few and often contradictory expression data on latrophilins in tumors and tumor cell lines are available. Their function in tumorigenesis has not been evaluated yet.

Most tumor cell lines and tumors show rather low **latrophilin-1** mRNA levels.[1]

Some tumor cell lines and tumors of different origin are **latrophilin-2** mRNA positive.[1] Latrophilin-2 protein is present to a small extent in some tumors.[2] Breast and lung tumors were judged as latrophilin-2 negative, which contrasts to the results of the above cited scientific articles.

A gene identification study, centred on a region of overlapping loss of heterozygosity in breast tumors within band 1p31.1, led to the characterisation of latrophilin-2.[8] A number of breast tumor cell lines apparently overexpressed the gene whilst others showed very low levels of transcription.[9]

The strong expression of latrophilin-2 in normal lung is reduced in half of the matched primary nonsmall cell lung carcinomas (NSCLC).[3] Over-representation of latrophilin-2 was not registered in any of these tumors, nor in any lung cancer cell line tested. Latrophilin-2 mRNAs are alternatively spliced to varying degrees with shifts in the major gene product to truncated or altered forms in some lines. Overall, the expression level and pattern of altered forms of latrophilin-2 varied between tumors and correlated to normal tissues.

The BioGPS database did not reveal or only rarely revealed expression of **latrophilin-3** in tumor cell lines and tumors whereas the HPA database suggests that latrophilin-3 is present in many tumor entities and most tumor cell lines.[1,2]

Normal smooth muscle and cardiac myocytes express high **ETL** mRNA levels. Most of the tumor cell lines are ETL-negative, except the mesenchymal cell line HT-1080.[1]

However, the tested cell line panel did not contain cell lines derived from muscle tumors such as leiomyosarcomas, thus a statement on ETL in relevant tumors is not possible.

## GROUP II: EGF-LIKE (CD97, EMR1-4)

CD97 is the only member of EGF-like adhesion-GPCRs that shows a broad, not cell-lineage specific expression.[1] Beside cells of the hematopoetic system, epithelial and muscle cells are CD97-positive.[10,11] Accordingly, most tumor cell lines as well as several tumor entities express CD97 at high levels.[1,2]

### Glycosylation of CD97 in Tumors

Depending on the cell type and transformation status of the cells, CD97 is completely or partly N-glycosylated or naked.[12] During tumor transformation, not only the CD97 protein expression level but also the degree of CD97 N-glycosylation varies as shown for leiomyomas and leiomyosarcomas.[11] However, tumorigenesis is not accompanied necessarily by N-glycosylation of naked CD97 in such tumors.[11]

The selection of the CD97 antibody influences the result in studies focused on the correlation between CD97 and histopathological subtypes, diagnosis, progression, or prognosis of tumors as shown for various tumor entities[13] because CD97 antibodies binding within the first EGF-like domain detect only N-glycosylated CD97. First studies on CD97 in tumors were carried out exclusively with antibodies to the EGF domains,[14,15] thus, the data are representative for N-glycosylated CD97 only. The number of CD97-positive tumors increased dramatically if antibodies detecting glycosylation nondependent epitopes were used for immunohistology.[13] Interestingly, binding to CD55, a ligand of CD97 known as decay accelerating factor (DAF),[16] depends greatly on N-glycosylation within the EGF-like domains of CD97 that has thus functional consequences.[12]

### CD97 in Carcinomas

CD97 is a dedifferentiation marker in the thyroid.[14,17] Papillary and follicular carcinomas were CD97-negative whereas most of the fatal anaplastic thyroid carcinomas expressed the molecule. However, in these studies N-glycosylation-dependent monoclonal antibodies were used.

CD97 is overexpressed in colorectal adenocarcinomas compared to the corresponding normal tissue.[1,18] Carcinomas with more strongly CD97-stained scattered tumor cells at the invasion front showed a poorer clinical stage as well as increased lymph vessel invasion compared to cases with uniform CD97 staining.[18] Gastric, pancreatic and esophageal carcinomas were also CD97-positive.[15,19]

Summarizing these studies, until now small patient groups have been examined, clinical outcome data correlated to CD97 expression are missing and anti-CD97 monoclonal antibodies recognizing glycosylation-dependent epitopes were often used. These facts undermine a statement on the relevance and function of CD97 in tumorigenesis.

## Experimental Studies

Injecting HT1080/Tet$^{off}$ cells overexpressing stably human CD97 in *scid* mice, led to a faster tumor induction.[20] In vitro, CD97 expression levels correlated with the migratory and invasive capacity of colorectal tumor cell lines and HT1080/Tet$^{off}$ cells, stably overexpressing CD97.[18,20] Interestingly, a CD97-dependent increase in random single cell migration is disrupted by truncating the seven-span transmembrane part (TM7) to TM1, suggesting for the first time signal transduction through an EGF-like adhesion-GPCR.[20] In addition, CD97 increased the secretion of chemokines and matrix metalloproteinases. In silico simulations demonstrated that these CD97-induced effects can increase the invasive capacity of tumors and cause the appearance of scattered tumor cells at the invasion front.[20]

## Soluble CD97 in Tumorigenesis

The CD97 α-chain has been shown to be shed from the membrane of CD97-expressing cells. It is very likely identical to soluble CD97 (sCD97), detected in synovial fluid of rheumatoid arthritis patients[21] but not in sera of tumor patients (own unpublished results). Thus, it is unclear whether the release of the CD97 α-chain happens during tumorigenesis in vivo.

In experimental studies, the CD97 α-chain promotes angiogenesis as demonstrated with purified protein in a directed in vivo angiogenesis assay and by enhanced vascularization of developing tumors expressing CD97. The CD97 α-chain acts by binding endothelial cells via interactions with CD97 ligands, glycosaminoglycans and integrins $\alpha_5\beta_1$ and $\alpha_v\beta_3$.[22] The involvement of the third known ligand of CD97, CD55,[16] in tumorigenesis has not been demonstrated yet.

In normal cells EMR1, -2 and -3 proteins are restricted to the myeloid system. In a few tumor cell lines and tumors **EMR1** and **EMR2** mRNAs are moderately present.[1] The pattern is not lineage specific and not restricted to cell lines of hematopoetic origin. Only one scientific article has been published on EMR2 in tumors: EMR2 protein was not present in colorectal tumor cell lines although many different EMR2 mRNA splice variants were found.[23] Correspondingly, EMR2 was only rarely expressed in colorectal adenocarcinomas.[23] This contrasts to the data obtained in colorectal carcinoma for CD97 showing high structural homology to EMR2.[24]

Nonhematopoetic tumor cell lines are **EMR3** mRNA negative.[1] EMR3 is present in some tumor entities as breast, colorectal, liver and testis but nearly always at low levels.[2]

**EMR4** protein is not present in human.

## GROUP III: IgG-LIKE (GPR123, GPR124, GPR125)

**GPR123** mRNA, conserved between the vertebrates, shows central nervous system specific expression,[25] no other data are available.

**GPR124** was originally identified as a tumor endothelial marker that is upregulated during tumor angiogenesis.[26] GPR124 is shed from endothelial cells and further proteolytic processing creates a protein subunit that mediates endothelial survival and subsequent tumor angiogenesis, via interactions with glycosaminoglycans and the integrin $\alpha_v\beta_3$.[27] GPR124 itself is not or only slightly present in tumors or tumor cell lines.[1,2]

**GPR125** is described to be expressed specifically in the choroid plexus,[28] but HPA data revealed nonlineage specific protein expression in normal tissues.[2] Some tumor cell lines express slight to moderate GPR125 mRNA levels but most of the tumors are GPR125 mRNA negative.[1]

## GROUP IV: CELSR-LIKE (CELSR 1-3)

Only few data are available on the expression of CELSRs in cancer. **CELSR1** mRNA was found to be expressed in gastrointestinal tumors by comparative integromics on noncanonical Wnt or planar cell polarity signalling molecules[29] but mRNA levels in primary tumors are rather low.[1]

As shown in the HPA database, **CELSR2** (Flamingo-1) is expressed at higher levels in nearly all examined tumor entities and tumor cell lines, but these data could not be confirmed by BioGPS mRNA profiles.[1,2]

**CELSR3** mRNA is present in some tumor cell lines and at low levels in several tumor entities.[1,2]

## GROUP V: GPR133 AND 144

BioGPS database data revealed that only a few tumor cell lines are **GPR133** mRNA positive. Tumor cell lines are **GPR144** mRNA negative or only slightly positive.[1]

## GROUP VI: GPR110, GPR111, GPR113, GPR115, GPR116

Many tumors of several entities are **GPR110** mRNA positive, but by contrast, most tumor cell lines are GPR110 mRNA negative.[1]

For the adhesion-GPRs **GPR111** and **GPR113** no data related to cancer are available.

**GPR115** protein shows moderate to high expression in normal tissues, most tumor entities and tumor cell lines.[2]

**GPR116** shows specific high expression of its mRNA in normal fetal and adult lung and moderate levels in some other tissues such as the kidney, thyroid and adrenal gland.[1] Lung and kidney tumors have especially elevated GPR116 mRNA levels. Most tumor cell lines are GPR116 mRNA negative or express low levels, except for the breast tumor cell line MDAMB231.[1]

## GROUP VII: BAI-LIKE (BAI 1-3)

Brain-specific angiogenesis inhibitors (BAIs) show nearly brain restricted expression and are involved in the regulation of brain tumor progression. The decreased level of the three BAI genes in glioma tissues has been discussed as one of the molecular markers for the prediction of high-grade gliomas.[30,31]

**Table 1.** Adhesion-GPCR: Available Database Data

| Group | | Receptor | Other Names | HPA (Protein Data); www.proteinatlas.org | | | AGCOH | BioGPS (mRNA Data); biogps.gnf.org | | |
|---|---|---|---|---|---|---|---|---|---|---|
| | | | | Normal Tissues | Tumor Cell Lines | Tumors | | Normal Tissues | Tumor Cell Lines | Tumors |
| I | latrophilin-like | 1 latrophilin 1 (LPHN1) | CIRL-1, lectomedin-2 | | | | | x | x | x |
| | | 2 latrophilin 2 (LPHN2) | Lectomedin-1, latrophilin homolog 1 (LPHH1), CIRL-2 | x | x | x | x | | | x |
| | | 3 latrophilin 3 (LPHN3) | Lectomedin-3, CIRL-3 | x | x | x | | x | x | x |
| | | 4 ETL | | | | | x | x | x | |
| II | EGF-like | 5 CD97 | | x | x | x | x | x | x | x |
| | | 6 EMR1 | | | | | x | x | x | x |
| | | 7 EMR2 | | | | | x | x | x | x |
| | | 8 EMR3 | | x | x | x | x | x | x | |
| | | 9 EMR4 | | protein not present in human | | | | | | |
| III | IgG-like | 10 GPR123 | | | | | x | | | |
| | | 11 GPR124 | | x | x | x | x | | x | |
| | | 12 GPR125 | | x | x | x | x | x | x | x |
| IV | CELSR-like | 13 CELSR1 | EGF LAG seven-pass G-type receptor | | | | x | x | x | x |
| | | 14 CELSR2 | | x | x | x | | x | x | x |
| | | 15 CELSR3 | | | | | x | x | x | x |

*continued on next page*

than release of $Ca_i^{2+}$. Indeed, when $LTX^{N4C}$ binds to any receptor, it might interact with other proteins, like ion channels, and thus cause exocytosis irrespective of latrophilin. These arguments have been remarkably answered by our recent finding that a recombinant single-chain antibody against latrophilin 1, isolated from a phage display library, can cause burst-like neurotransmitter exocytosis similar to that induced by $LTX^{N4C}$ (paper in preparation). This ultimately proves that stimulation of neuronal latrophilin by agonists sends an exocytotic signal via a G protein pathway.

Neuronal studies also demonstrate that the main signals sent by latrophilin are relatively fast and reach a maximum within several seconds or minutes. Furthermore, this signalling retains all its characteristics even in synaptosomes or neuromuscular junctions, subcellular systems consisting of severed nerve terminals and lacking the neuronal somata. This leads to an important conclusion that, consistent with $Ca^{2+}$ signalling, latrophilin acts locally, within presynaptic nerve terminals and does not necessarily send signals to the cell body and the nucleus.[52,55]

Of course, it is also possible that latrophilin can link to other signalling pathways, especially considering the promiscuous reassociation of its NTF with CTF's from other adhesion-GPCRs. Therefore, in-depth studies of both G protein-coupled and any alternative mechanisms are required.

## LIGANDS AND INTERACTING PARTNERS OF LATROPHILINS

### Extracellular Ligands

*α-Latrotoxin*

The main exogenous ligand of latrophilin 1 is α-latrotoxin. The toxin stimulates its receptor and thus can be classed as an agonist. The interaction of α-latrotoxin with latrophilin was tested using various truncated constructs of latrophilin.[25] This study demonstrated that a very large fragment of NTF containing HRM, Stalk and GPS (390 amino acids) may be necessary to bind α-latrotoxin strongly, suggesting a multi-point interaction between toxin and latrophilin, with low-affinity binding at each point. In particular, HRM, a putative ligand-binding region of hormone receptors, alone was unable to mediate toxin binding. It may be possible that some peptide within the toxin molecule mimics a natural ligand of HRM but interacts with this domain only weakly. The association of toxin with NTF at additional sites may provide sufficient retention of the hormone-mimetic toxin peptide in contact with HRM, leading to receptor activation.

Black widow spider venom also contains another component (possibly ε-latroinsectotoxin) that kills *C. elegans* worms on injection.[56] Knockout and RNAi studies have shown that the toxic effects of the venom is mediated by the LAT-1 orthologue of latrophilins in *C. elegans*, but not by LAT-2.[56]

*Cyclooctadepsipeptides*

Latrophilin orthologues from the parasitic nematode *Haemonchus contortus* (HC110-R) and *C. elegans* (LAT-1) were thought to bind the anthelmintic cyclical depsipeptides, PF1022A (a natural secondary metabolite of the fungus *Mycelia sterilia*) and its semisynthetic derivative emodepside.[57] Electrophysiological studies revealed that

Table 1. Continued

| Group | | Receptor | Other Names | HPA (Protein Data); www.proteinatlas.org | | | AGCOH | BioGPS (mRNA Data); biogps.gnf.org | | |
|---|---|---|---|---|---|---|---|---|---|---|
| | | | | Normal Tissues | Tumor Cell Lines | Tumors | | Normal Tissues | Tumor Cell Lines | Tumors |
| V | 16 | GPR133 | | | | | | x | x | |
| | 17 | GPR144 | | | | | | x | x | |
| VI | 18 | GPR110 | | | | | | x | x | x |
| | 19 | GPR111 | | | | | | x | | |
| | 20 | GPR113 | | | | | no data available | | | |
| | 21 | GPR115 | | x | x | x | | x | | |
| | 22 | GPR116 | | | | | | x | x | x |
| VII BAI-like | 23 | BAI1 | | | | | | x | x | x |
| | 24 | BAI2 | | | | | | x | x | x |
| | 25 | BAI3 | | x | x | x | | x | x | x |
| VIII | 26 | GPR56 | TM7XN1 | | | | | x | x | x |
| | 27 | GPR64 | HE6 | x | x | x | | x | x | |
| | 28 | GPR97 | | | | | | x | | |
| | 29 | GPR112 | | | | | no data available | | | |
| | 30 | GPR114 | | x | x | x | | x | | |
| | 31 | GPR126 | | | | | | x | x | x |
| | 32 | GPR128 | | | | | | x | | |
| un-grouped | 33 | VLGR1 | MASS1 | | | | | x | x | x |

Abbreviations: AGCOH: The Atlas of Genetics and Cytogenetics in Oncology and Haematology (http://atlasgeneticsoncology.org); HPA: Human protein atlas.

**BAI1** is expressed in neuronal cells of the cerebral cortex but not in astrocytes.[32] It is not clear, whether BAI1 is produced by the tumor cells themselves or by cells infiltrating and surrounding the tumor, because most tumor cell lines and tumors are BAI1 mRNA negative.[1] BAI1 was initially identified as a p53-regulated gene whose protein product could inhibit angiogenesis.[33] The proteolytically cleaved BAI1 extracellular domain inhibits endothelial cell proliferation by binding $\alpha_v\beta_5$ integrin via its thrombospondin type 1 repeats.[34] The BAI1 extracellular domain can also inhibit in vivo angiogenesis and tumor xenograft growth in mice.[35] The transfer of the BAI1 gene to the mouse renal cell carcinoma cell line Renca suppresses tumor growth via the inhibition of angiogenesis.[36] Taken together, BAI1 seems to regulate vascularisation of tumors and is thus a gene therapy candidate for the treatment of (brain) tumors, especially human glioblastomas.[37]

**BAI2** mRNA is not present in most tumor cell lines but in some tumors of different entities.[1]

Most tumor cell lines and tumors are **BAI3** mRNA negative.[1] The data contrasts to that of BAI3 protein: BAI3 is moderately to highly expressed in most tumor entities.[2]

## GROUP VIII: MISCELLANEOUS (GPR56, GPR64, GPR97, GPR112, GPR114, GPR126)

**GPR56** mRNA is expressed in various normal human tissues.[38] It plays a role in brain development.[39,40]

The data of GPR56 in tumors are contradictory and not consistent. Tumor cell lines show varying levels of GPR56 mRNA,[38] although BioGPS identified only few tumor cell lines, mainly melanocytic, as GPR56 mRNA positive.[1] Some pancreatic tumor cell lines showed high levels of GPR56 mRNA but the GPR56 protein was low or undetectable.[41]

GPR56 mRNA was expressed in poorly and intermediately metastasizing melanoma cell lines whereas it was downregulated markedly in the highly metastatic.[38] The data were confirmed in tumors from cell lines derived from poorly metastatic A375eco melanoma cells injected into immunodeficient mice: highly metastatic cell lines showed lower GPR56 mRNA compared to the A375eco parental cells.[42,43] Reduction of GPR56 in A375eco cells enhanced tumor growth and metastasis. GPR56 interacted with a ubiquitously expressed crosslinking enzyme, tissue transglutaminase 2 (TG2), located in the extracellular matrix. TG2 itself played suppressive roles in tumor progression[44] and therefore it might contribute to GPR56-mediated suppression of melanoma metastasis in the used mouse model.[42] In fact, the growth inhibition by GPR56 only occurs in vivo, suggesting that the function of GPR56 involves a factor in the tumor microenvironment.[43]

These data, all derived from melanoma cell lines, contrast clearly to those obtained from other tumor entities. In many human glioblastomas GPR56 was upregulated and seemed to promote cellular adhesion signalling.[45] Higher GPR56 expression was correlated with cellular transformation phenotypes of several cancer tissues compared with their normal counterparts, implying a potential oncogenic function of GPR56.[46] GPR56 silencing resulted in apoptosis induction and reduced anchorage-independent growth of cancer cells in vitro. Significant tumor responses, including regression, were induced in vivo after GPR56 silencing in xenograft models.[46] Recently it was shown that splice variants of GPR56 mRNA regulate differentially the activity of transcription factors associated with tumorigenesis[47] perhaps explaining in part the contradictious results on GPR56 in tumors.

**GPR64 (HE6)** associated with apical membranes of efferent and epididymal duct epithelia[48] is absent in most other normal tissues and tumors.[1,2] Database data on tumor cell lines are contradictory: GPR64 mRNA is missing according to BioGPS data whereas many tumor cell lines are GPR64 protein positive in the HPA database.[1,2]

Only few data on **GPR97** mRNA on normal cells are available.[1] GPR97 seems to be restricted to cells of hematopoietic origin.

For **GPR112** no data are available.

**GPR114** shows mild to strong expression in tumors of several origins.[2] Accordingly, many tumor cell lines are GPR114 protein positive.[2]

Nearly half of the examined tumor cell lines of different origin showed elevated **GPR126** mRNA levels.[1] Consequently, some tumors of different tumor entities are strongly GPR126 mRNA positive although the levels varied between the patients.[1]

## UNGROUPED (GPR128, VLGR1)

For **GPR128** in tumors no data are available.

Tumor cell lines and primary tumors are **VLGR1** (MASS1) mRNA negative.[1]

## CONCLUSION

Adhesion molecules, mediating cell-cell and cell-matrix interactions, are regulated during tumor development and progression. The inhibition of adhesion, migration, invasion or angiogenesis by blocking adhesion receptors is a therapeutic approach to improve tumor control.

Adhesion-GPCRs are suggested to have, as the name implies, an adhesive function in tumors. However, the amount of information on the expression of adhesion-GPCRs in tumor cell lines and tumors is rather limited. Most of the adhesion-GPCRs are still poorly studied orphans with unknown functions.

**Summary of Main Findings:**

1. Comprehensive scientific data on adhesion-GPCRs in tumors are available for a few class members only, such as latrophilin-2, CD97, GPR124, BAI1 and GPR56. These adhesion-GPCRs are involved (i) in tumor angiogenesis as a tumor angiogenesis inhibitor (BAIs) or as a promoter (CD97) or as a marker of tumor endothelial cells (GPR124) and (ii) in the interaction of the tumor cell with its microenvironment (CD97). (iii) GPR56 acts as a tumor growth and metastasis suppressor in melanomas.
2. Some of the receptors such as GPR97 and GPR110 to GPR116 have been found during searches in the human genome databases[49] and have not been explored outside database screening until now. Even in accessible databases any data on GPR112 and GPR113 are missing.

3. Many of the adhesion-GPCR mRNAs are present in tumor cell lines and tumors[1] suggesting the presence of the encoded proteins in malignant tumors and a function of these receptors in tumorigenesis. For example, high expression of GPR115 and GPR126 mRNA was shown in the databases. However, no scientific articles on these adhesion-GPCRs in tumors have been published until now.
4. The cellular source of adhesion-GPCRs involved in cancer may not be just the tumor cell. In some cases, such as for BAIs the adhesion-GPCRs are partly expressed in cells of the tumor microenvironment regulating tumor growth.
5. For latrophilin-3, BAI3 and GPR64 mRNA data in the BioGPS and protein data from the HPA are contradictory, thus, evaluation was not possible.[1,2]
6. Splice variants are very frequent for the majority of adhesion-GPCRs. Alternative splicing of these adhesion-GPCRs may influence how these receptors enable cellular interactions during tumorigenesis. Only data on different splice variants of latrophilin-2 mRNA in malignant compared to the corresponding normal tissues are available.
7. Different glycosylation and thus different ligand binding of an adhesion-GPCR in tumors compared to the corresponding normal tissue has been demonstrated for CD97. Whether this will be a general principle for adhesion-GPCRs has to be investigated.

Taken together, there is evidence for the expression and several functions of adhesion-GPCRs in tumors but these data are as heterogeneous as the family of adhesion-GPCRs. We are at the very beginning in our understanding of adhesion-GPCRs in tumorigenesis.

## REFERENCES

1. BioGPS; http://biogps.gnf; 2009.
2. Human Protein Atlas (HPA); http://www.proteinatlas.org; 2009.
3. Atlas of Genetics and Cytogenetics in Oncology and Haematology (AGCOH); http://atlasgeneticsoncology.org; 2009.
4. Bjarnadottir TK, Fredriksson R, Schioth HB. The adhesion-GPCRs: a unique family of G protein-coupled receptors with important roles in both central and peripheral tissues. Cell Mol Life Sci 2007; 64:2104-2119.
5. Ichtchenko K, Bittner MA, Krasnoperov V et al. A novel ubiquitously expressed alpha-latrotoxin receptor is a member of the CIRL family of G-protein-coupled receptors. J Biol Chem 1999; 274:5491-5498.
6. Matsushita H, Lelianova VG, Ushkaryov YA. The latrophilin family: multiply spliced G protein-coupled receptors with differential tissue distribution. FEBS Lett 1999; 443:348-352.
7. Sudhof TC. Alpha-latrotoxin and its receptors: neurexins and CIRL/latrophilins. Annu Rev Neurosci 2001; 24:933-962.
8. White GR, Varley JM, Heighway J. Isolation and characterization of a human homologue of the latrophilin gene from a region of 1p31.1 implicated in breast cancer. Oncogene 1998; 17:3513-3519.
9. White GR, Varley JM, Heighway J. Genomic structure and expression profile of LPHH1, a 7TM gene variably expressed in breast cancer cell lines. Biochim Biophys Acta 2000; 1491:75-92.
10. Jaspars LH, Vos W, Aust G et al. Tissue distribution of the human CD97 EGF-TM7 receptor. Tissue Antigens 2001; 57:325-331.
11. Aust G, Wandel E, Boltze C et al. Diversity of CD97 in smooth muscle cells (SMCs). Cell Tissue Res 2006; 323:1-9.
12. Wobus M, Vogel B, Schmücking E et al. N-glycosylation of CD97 within the EGF domains is crucial for epitope accessibility in normal and malignant cells as well as CD55 ligand binding. Int J Cancer 2004; 112:815-822.

13. Aust G. Immunohistochemical detection of CD97 in colorectal carcinomas. In: Hayat MA, ed. Immunohistochemistry and in situ hybridization of human carcinomas. Burlington, San Diego, London: Elsevier Academic Press, 2005:201-206.
14. Aust G, Eichler W, Laue S et al. CD97: A dedifferentiation marker in human thyroid carcinomas. Cancer Res 1997; 57:1798-1806.
15. Boltze C, Schneider-Stock R, Aust G et al. CD97, CD95 and Fas-L clearly discriminate between chronic pancreatitis and pancreatic ductal adenocarcinoma in perioperative evaluation of cryocut sections. Pathol Int 2002; 52:83-88.
16. Hamann J, Vogel B, van Schijndel GM et al. The seven-span transmembrane receptor CD97 has a cellular ligand (CD55, DAF). J Exp Med 1996, 184.1185-1189.
17. Hoang-Vu C, Bull K, Schwarz I et al. Regulation of CD97 protein in thyroid carcinoma. J Clin Endocrinol Metab 1999; 84:1104-1109.
18. Steinert M, Wobus M, Boltze C et al. Expression and regulation of CD97 in colorectal carcinoma cell lines and tumor tissues. Am J Pathol 2002; 161:1657-1667.
19. Aust G, Steinert M, Schütz A et al. CD97, but not its closely related EGF-TM7 family member EMR2, is expressed on gastric, pancreatic and esophageal carcinomas. Am J Clin Pathol 2002; 118:699-707.
20. Galle J, Sittig D, Hanisch I et al. Individual cell—based models of tumor—environment interactions. Multiple effects of CD97 on tumor invasion. Am J Pathol 2006; 169:1802-1811.
21. Hamann J, Wishaupt JO, van Lier RA et al. Expression of the activation antigen CD97 and its ligand CD55 in rheumatoid synovial tissue. Arthritis Rheum 1999; 42:650-658.
22. Wang T, Ward Y, Tian L et al. CD97, an adhesion receptor on inflammatory cells, stimulates angiogenesis through binding integrin counter receptors on endothelial cells. Blood 2004; 105:2836-2844.
23. Aust G, Hamann J, Schilling N et al. Detection of alternatively spliced EMR2 mRNAs in colorectal tumor cell lines but rare expression of the molecule in the corresponding adenocarcinomas. Virchows Arch 2003; 443:32-37.
24. Lin HH, Stacey M, Hamann J et al. Human EMR2, a novel EGF-TM7 molecule on chromosome 19p13.1, is closely related to CD97. Genomics 2000; 67:188-200.
25. Lagerstrom MC, Rabe N, Haitina T et al. The evolutionary history and tissue mapping of GPR123: specific CNS expression pattern predominantly in thalamic nuclei and regions containing large pyramidal cells. J Neurochem 2007; 100:1129-1142.
26. Carson-Walter EB, Watkins DN, Nanda A et al. Cell surface tumor endothelial markers are conserved in mice and humans. Cancer Res 2001; 61:6649-6655.
27. Vallon M, Essler M. Proteolytically processed soluble tumor endothelial marker (TEM) 5 mediates endothelial cell survival during angiogenesis by linking integrin alpha(v)beta3 to glycosaminoglycans. J Biol Chem 2006; 281:34179-34188.
28. Pickering C, Hagglund M, Szmydynger-Chodobska J et al. The Adhesion GPCR GPR125 is specifically expressed in the choroid plexus and is upregulated following brain injury. BMC Neurosci 2008; 9:97.
29. Katoh M, Katoh M. Comparative integromics on noncanonical WNT or planar cell polarity signaling molecules: transcriptional mechanism of PTK7 in colorectal cancer and that of SEMA6A in undifferentiated ES cells. Int J Mol Med 2007; 20:405-409.
30. Kaur B, Brat DJ, Calkins CC et al. Brain angiogenesis inhibitor 1 is differentially expressed in normal brain and glioblastoma independently of p53 expression. Am J Pathol 2003; 162:19-27.
31. Kee HJ, Ahn KY, Choi KC et al. Expression of brain-specific angiogenesis inhibitor 3 (BAI3) in normal brain and implications for BAI3 in ischemia-induced brain angiogenesis and malignant glioma. FEBS Lett 2004; 569:307-316.
32. Mori K, Kanemura Y, Fujikawa H et al. Brain-specific angiogenesis inhibitor 1 (BAI1) is expressed in human cerebral neuronal cells. Neurosci Res 2002; 43:69-74.
33. Nishimori H, Shiratsuchi T, Urano T et al. A novel brain-specific p53-target gene, BAI1, containing thrombospondin type 1 repeats inhibits experimental angiogenesis. Oncogene 1997; 15:2145-2150.
34. Koh JT, Kook H, Kee HJ et al. Extracellular fragment of brain-specific angiogenesis inhibitor 1 suppresses endothelial cell proliferation by blocking alphavbeta5 integrin. Exp Cell Res 2004; 294:172-184.
35. Kaur B, Brat DJ, Devi NS et al. Vasculostatin, a proteolytic fragment of brain angiogenesis inhibitor 1, is an antiangiogenic and antitumorigenic factor. Oncogene 2005; 24:3632-3642.
36. Kudo S, Konda R, Obara W et al. Inhibition of tumor growth through suppression of angiogenesis by brain-specific angiogenesis inhibitor 1 gene transfer in murine renal cell carcinoma. Oncol Rep 2007; 18:785-791.
37. Kang X, Xiao X, Harata M et al. K. Antiangiogenic activity of BAI1 in vivo: implications for gene therapy of human glioblastomas. Cancer Gene Ther 2006; 13:385-392.

38. Zendman AJ, Cornelissen IM, Weidle UH et al. TM7XN1, a novel human EGF-TM7-like cDNA, detected with mRNA differential display using human melanoma cell lines with different metastatic potential. FEBS Lett 1999; 446:292-298.
39. Jin Z, Tietjen I, Bu L et al. Disease-associated mutations affect GPR56 protein trafficking and cell surface expression. Hum Mol Genet 2007; 16:1972-1985.
40. Piao X, Hill RS, Bodell A et al. G protein-coupled receptor-dependent development of human frontal cortex. Science 2004; 303:2033-2036.
41. Huang Y, Fan J, Yang J et al. Characterization of GPR56 protein and its suppressed expression in human pancreatic cancer cells. Mol Cell Biochem 2008; 308:133-139.
42. Xu L, Begum S, Hearn JD et al. GPR56, an atypical G protein-coupled receptor, binds tissue transglutaminase, TG2 and inhibits melanoma tumor growth and metastasis. Proc Natl Acad Sci USA 2006; 103:9023-9028.
43. Xu, L, Hynes Ro. GPR56 and TgG2 possible roles in suppression of tumor growth by the microenvironment. Cell Cycle 2007; 6:760-765.
44. Jones RA, Kotsakis P, Johnson TS et al. Matrix changes induced by transglutaminase 2 lead to inhibition of angiogenesis and tumor growth. Cell Death Differ 2006; 13:1442-1453.
45. Shashidhar S, Lorente G, Nagavarapu U et al. GPR56 is a GPCR that is overexpressed in gliomas and functions in tumor cell adhesion. Oncogene 2005; 24:1673-1682.
46. Ke N, Sundaram R, Liu G et al. Orphan G protein-coupled receptor GPR56 plays a role in cell transformation and tumorigenesis involving the cell adhesion pathway. Mol Cancer Ther 2007; 6:1840-1850.
47. Kim JE, Han JM, Park CR et al. Splicing variants of the orphan G-protein-coupled receptor GPR56 regulate the activity of transcription factors associated with tumorigenesis. J Cancer Res Clin Oncol 2009.
48. Obermann H, Samalecos A, Osterhoff C et al. HE6, a two-subunit heptahelical receptor associated with apical membranes of efferent and epididymal duct epithelia. Mol Reprod Dev 2003; 64:13-26.
49. Fredriksson R, Lagerstrom MC, Hoglund PJ et al. Novel human G protein-coupled receptors with long N-terminals containing GPS domains and Ser/Thr-rich regions. FEBS Lett 2002; 531:407-414.

# CHAPTER 10

# IMMUNITY AND ADHESION-GPCRs

Simon Yona, Hsi-Hsien Lin and Martin Stacey*

**Abstract:** Adhesion-GPCRs are unusual, owing to their unique structure, comprising a large and complex extracellular domain composed of various common protein modules. Adhesion-GPCR family members are expressed ubiquitously; however the expression of each receptor is highly regulated and often restricted to specific cell types. The EGF-TM7 adhesion-GPCR subfamily members are predominantly expressed by leukocytes and involved in coordinating both the innate and acquired immune responses. Here we highlight some immunological insights in relation to EGF-TM7 proteins and other members of the adhesion-GPCR family.

## INTRODUCTION: ADHESION-GPCRs IN IMMUNOLOGY

Inflammation is characterised by the recruitment of blood borne phagocytes to sites of injury or infection. This response is coordinated by an assortment of membrane bound receptors, including lectins, TLRs, selectins, GPCRs, integrins and the Ig superfamily, found on leukocytes.[1] Interestingly, the expression pattern of Epidermal Growth Factor (EGF)-seven transmembrane (TM7) receptors, the first described adhesion-GPCRs, are predominantly leukocyte-restricted. The EGF-TM7 family comprises of CD97, EMR1 (F4/80 receptor) EMR2, EMR3 and EMR4, all the members of this family posses extracellular domains which contain a GPS motif and multiple EGF-like repeat units, that undergo alternative splicing thereby producing multiple receptor isoforms[2] (Fig. 1). It is now apparent that this family is involved in many aspects of leukocyte development and activation,[3-10] further details may be found in this volume by Lin et al and Hamann et al. More recently, other adhesion-GPCR members have been implicated in the immune response, including BAI1 and GPR56.

*Corresponding Author: Martin Stacey—Department of Immunology, Leeds University, Leeds, UK.
Email: m.stacey@leeds.ac.uk

*Adhesion-GPCRs: Structure to Function*, edited by Simon Yona and Martin Stacey.
©2010 Landes Bioscience and Springer Science+Business Media.

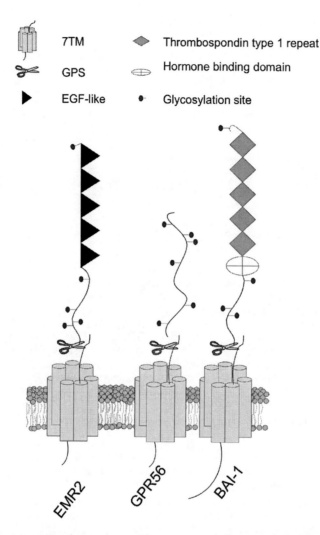

**Figure 1.** The structure of adhesion-GPCRs. Adhesion-GPCRs are composed of a large N-terminal region, which often possess diverse arrangements of protein modules coupled to a seven-span transmembrane moiety. The majority of adhesion-GPCRs undergo an autocatalytic cleavage event (depicted by scissors) within the ER at a conserved G protein-coupled proteolytic site (GPS). Noncovalent association of the TM7 and extracellular subunit results in the formation of a heterodimer at the cell surface. Protein domains and N-linked glycosylation sites were predicted using SMART and NetNGlyc 1.0 algorithms respectively.

## THE EGF-TM7 FAMILY

Historically, the first described member of adhesion-GPCR family was the macrophage specific antigen recognised by the monoclonal antibody F4/80.[11] Since the early 1980s the F4/80 receptor (mouse EMR1) has provided an excellent marker for a number of murine macrophage subpopulations such as microglia, Kupffer cells, splenic red pulp macrophages as well as in Langerhans cells. The F4/80 receptor is also expressed, albeit at lower levels, on alveolar macrophages, dendritic cells (DCs) and eosinophils. The function of the F4/80 receptor remained elusive until the generation of the *F4/80 receptor* null mouse, which demonstrated that the receptor was critical for the induction of peripheral tolerance in a number of in vivo models.[12] Curiously, the protein expression profile of the human F4/80 receptor orthologue, EMR1, is restricted to eosinophils.[13] Nevertheless *EMR1* mRNA is found at high levels in CD14+ monocytes (http://symatlas.gnf.org/SymAtlas), suggesting that this protein undergoes tight translational control.

Unlike the highly restricted nature of EMR1 the other EGF-TM7 proteins, CD97, EMR2, EMR3 and EMR4, are expressed largely by myeloid leukocytes, including PMNs (polymorphonuclear cells), monocytes, macrophages and dendritic cells.[3,7,14] At present the best characterised EGF-TM7 receptors are CD97 and EMR2 genes which are believed to have arisen through recent gene duplication and conversion events.[15] Although highly homologous differences in expression profiles and ligand binding have been demonstrated for these two proteins. CD97 has a more promiscuous expression pattern, being found on activated T cells and B cells, smooth muscle cells as well as myeloid leukocytes. Furthermore, although the extracellular EGF domains of CD97 and EMR2 are 97% identical, significant differential ligand binding activity has been demonstrated, for example, the receptor isoforms possessing EGF domains 1, 2, 5 differ by a mere three amino acids; however, CD97 binds to its ligand, the complement control protein, Decay Accelerating Factor (DAF/CD55) with at least one order of magnitude higher affinity. The exact immune significance of the CD97-CD55 interaction has not yet been fully dissected; however during the adaptive immune response, CD97 is able costimulate T cells by binding to DAF resulting in T-cell proliferation, IL-10 production and enhanced expression of activation markers independent of DAF's role in complement activity.[16] The deletion of either DAF (ligand) or CD97 (receptor) genes ameliorates arthritis in rodent models, supporting a role for the CD97-DAF complex during the immune response.[17]

## GONE FISHING

A major hindrance to the characterisation of adhesion-GPCRs to date has been the lack of defined ligands. One technique that has proved particularly successful in de-orphanising adhesion-GPCRs, is the use of multivalent, high avidity probes. Recombinant soluble extracellular regions of adhesion-GPCRs were engineered to contain an Fc-fragment and a C-terminal biotinylation signal for purification and specific biotinylation. Subsequent coupling to fluorescent/magnetic led to the identification and characterization of a number of low affinity cellular ligands for myeloid cell surface receptors.[18] Using this technique the largest isoforms of both CD97 and EMR2 containing all five EGF domains have been shown to bind the extracellular matrix component chondroitin sulphate.[5,8] As chondroitin sulphate synthesis is tightly regulated in a tissue and cell specific fashion, during wound repair, inflammation and tumourogensis.[19,20] It has been proposed that chondroitin sulphate heterogeneity may regulate the migration or differentiation of EMR2/CD97 expressing myeloid cells during wound repair, inflammation and tumourogensis. Congruent to the potential roles of EMR2 and CD97 in inflammation, CD97 binds to chondroitin sulphate in the synovial tissue from rheumatoid arthritis patients.[21] In addition, EMR2 is highly expressed on peripheral PMNs from patients with systemic inflammatory response syndrome,[22] a condition defined by a marked increase in neutrophil number and hyperactivation which often results in multiple organ failure and death. Indeed in vitro studies have demonstrated that PMNs ligated with the EMR2 specific activating antibody, 2A1, exhibit enhanced adhesion, migration and generation antimicrobial mediators towards a number of pro-inflammatory mediators[10] (Fig. 2A). Furthermore, CD97 has been implicated in PMN migration, in mouse models of colitis, in the clearance

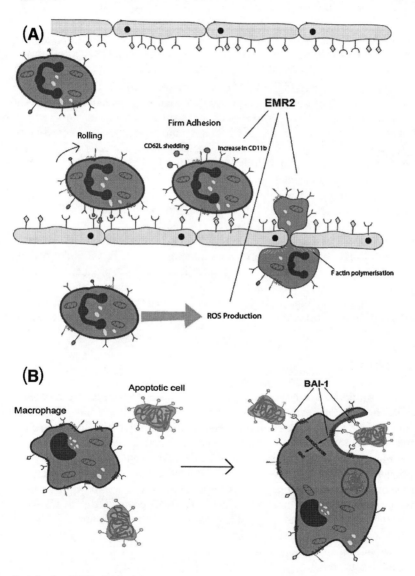

**Figure 2.** Adhesion-GPCRs in the immune response. The adhesion-GPCRs are involved in number of functional roles within the immune system. A) Ligation of EMR2 with an activating antibody regulates PMN activation. PMN migration and activation is critical during the initial phase of the innate immune response. EMR2 ligation increases PMN adhesion and migration to the site of injury or infection. In addition, EMR2 activation augments the production/degranulation of antimicrobial products such as superoxide and myeloperoxidase. B) BAI1 functions as a phagocytic receptor for apoptotic cells. The resolution phase of the inflammatory response requires the clearance of apoptotic cells, facilitating the restoration of tissue integrity and homeostasis. BAI1 directly engages phosphatidylserine on the surface of apoptotic cells and signals upstream of the ELMO–Dock180 module, which promotes the internalisation of apoptotic cells.

of *Streptococcus pneumonia* and in joint damage during collagen-induced arthritis.[4,6,23,24] The administration of blocking CD97 antibodies reduced the neutrophil-dependent mobilization of haematopoietic and progenitor cells.[9] Curiously, *CD97* null mice are less susceptible to *Listeria monocytogenes* infection.[25]

The immune roles of the remaining EGF-TM7 members EMR3 and EMR4[14,26] are less well-characterised. EMR3 expression is highly expressed in leukaemias (NCBI-Unigene database) and antibody studies have shown it to be a marker of fully differentiated PMNs. Unlike the other EGF-TM7 receptors, EMR3 is only up-regulated during late granulopoiesis. Interestingly *EMR4* is a rapidly evolving gene. It is presumed to be pseudogene in humans due to a premature stop codon within the extracellular region; however, it is found as an intact full length coding gene in apes and lower species.[27] Indeed, mouse studies have shown that the specific delivery of antigen to CD8⁻ DCs via the intact murine EMR4 receptor dramatically enhances antibody production, suggesting a function role in DC biology.[28] Together the above findings provide strong evidence for the involvement of EGF-TM7 receptors within the immune response and show their potential as therapeutic targets for infectious and inflammatory diseases.

## BAI1

In addition to the activation phase of an inflammatory response, the subsequent highly coordinated resolution phase requires the efficient clearance of apoptotic bodies, thereby facilitating the restoration of tissue integrity and homeostasis. The adhesion-GPCR brain angiogenesis inhibitor 1 (BAI1), was previously thought to be restricted to the CNS, recent data has shown BAI1 to mediate the clearance of apoptotic bodies (see Park et al, in this volume). Recent reports have implicated the adhesion-GPCR BAI1, in this resolution phase. BAI1, an adhesion-GPCR expressed on macrophages, posses tandem thrombospondin repeats coupled to a TM7 region via a GPS containing stalk region (Fig. 1). These extracellular thrombospondin repeats bind phosphatidylserine found on the outer leaflet of the plasma membrane on apoptotic cells, an established "eat me" signal.[29] This binding, mediates TM7-dependent signalling via the interaction of guanine nucleotide exchange factors ELMO (Engulfment and cell Motility) and DOCK180 (Dedicator of Cytokinesis protein) and the small GTPase Rac. Subsequent actin polymerisation at the site of binding results in the phagocytosis and the clearance of apoptotic bodies (Fig. 2B).

## GPR56

Although the adhesion-GPCR GPR56 has been implicated in cancer progression and brain development (see Xu et al and Piao et al, this volume), recent antibody studies have identified GPR56 as a novel marker for the inflammatory CD56$^{dull}$ CD16⁺ Natural Killer (NK) cells found in peripheral blood and in inflamed tissues, allowing for the discrimination of NK subsets.[30] The exact role of GPR56 during inflammation and the immune response has yet to be elucidated. However, when NK cells were treated by IL-2, IL-15 or IL-18 GPR56 expression is down-regulated, with the most effective reduction observed following IL-18 treatment. IL-18 is capable of inducing CD56$^{dull}$ CD16⁺ NK cells to express the chemokine receptor CCR7 de novo, which in turn these cells might mediate the chemotactic migration of NK cells. It is therefore possible that GPR56 plays a role in NK cell migration/retention within tissues, a cellular function that has already been shown for GPR56 in the context of tumour development.

Microarray and EST data (http://symatlas.gnf.org/SymAtlas/, NCBI-Unigene database) illustrate the high expression of other adhesion-GPCRs on leukocytes including *GPR97* in immune related sites such as bone marrow, spleen and whole blood suggesting potential roles within the immune system.

## CONCLUSION

The restricted expression pattern profile of adhesion-GPCRs promotes their particular functional roles within a number of physiological systems (see other chapters in this volume). In recent years, the EGF-TM7 family members have been implicated in a number of immune conditions from the initiation to the resolution of the immune response. Future research in these areas will hopefully increase our understanding of these receptors during the immune response and should provide future therapeutic targets for a range of human diseases.

## ACKNOWLEDGEMENTS

S Yona is the recipient of a FEBS international Fellowship, HH Lin is funded by National Science Council, Taiwan and M Stacey receives funding from Yorkshire Cancer Research and Breast Cancer Campaign, UK.

## REFERENCES

1. Taylor PR, Martinez-Pomares L, Stacey M et al. Macrophage receptors and immune recognition. Annu Rev Immunol 2005; 23:901-944.
2. Yona S, Lin HH, Siu WO et al. Adhesion-GPCRs: emerging roles for novel receptors. Trends Biochem Sci 2008; 33(10):491-500.
3. Chang GW, Davies JQ, Stacey M et al. CD312, the human adhesion-GPCR EMR2, is differentially expressed during differentiation, maturation and activation of myeloid cells. Biochem Biophys Res Commun 2007; 353(1):133-138.
4. Hamann J WJ, van Lier RA, Smeets TJ et al. Expression of the activation antigen CD97 and its ligand CD55 in rheumatoid synovial tissue. Arthritis Rheum 1999; 42(4):650-658.
5. Kwakkenbos MJ, Pouwels W, Matmati M et al. Expression of the largest CD97 and EMR2 isoforms on leukocytes facilitates a specific interaction with chondroitin sulfate on B-cells. J Leukoc Biol 2005; 77(1):112-119.
6. Leemans JC, te Velde AA, Florquin S et al. The Epidermal Growth Factor-Seven Transmembrane (EGF-TM7) Receptor CD97 Is Required for Neutrophil Migration and Host Defense. J Immunol 2004; 172(2):1125-1131.
7. Matmati M, Pouwels W, van Bruggen R et al. The human EGF-TM7 receptor EMR3 is a marker for mature granulocytes. J Leukoc Biol 2007; 81(2):440-448.
8. Stacey M, Chang GW, Davies JQ et al. The epidermal growth factor-like domains of the human EMR2 receptor mediate cell attachment through chondroitin sulphate glycosaminoglycans. Blood 2003; 102(8):2916-2924.
9. van Pel M, Hagoort H, Kwakkenbos MJ et al. Differential role of CD97 in interleukin-8-induced and granulocyte-colony stimulating factor-induced hematopoietic stem and progenitor cell mobilization. Haematologica 2008; 93(4):601-604.
10. Yona S, Lin HH, Dri P et al. Ligation of the adhesion-GPCR EMR2 regulates human neutrophil function. FASEB J 2008; 22(3):741-751.

11. Austyn JM, Gordon S. F4/80, a monoclonal antibody directed specifically against the mouse macrophage. Eur J Immunol 1981; 11(10):805-815.
12. Lin HH, Faunce D, Stacey M et al. The macrophage receptor F4/80 is involved in the induction of CD8⁺ regulatory T-cells in peripheral tolerance. J Exp Med 2005; 201(10):1615-1625.
13. Hamann J, Koning N, Pouwels W et al. EMR1, the human homolog of F4/80, is an eosinophil-specific receptor. Eur J Immunol 2007; 37(10):2797-2802.
14. Stacey M, Chang GW, Sanos SL et al. EMR4, a novel epidermal growth factor (EGF)-TM7 molecule up-regulated in activated mouse macrophages, binds to a putative cellular ligand on B lymphoma cell line A20. J Biol Chem 2002; 277(32):29283-29293.
15. Kwakkenbos MJ, Matmati M, Madsen O et al. An unusual mode of concerted evolution of the EGF-TM7 receptor chimera EMR2. FASEB J 2006; 20(14):2582-2584.
16. Capasso M, Durrant LG, Stacey M et al. Costimulation via CD55 on human CD4⁺ T-cells mediated by CD97. J Immunol 2006; 177(2):1070-1077.
17. Hoek RM, de Launay D, Kop EN et al. Deletion of either CD55 or CD97 ameliorates arthritis in mouse models. Arthritis Rheum 62(4):1036-1042.
18. Lin HH, Chang GW, Huang YS et al. Multivalent protein probes for the identification and characterization of cognate cellular ligands for myeloid cell surface receptors. Methods Mol Biol 2009; 531:89-101.
19. Prydz K, Dalen KT. Synthesis and sorting of proteoglycans. J Cell Sci 2000; 113 Pt 2:193-205.
20. Taylor KR, Gallo RL. Glycosaminoglycans and their proteoglycans: host-associated molecular patterns for initiation and modulation of inflammation. FASEB J 2006; 20(1):9-22.
21. Kop EN, Kwakkenbos MJ, Teske GJ et al. Identification of the epidermal growth factor-TM7 receptor EMR2 and its ligand dermatan sulfate in rheumatoid synovial tissue. Arthritis Rheum 2005; 52(2):442-450.
22. Yona S, Lin HH, Dri P et al. Ligation of the adhesion-GPCR EMR2 regulates human neutrophil function. FASEB J 2007.
23. Visser L, de Vos AF, Hamann J et al. Expression of the EGF-TM7 receptor CD97 and its ligand CD55 (DAF) in multiple sclerosis. J Neuroimmunol 2002; 132(1-2):156-163.
24. Kop EN, Adriaansen J, Smeets TJ et al. CD97 neutralisation increases resistance to collagen-induced arthritis in mice. Arthritis Res Ther 2006; 8(5):R155.
25. Wang T, Tian L, Haino M et al. Improved antibacterial host defense and altered peripheral granulocyte homeostasis in mice lacking the adhesion class G protein receptor CD97. Infect Immun 2007; 75(3):1144-1153.
26. Stacey M, Lin HH, Hilyard KL et al. Human Epidermal Growth Factor (EGF) module-containing mucin-like hormone receptor 3 is a new member of the EGF-TM7 family that recognizes a ligand on human macrophages and activated neutrophils. J Biol Chem 2001; 276(22):18863-18870.
27. Hamann J, Kwakkenbos MJ, de Jong EC et al. Inactivation of the EGF-TM7 receptor EMR4 after the Pan-Homo divergence. Eur J Immunol 2003; 33(5):1365-1371.
28. Corbett AJ, Caminschi I, McKenzie BS et al. Antigen delivery via two molecules on the CD8⁻ dendritic cell subset induces humoral immunity in the absence of conventional "danger". Eur J Immunol 2005; 35(10):2815-2825.
29. Park D, Tosello-Trampont AC, Elliott MR et al. BAI1 is an engulfment receptor for apoptotic cells upstream of the ELMO/Dock180/Rac module. Nature 2007; 450(7168):430-434.
30. Della Chiesa M, Falco M et al. GPR56 as a novel marker identifying the CD56dull CD16⁺ NK cell subset both in blood stream and in inflamed peripheral tissues. Int Immunol 2010; 22(2):91-100.

# CHAPTER 11

# CD97 IN LEUKOCYTE TRAFFICKING

Jörg Hamann,* Henrike Veninga, Dorien M. de Groot,
Lizette Visser, Claudia L. Hofstra, Paul P. Tak, Jon D. Laman,
Annemieke M. Boots and Hans van Eenennaam

**Abstract:** CD97 is a member of the EGF-TM7 family of adhesion G protein-coupled receptors (GPCRs) broadly expressed on leukocytes. CD97 interacts with several cellular ligands via its N-terminal epidermal growth factor (EGF)-like domains. To understand the biological function of CD97, monoclonal antibodies (mAbs) specific for individual EGF domains have been applied in a variety of in vivo models in mice, which represent different aspects of innate and adaptive immunity. Targeting CD97 by mAbs inhibited the accumulation of neutrophilic granulocytes at sites of inflammation thereby affecting antibacterial host defense, inflammatory disorders and stem cell mobilization from bone marrow. Interestingly, targeting CD97 did not impact antigen-specific (adaptive response) models such as delayed type hypersensitivity (DTH) or experimental autoimmune encephalomyelitis (EAE). However, collagen-induced arthritis (CIA), a model for rheumatoid arthritis, was significantly ameliorated suggesting therapeutic value of CD97 targeting. CD97-deficient mice are essentially normal at steady state except for a mild granulocytosis, which increases under inflammatory conditions. Comparison of the consequences of antibody treatment and gene targeting implies that CD97 mAbs actively inhibit the innate response presumably at the level of granulocyte or macrophage recruitment to sites of inflammation. Based on the collected data, we propose that the CD97 mAbs either activate CD97-mediated signal transduction via a yet unknown mechanism or act by inducing CD97 internalization, making CD97 unavailable for binding to its ligands and thereby blocking recruitment of neutrophils and possibly macrophages.

---

*Corresponding Author: Jörg Hamann—Department of Experimental Immunology, Academic Medical Center, University of Amsterdam, Meibergdreef 9, 1105 AZ Amsterdam, The Netherlands.
Email: j.hamann@amc.uva.nl

*Adhesion-GPCRs: Structure to Function*, edited by Simon Yona and Martin Stacey.
©2010 Landes Bioscience and Springer Science+Business Media.

## CD97 IS A PROTOTYPICAL EGF-TM7 RECEPTOR

CD97[1] is a prototypical member of the EGF-TM7 family of adhesion-GPCRs.[2-4] Highly conserved during mammalian evolution,[5] CD97 is broadly present on almost all types of leukocytes, with highest expression levels found on myeloid cells.[1,6] In addition, CD97 is expressed by normal and malignant epithelial and muscle cells.[7-10]

Like all members of EGF-TM7 family, CD97 possesses tandemly arranged EGF-like domains.[11] Due to alternative splicing, three CD97 isoforms with differently ordered EGF-like domains exist in man and mouse (Fig. 1A and B).[12-14] The first two EGF domains found in all isoforms interact with CD55 (decay-accelerating factor), a regulator of the complement cascade.[13-18] The second last EGF domain, which is present only in the largest isoform, binds chondroitin sulfate B (dermatan sulfate), a glycosaminoglycan that is found abundantly on cell surfaces and extracellular matrix.[19,20] Finally, an Arg-Gly-Asp (RGD) motif in the stalk region of human CD97 is bound by the integrin $\alpha5\beta1$ (very late antigen-5) and possibly $\alpha v\beta3$.[21] While the interaction with CD55 and chondroitin sulfate B is evolutionary conserved, the integrin-binding RGD motif is only present in hominoids.[5]

## CD97 ANTIBODY TREATMENT INHIBITS GRANULOCYTE TRAFFICKING

Based on its molecular structure and ligand interactions, a role for CD97 in leukocyte trafficking has been proposed.[15] To investigate the biological function of CD97 in a ligand-specific context, antibodies directed against specific EGF domains were generated in Armenian hamsters (Fig. 1C). Two CD97 mAbs were isolated: 1B2, blocking the CD97-CD55 interaction and 1C5, interfering with the supposed binding site for chondroitin sulfate. The functional consequences of targeting CD97 and blocking its ligands interactions were studied using several mouse inflammation models each representing different aspects of innate and antigen-dependent immunity.

An initial study revealed reduced migration of radioactively labeled neutrophils to the inflamed colon in dextran sulfate sodium (DSS)-treated mice.[18] The consequences of this defect in neutrophil migration for host defense were demonstrated in an acute mouse model of pneumococcal pneumonia provoked by intranasal infection with *Streptococcus pneumoniae*.[18] Mice treated with CD97 mAbs 1B2 or 1C5 displayed a reduced granulocytic inflammatory infiltrate at 20 hours after inoculation. This was associated with a significantly enhanced outgrowth of bacteria in the lungs at 44 hours and a strongly diminished survival ($p < 0.0001$). A similar reduction in recruitment of granulocytes to the lungs was observed when mice were challenged with lipopolysaccharide (LPS). As shown in Figure 2A, the 1B2 mAb was able to inhibit neutrophil migration to the lungs after 24 hours to a similar extent as a mAb to CXCR2, a chemokine receptor demonstrated to be relevant to neutrophil recruitment. Next, we studied whether 1B2 could block interleukin (IL)-8-induced stem cell mobilization from bone marrow, which has been described to act through a neutrophil-dependent mechanism. Complete inhibition of mobilization was found in mice treated with 1B2.[23] Cell sorting and subsequent culture experiments indicated that CD97 is not expressed on colony-forming stem cells thus indicating that the absence of stem cell mobilization after CD97 mAb administration is due to its effect on neutrophil function.

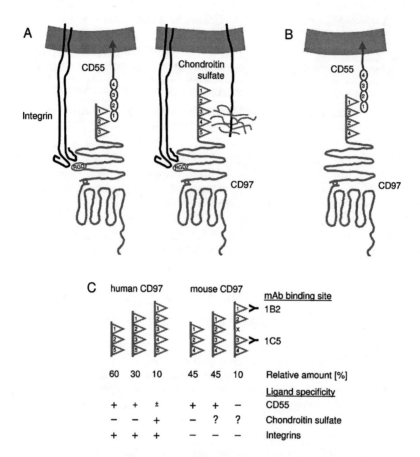

**Figure 1.** A) Cartoon representation of human CD97 interacting with its cellular ligands. At the cell surface, CD97 is expressed as a noncovalently associated dimer consisting of an extracellular α and a membrane-spanning β chain. The two chains result from autocatalytic processing of a CD97 propeptide.[22] Alternative splicing generates isoforms with three, four or five EGF domains.[12] Shown here are the smallest and the largest isoform. While EGF domain 1 and 2 interact with CD55, EGF domain 4, which only is present in the largest isoform, binds chondroitin sulfate B. Integrins bind a RGD motif in the stalk region of human CD97. B) Mouse CD97 has a structure similar to human CD97 albeit that the maximal number of EGF domains is four.[13,14] Shown here is the middle isoform. In the largest isoform, the EGF domains 2 and 3 are separated by 45 amino acids. C) Characteristics of human and mouse CD97 isoforms. Depicted is the composition of the EGF domain region, the relative amount of transcripts present in leukocytes and the ligand specificity. In humans, affinity for CD55 correlates inversely with the number of EGF domains. An interaction of EGF domain 3 of mouse CD97 (the homolog of EGF domain 4 in humans) with chondroitin sulfate B still needs to be proven. The binding site of mAbs recognizing specific EGF domains in mouse CD97 is indicated.

In an effort to determine whether (1) CD97 is important in the recruitment of granulocytes to other compartments such as the skin and (2) CD97 affects the inflammation in a model, which has been shown to represent aspects of both innate and adaptive immunity, 1B2 mAb was tested in the oxazolone DTH model. In this model, macrophages and granulocytes are recruited into the skin upon challenge with oxazolone and based on the efficient inhibition by the immunosuppressant drugs fingolimod (FTY720) and

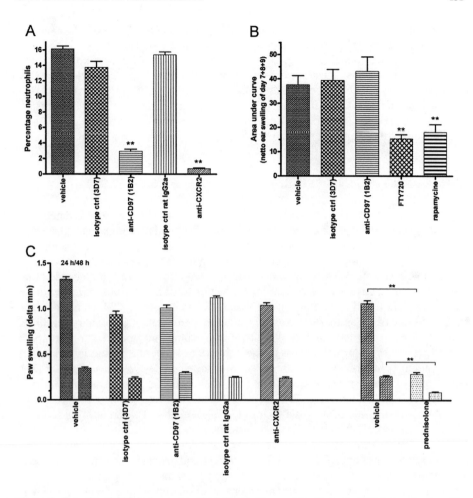

**Figure 2.** Targeting CD97 in lung and skin inflammation models. A) In the LPS-induced pulmonitis model, 4 mice per group were treated with 25 mg/kg antibody 2 hours prior to intraperitoneal (ip) bolus injection of 1 mg/kg LPS. 24 hours after LPS injection, lungs were harvested and infiltrated neutrophils (Ly6G and CD11b positive) were quantified by flow cytometry. B) In the oxazolone DTH model, 10 mice per group were sensitized with oxazolone on day 0 on the skin of the belly and 1B2 (25 mg/kg) was administered every other day from day 0 on ip. On day 6, mice were challenged on the skin of the ear with oxazolone and ear thickness was determined on day 7, 8 and 9. Rapamycine (1.5 mg/kg) and FTY720 (1 mg/kg) were used as positive and the subclass-matched irrelevant antibody 3D7 was used as a negative control. C) In the tetanus toxoid DTH model, 8 mice per group were sensitized with tetanus toxoid (3.75 LF in 50% DDA) 7 days prior to challenge. At day 7, 4 hours before tetanus toxoid challenge (2.5 LF in 0.1% alum) in the paw, 25 mg/kg antibody or 3 mg/kg prednisolone was given ip. Paw swelling was determined on day 8 (24 hours after challenge, left bars) and 9 (48 hours, right bars) and compared to nonchallenged paws and expressed as delta mm. Data are mean ± SEM. ** = $p < 0.01$.

rapamycin it has been suggested that also T-cell recruitment and/or activation are affected (see Fig. 2B). 1B2 was not able to inhibit the recruitment of granulocytes to the ear of the mice as was determined by measuring the skin thickness of the ear, suggesting that either 1B2 is not effective in preventing migration of granulocytes to skin or that the

T-cell component of this inflammation counteracts a potential effect. Also in a second more antigen-driven DTH model, induced and challenged with tetanus toxoid, 1B2 was not able to suppress the DTH (Fig. 2C). The data imply that 1B2 does not affect the antigen-specific response and associated recruitment of inflammatory cells. Notably, also anti-CXCR2 was not able to suppress tetanus toxoid DTH, thus implying that this DTH reaction is not dependent on neutrophil infiltration. Taken together, targeting CD97 using antibodies in the different inflammation models suggests that CD97 plays an important role in neutrophil recruitment and function.

## CD97 TARGETING IN ANTIGEN-DRIVEN DISEASE MODELS

The role and therapeutic potential of CD97 in arthritis has been studied in detail. We previously suggested that the colocalization of CD97+ intimal macrophages with CD55+ fibroblast-like synoviocytes and/or chondroitin sulfate-carrying structures underneath the lining in synovia might be involved in sustaining the chronic synovial inflammation in rheumatoid arthritis.[24] To study the relevance of targeting CD97 and blockade of its interaction with its ligands, we assessed the effect of 1B2 and 1C5 mAbs in the mouse CIA model. We found that treatment of DBA/J1 mice developing CIA with mAb 1B2 starting from day 21 on (early disease) resulted in significant suppression of arthritis scores ($p < 0.05$) and less joint swelling ($p < 0.05$) compared to the control groups.[25] Radiological and histological analysis showed inhibition of bone destruction, inflammation and granulocyte infiltration. When treatment was started on day 35 (late disease), CD97 mAb application had similar effects albeit less pronounced. In subsequent experiments, we demonstrated that targeting CD97 with 1B2 is as effective in therapeutic treatment of CIA as targeting tumor necrosis factor (TNF) (reduction of mean arthritis score of 60%).[26] Radiological analysis and immunohistochemistry showed that 1B2 and anti-TNF had similar favorable effects on bone destruction. In addition, using the 1C5 mAb, which targets only 55% of total CD97, we found a strong trend towards reduction of the mean arthritis score of 40% ($p = 0.057$). These results support the notion that the interaction between CD97 and its ligand(s) contributes to synovial inflammation and bone destruction in arthritis.

In addition, we analyzed EAE, a mouse model for multiple sclerosis. We had previously shown that CD97 is abundantly expressed in multiple sclerosis lesions by different cell types including macrophages/microglia and T cells.[27] In addition, CD55 is expressed by endothelial cells and by macrophages/microglia in active lesions. In a standard EAE model, SJL/J mice were treated with up to 1 mg 1B2 from two days before immunization with proteolipid protein (PLP) peptide fragment 139-151 and then every other day until day 12 after immunization. Using this intense antibody regimen, EAE day of onset, incidence and severity were not significantly affected by 1B2 treatment (Fig. 3). The somewhat higher EAE scores in the antibody-treated groups may be due to biological variation in a complex disease model.

Table 1 summarizes the results of antibody treatments in different mouse immunological models, each representing different aspects of innate and adaptive immunity. In all cases where mAbs 1B2 and 1C5 were studied in parallel, 1C5 was less effective than 1B2, which can be explained by its restricted binding to only the larger isoforms of mouse CD97 (only present in about half of all CD97 molecules). Taken together, these results support the view that CD97 mAbs efficiently inhibit the

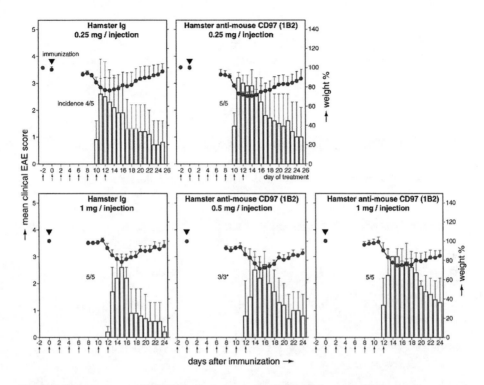

**Figure 3.** Early treatment with CD97 mAb does not affect EAE development. SJL/J mice (5 per group) were immunized subcutaneously with PLP$_{139\text{-}151}$ in complete Freund's adjuvant and additionally injected intravenously with heat-killed *Bordetella pertussis* bacteria. Treatment (arrows) with mAb 1B2 or hamster immunoglobulin control antibody ip was initiated two days before immunization (arrowheads) and applied every other day thereafter until day 12 after immunization. In two separate experiments mice received 0.25, 0.5 or 1 mg 1B2. Mice were weighed and scored for clinical signs of EAE daily. Bars represent the mean clinical scores, lines the mean weight, both ± SD. *, two mice died before the onset of EAE.

recruitment of granulocytes to sites of inflammation. This is especially evident when CD97 antibodies are tested in models, which allow studying the migration of neutrophils in isolation, e.g., LPS- or thioglycollate-induced neutrophil recruitment. However, when CD97 is targeted in more complex models, which represent in addition to innate immunity also aspects of adaptive immunity, i.e., DTH models, no effect of CD97 mAbs is found suggesting that CD97 targeting does not interfere with T-cell recruitment or recall responses. This notion is further substantiated by the lack of effect of 1B2 in a T-cell-driven EAE model, in which disease can be blocked by FTY720 treatment.[28] Interestingly, the 60% inhibitory effect of 1B2 treatment in established CIA, which largely depends on IL-1β, may suggest a role for granulocytes in this model. Alternatively, other cell types such as macrophages may be affected here by CD97 blockade. In addition, the CIA model in contrast to the EAE models has been demonstrated to be only partly dependent on T-cell recruitment and/or activation.[29] Collectively, our data show that CD97 targeting predominantly affects the innate response (neutrophils and possibly macrophages) and not the adaptive response. Clearly, additional research is needed to further substantiate this notion.

**Table 1.** Consequences of CD97 antibody treatment in immune responses in mice

| Model | Mouse Strain | Causing Agent | CD97 mAb | Effects of Antibody Treatment | Reference |
|---|---|---|---|---|---|
| *Innate* | | | | | |
| Colitis | BALB/c | DSS | 1B2 | Granulocyte infiltration ↓ | 18 |
| Peritonitis | C57BL/6J | Thioglycollate | 1B2 and 1C5 | Granulocyte infiltration ↓ | 30 |
| Pulmonitis | C57BL/6J | LPS | 1B2 | Granulocyte infiltration ↓ | De Groot et al, this chapter |
| Pneumonia | BALB/c | *S. pneumoniae* | 1B2 and 1C5 | Granulocyte infiltration ↓ Bacterial clearance ↓ Survival ↓ | 18 |
| Stem cell mobilization | BALB/c | IL-8 | 1B2 | Mobilization of hematopoietic stem and progenitor cells ↓ | 23 |
| *Antigen-dependent* | | | | | |
| DTH | BALB/c | Oxazolone | 1B2 | No effect | Hofstra et al, this chapter |
| DTH | BALB/c | Tetanus toxoid | 1B2 | No effect | De Groot et al, this chapter |
| CIA | DBA/J1 | Bovine collagen | 1B2 and 1C5 | Disease activity ↓ Bone destruction ↓ | 25,26 |
| EAE | SJL/J | $PLP_{139-151}$ | 1B2 | No effect | Visser and Laman, this chapter |

## ANTIBODY TREATMENT VERSUS GENE TARGETING

A remarkable finding of the experiments described above was the ability of two different CD97 mAbs to inhibit granulocyte trafficking. While 1B2 interferes with CD55 binding in vitro, 1C5 binds at the predicted interaction site for chondroitin sulfate B. We concluded that either different CD97-ligand interactions equally contribute to granulocyte trafficking, that the antibodies used in the in vivo experiments did not act as antagonists or that the antibodies exert another effect. Importantly, the CD97-specific effect of the antibody activity was confirmed by demonstrating that granulocyte migration into the inflamed peritoneum was blocked in wild-type but not in knockout mice.[30]

CD97-deficient mice, which were independently developed by the laboratory of Kathleen Kelly and by us, showed a mild granulocytosis at steady state that increased under inflammatory conditions.[30,31] Recruitment of granulocytes to sites of inflammation was unaffected and the immune response to *S. pneumoniae*-induced pneumonia in our CD97 knockout mouse was the same as in wild-type mice.[30] In contrast, Kelly and coworkers observed an improved immune response in their CD97 knockout mouse in an acute infection with *Listeria monocytogenes* that possibly was due to the granulocytosis.[31] Future studies will have to elucidate whether this is related to the different pathogens used or is caused by other factors.

## IN VIVO STUDIES START TO UNVEIL THE CD97 MECHANISM OF ACTION

One explanation for the differences observed between CD97 gene targeting and antibody treatment is that the CD97 knockout mice do not show the granulocyte migration defect due to compensatory mechanisms that develop during ontogeny. Alternatively, CD97 antibodies might actively induce an inhibitory effect that disturbs granulocyte and possible macrophage transmigration, which is not perturbed by the absence of the molecule.[30] An agonistic effect of CD97 antibodies could involve regulation of the interaction between the $\alpha$ and the $\beta$ chain or interference with receptor dimerization, which is a characteristic of EGF-TM7 receptors.[32] However, up to now no such signaling event could be demonstrated despite many efforts studying known GPCR mediators (Garritsen, Van Elsas and Van Puijenbroek, unpublished data).

Another putative mechanism of action for targeting CD97 with antibodies may be formed by the depletion of soluble CD97 from the circulation. The $\alpha$ chain of CD97 has been shown to be shed from the membrane of CD97-expressing cells and to be increased in the sera of CIA mice and of patients suffering from rheumatoid arthritis.[26,33] Up to now, two functions have been ascribed to the $\alpha$ chain: chemoattractant action and costimulation of T cells.[21,34] Since CD97 is implied as a contributor to granulocyte recruitment, chemotactic properties of the $\alpha$ chain are especially relevant. Soluble CD97 was shown to induce chemotaxis of human umbilical vein endothelial cells (HUVEC) in an integrin-dependent and thus human-specific manner.[21] In addition, Spendlove and coworkers demonstrated that soluble CD97 can crosslink CD55 molecules on T cells in vitro, thereby in the presence of anti-CD3 stimulus increasing T-cell proliferation, upregulating CD69 and CD25 and inducing the secretion of IL-10 and granulocyte-macrophage colony-stimulating factor (GM-CSF).[34] However, our observation that targeting CD97 is probably not effective in in vivo models, which have been shown to be highly dependent on T-cell recruitment and activation, raises doubt on whether this mechanism translates to the in vivo situation. In addition, we recently demonstrated that soluble CD97 was also increased in CIA mice effectively treated with IL1$\beta$ antibodies,[26] suggesting that the increased soluble CD97 is more an epiphenomenon rather than a causative factor in this disease model.

A third putative mechanism of action is suggested by our own work, which showed that binding of 1B2 to CD97-positive cells induces internalization of the CD97 receptor.[26] Although quite common for classical GPCRs, this is the first time this has been observed for an EGF-TM7 receptor family member. Upon binding of 1B2, CD97 internalizes and thereby would become unavailable for binding to its ligands. This latter mechanism would also explain why 1C5 and 1B2, although they block the binding to two different ligands, are both able to inhibit CD97 function. The relative abundance of binding (1B2 binding all CD97 molecules and 1C5 binding only about half of all CD97 molecules) would explain why 1B2 effects are reproducibly statistically significant and the inhibition of 1C5 does not reach statistical significance.

## CONCLUSION

Although studies in mice provided first insights into the physiological functions of CD97, we are still missing crucial knowledge for a comprehensive biological understanding of this adhesion-GPCR. Interestingly, new experiments recently confirmed the interaction between CD97 and CD55 in vivo. We found that CD97 expression in CD55 knockout mice is significantly increased on all leukocytes and that this phenotype is reversible after transfer of CD55-deficient leukocytes into wild-type mice (Veninga et al, work in progress). Elucidating the molecular processes engaged by CD97 and the working mechanism of CD97 antibodies may lead to the development of novel therapeutics for the treatment of inflammatory disorders.

## ACKNOWLEDGEMENTS

We like to thank all colleagues and collaborators who have contributed to the studies discussed here. The work described here has been supported by the Netherlands Organization for Scientific Research (NWO), the Dutch Arthritis Association, the Dutch MS Research Foundation and Organon (Schering-Plough).

## REFERENCES

1. Eichler W, Aust G, Hamann D. Characterization of an early activation-dependent antigen on lymphocytes defined by the monoclonal antibody BL-Ac(F2). Scand J Immunol 1994; 39:111-115.
2. Kwakkenbos MJ, Kop EN, Stacey M et al. The EGF-TM7 family: a postgenomic view. Immunogenetics 2004; 55:655-666.
3. Lagerström MC, Schiöth HB. Structural diversity of G protein-coupled receptors and significance for drug discovery. Nat Rev Drug Discov 2008; 7:339-357.
4. Yona S., Lin HH, Siu WO et al. Adhesion-GPCRs: emerging roles for novel receptors. Trends Biochem Sci 2008; 33:491-500.
5. Kwakkenbos MJ, Matmati M, Madsen O et al. An unusual mode of concerted evolution of the EGF-TM7 receptor chimera EMR2. FASEB J 2006; 20:2582-2584.
6. Kop EN, Matmati M, Pouwels W et al. Differential expression of CD97 on human lymphocyte subsets and limited effect of CD97 antibodies on allogeneic T-cell stimulation. Immunol Lett 2009; 123:160-168.
7. Jaspars LH, Vos W, Aust G et al. Tissue distribution of the human CD97 EGF-TM7 receptor. Tissue Antigens 2001; 57:325-331.
8. Aust G, Eichler W, Laue S et al. CD97: a dedifferentiation marker in human thyroid carcinomas. Cancer Res 1997; 57:1798-1806.
9. Aust G, Steinert M, Schütz A et al. CD97, but not its closely related EGF-TM7 family member EMR2, is expressed on gastric, pancreatic and esophageal carcinomas. Am J Clin Pathol 2002; 118:699-707.
10. Aust G, Wandel E, Boltze C et al. Diversity of CD97 in smooth muscle cells. Cell Tissue Res 2006; 324:139-147.
11. Hamann J, Eichler W, Hamann D et al. Expression cloning and chromosomal mapping of the leukocyte activation antigen CD97, a new seven-span transmembrane molecule of the secretion receptor superfamily with an unusual extracellular domain. J Immunol 1995; 155:1942-1950.

12. Gray JX, Haino M, Roth MJ et al. CD97 is a processed, seven-transmembrane, heterodimeric receptor associated with inflammation. J Immunol 1996; 157:5438-5447.
13. Qian YM, Haino M, Kelly K et al. Structural characterization of mouse CD97 and study of its specific interaction with murine decay-accelerating factor (DAF, CD55). Immunol 1999; 98:303-311.
14. Hamann J, van Zeventer C, Bijl A et al. Molecular cloning and characterization of mouse CD97. Int Immunol 2000; 12:439-448.
15. Hamann J, Vogel B, van Schijndel GMW et al. The seven-span transmembrane receptor CD97 has a cellular ligand (CD55, DAF). J Exp Med 1996; 184:1185-1189.
16. Hamann J, Stortelers C, Kiss-Toth E et al. Characterization of the CD55 (DAF)-binding site on the seven-span transmembrane receptor CD97. Eur J Immunol 1998; 28:1701-1707.
17. Lin HH, Stacey M, Saxby C et al. Molecular analysis of the epidermal growth factor-like short consensus repeat domain-mediated protein-protein interactions: dissection of the CD97-CD55 complex. J Biol Chem 2001; 276:24160-24169.
18. Leemans JC, te Velde AA, Florquin S et al. The epidermal growth factor-seven transmembrane (EGF-TM7) receptor CD97 is required for neutrophil migration and host defense. J Immunol 2004; 172:1125-1131.
19. Stacey M, Chang GW, Davies JQ et al. The epidermal growth factor-like domains of human EMR2 receptor mediates cell attachment through chondroitin sulphate glycosaminoglycans. Blood 2003; 102:2916-2924.
20. Kwakkenbos MJ, Pouwels W, Matmati M et al. Expression of the largest CD97 and EMR2 isoforms on leukocytes facilitates a specific interaction with chondroitin sulfate on B-cells. J Leukoc Biol 2005; 77:112-119.
21. Wang T, Ward Y, Tian LH et al. CD97, an adhesion receptor on inflammatory cells, stimulates angiogenesis through binding integrin counterreceptors on endothelial cells. Blood 2005; 105:2836-2844.
22. Lin HH, Chang GW, Davies JQ et al. Autocatalytic cleavage of the EMR2 receptor occurs at a conserved G protein-coupled receptor proteolytic site motif. J Biol Chem 2004; 279:31823-31832.
23. Van Pel M, Hagoort H, Kwakkenbos MJ et al. Differential role of CD97 in interleukin-8-induced and granulocyte-colony stimulating factor-induced hematopoietic stem and progenitor cell mobilization. Haematologica 2008; 93:601-604.
24. Kop EN, Kwakkenbos MJ, Teske GJD et al. Identification of the epidermal growth factor-TM7 receptor EMR2 and its ligand dermatan sulfate in rheumatoid synovial tissue. Arthritis Rheum 2005; 52:442-450.
25. Kop EN, Adriaansen J, Smeets TJ et al. CD97 neutralisation increases resistance to collagen-induced arthritis in mice. Arthritis Res Ther 2006; 8:R155.
26. De Groot DM, Vogel G, Dulos J et al. Therapeutic antibody targeting of CD97 in experimental arthritis; the role of antigen expression, shedding and internalization on the pharmacokinetics of anti-CD97 monoclonal antibody 1B2. J Immunol 2009; 183:4127-4134.
27. Visser L, de Vos AF, Hamann J et al. Expression of the EGF-TM7 receptor CD97 and its ligand CD55 (DAF) in multiple sclerosis. J Neuroimmunol 2002; 132:156-163.
28. Kataoka H, Sugahara K, Shimano K et al. FTY720, sphingosine 1-phosphate receptor modulator, ameliorates experimental autoimmune encephalomyelitis by inhibition of T-cell infiltration. Cell Mol Immunol 2005; 2:439-448.
29. Ijima K, Murakami M, Okamoto H et al. Successful gene therapy via intraarticular injection of adenovirus vector containing CTLA4IgG in a murine model of type II collagen-induced arthritis. Hum Gene Ther 2001; 12:1063-1077.
30. Veninga H, Becker S, Hoek RM et al. Analysis of CD97 expression and manipulation: antibody treatment but not gene targeting curtails granulocyte migration. J Immunol 2008; 181:6574-6583.
31. Wang T, Tian L, Haino M et al. Improved antibacterial host defense and altered peripheral granulocyte homeostasis in mice lacking the adhesion class G protein receptor CD97. Infect Immun 2007; 75:1144-1153.
32. Davies JQ, Chang GW, Yona S et al. The role of receptor oligomerization in modulating the expression and function of leukocyte adhesion-G protein-coupled receptors. J Biol Chem 2007; 282:27343-27353.
33. Hamann J, Wishaupt JO, van Lier RAW et al. Expression of the actvation antigen CD97 and its ligand in rheumatoid synovial tissue. Arthritis Rheum 1999; 42:650-658.
34. Capasso M, Durrant LG, Stacey M et al. Costimulation via CD55 on human CD4+ T-cells mediated by CD97. J Immunol 2006; 177:1070-1077.

# CHAPTER 12

# THE ROLE OF CD97 IN REGULATING ADAPTIVE T-CELL RESPONSES

## Ian Spendlove* and Ruhcha Sutavani

**Abstract:** CD97 was identified as an early activation marker on T cells, having low expression on naive T cells. This is a common feature of molecules that have a role in T-cell function. It was subsequently identified as a ligand for CD55, which has been previously identified as an innate regulator of complement. The interaction of this receptor-ligand pair has been shown to provide a potent costimulatory signal to human T cells, despite their modest affinity. Though both CD97 and CD55 are expressed on T cells as well as antigen presenting cells (APCs), their interaction is significant when CD97 on APCs interacts with CD55 on T cells. The converse interaction is poorly defined and may be less significant. A unique aspect of the interaction of CD97 with CD55 is the stimulation of naive T cells, leading to the induction of IL-10 producing cells that behave like Tr1 regulatory cells. This raises a number of questions regarding the dual functions of CD55; regulating complement and stimulating T cells via CD97 interaction and any potential overlap in the consequences of these dual roles.

## INTRODUCTION: STRUCTURE OF CD97 AND ITS INTERACTION WITH CD55

The characterisation of the CD97 molecule was begun by Eichler et al, who first described it as an early activation marker of lymphocytes with a size of 78-85 kDa. The molecule is expressed by monocytes, granulocytes and activated B and T cells. Although low levels of CD97 were detectable on resting lymphocytes, it was found to be rapidly upregulated (within 2-4 hours) following stimulation of peripheral blood mononuclear cells with concanavalin A and via CD3/CD28 costimulation.[1,2]

---

*Corresponding Author: Ian Spendlove—The University of Nottingham, Academic Clinical Oncology, The City Hospital, Hucknall Road, Nottingham, Ng5 1PB, UK. Email: ian.spendlove@nottingham.ac.uk

*Adhesion-GPCRs: Structure to Function,* edited by Simon Yona and Martin Stacey.
©2010 Landes Bioscience and Springer Science+Business Media.

# THE ROLE OF CD97 IN REGULATING ADAPTIVE T-CELL RESPONSES

The gene structure and isoforms of CD97 are discussed in other chapters of this book. The protein structure of CD97, shown interacting with CD55 (Fig. 1) is modelled on the crystal structure of EMR2, a closely related family member. EMR2 differs from CD97 in the 1, 2 and 5 EGF-domain isoform by only 3 amino acids.[3] The model shows the interaction of CD97 with CD55 based on NMR and crystallographic studies.[4] The 1, 2, 5 isoform of CD97 binds optimally to CD55 with low affinity (dissociation constant ($K_D$) of 86 μM) and exhibits a rapid off-rate. Although the 1, 2, 5 isoforms differs by only 3 amino acids in the EGF-like domains, there is greater variability in the stalk region of the molecules. While EMR2 is able to bind CD55 it does so with a $K_D$ of at least an order of magnitude less than CD97. This indicates that the EGF-like domains are responsible for the interaction with CD55 and that the interaction can be affected by very small changes in the primary sequence.[5]

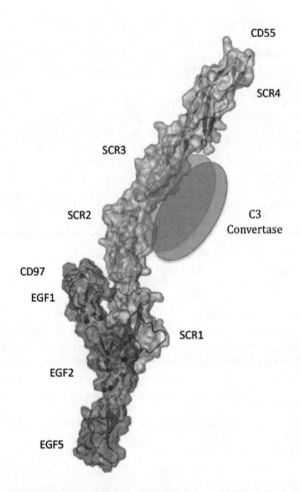

**Figure 1.** A model of CD97 (bottom left, dark grey) showing the EGF domains 1, 2 and 5 interacting with CD55 (light grey) showing SCR domains 1-4. The diagram also shows the proposed positioning of the C3 convertase on the opposing face to that of CD97. The model is based on crystallographic studies of CD55, CD97 supported by NMR studies of their interaction. Mutagenesis studies have revealed the proposed surfaces responsible for C3 convertase interaction (refs. 4, 25 and Lea 2004).

The cloning of mouse CD97 revealed that there was a phylogenetic restriction in binding to CD55. Mouse CD97 was able to bind both rat and mouse CD55 expressing erythrocytes and human CD97 only bound to human and primate erythrocytes.[6,7] It was shown that the same 1, 2, 5 isoform of mCD97 interacts with mCD55 and is required for binding of mouse erythrocytes and splenocytes to human embryonic kidney (HEK) 293 cells transfected with mouse CD97. However, the regulation of mCD97 on lymphocytes appears different to that of hCD97. Whereas activation of human T cells results in upregulation of CD97, interestingly, mCD97 levels appear to decline in splenocytes following stimulation, with levels return to normal by 48 hours poststimulation.[7]

## COSTIMULATION OF T CELLS

Ligand-receptor interactions are the main regulatory mechanism controlling effector functions of T cells. These interactions are tightly regulated by a host of mechanisms including environmental pro- and anti-inflammatory factors and cytokines that control receptor expression and therefore regulate the involvement of these receptors in T-cell activation.

There is also a wide range of receptor-ligand interactions known to influence T-cell activation and their regulation is tightly orchestrated during both primary and secondary stimulation cycles to generating an effective immune response. These include, the so-called "signal one", MHC-peptide-T-cell receptor (TCR-MHCp) interaction and a multitude of potential "signal two"s and their receptors that promote T-cell interaction with APC, including; CD28, CD80/86, CD27, CD70, HVEM, LIGHT, OX40-(L), 4-1BB-(L), CD30-(L), CD5, CD9, CD2, CD44, CD11a- ICAM1-3, some of which are present on both APC and T cell. Though "signal one" and "signal two" are both essential, costimulation via any one of the "signal two"s, alongside the "signal one", is able to promote cellular activation and drive cells through cycle resulting in proliferation.[8-11]

These costimulatory molecules fulfil multiple roles in the activation of T cells. Most are able to enhance the effect of TCR-MHCp interaction on T-cell activation and thus reduce the number of interacting TCR-MHCp required to stimulate T-cell activation. This predominantly occurs in the case of effector and memory T cells where costimulation enhances the reactivation of T cells.

The second role these costimulators and their ligands fulfil is the primary stimulation of naïve T cells to generate antigen specific effector cells. Signal one (TCR-MHCp) will drive proliferation of naïve T cells but these cells fail to induce Interleukin-2 (IL-2) production and apoptose after a short period of time. Signal two, co-engagement of CD28, stimulates IL-2 gene transcription. IL-2 acts as an autocrine growth factor, promoting both proliferation and differentiation of naïve T cells into an effector phenotype. Interestingly, only select costimulators fulfil this role for naive T cells, like CD28 and 4-1BB, while most others fail, indicating their primary role is in modulation of effector responses rather than naïve differentiation. In both settings however, costimulation alone is not enough to promote T-cell activation but also requires simultaneous TCR engagement.

The engagement of CD55 on effector T cells by CD97 provides a powerful costimulus in the case of both CD4 and CD8 T cells, when used in vitro in conjunction with anti-CD3 as a surrogate TCR-MHCp. This results in the upregulation of activation markers CD69 and CD25, increase in cell numbers entering S phase, cell proliferation (Fig. 2) and secretion of cytokines. However, CD97-CD55 engagement alone, at any concentration,

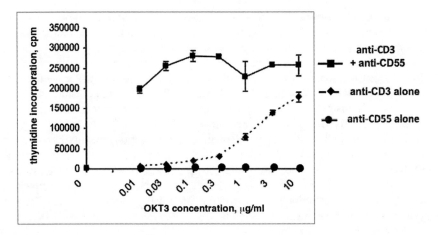

**Figure 2.** Purified CD4+ T cells were stimulated with a titration of plate bound OKT3 (anti-CD3) [diamonds], or in the presence of a constant concentration (5 μg/ml) of anti-CD55 antibody (791T/36) [squares], or with anti-CD55 antibody alone [circles]. Cells were stimulated for 3 days and tritiated thymidine incorporation used to measure proliferation. Similar results are seen with CD8 and naive T cells.

does not influence any of these events, indicating a dependency on TCR engagement[12] (and unpublished observations). This is supported by the observation that antibodies to CD55 block the CD97 interaction and abrogate the effects of costimulation.[4] It was also demonstrated that soluble CD55, when introduced in T-cell and peptide-pulsed monocyte assays, was able to inhibit the proliferative response of T cells and cytokine secretion by them.[13] It was recently demonstrated that antibodies to CD97 were able to diminish T-cell activation in vitro, although the antibody was to the stalk region common to all the EGF-TM7 family and not specifically CD97.[2]

## T-CELL SIGNALLING

When CD97 acts as a ligand for CD55, it initiates a signalling event within T cells that, when added to TCR signalling enhances proliferation and cytokine secretion. The paradigm for this type of costimulation is CD28, which initiates a complex series of signalling events resulting in T-cell growth, cytoskeletal reorganisation, proliferation and cytokine release (Fig. 3). The main signalling pathways that are activated cause increase in intracellular calcium levels, activation of the mitogen activated protein kinases (MAPKs), activation of phosphoinositol-3-kinase (PI3K) and translocation of the transcription factor NFκB to the nucleus. However, these pathways are by no means distinct and there is much overlap of the TCR and CD28 signals.

Cross linking of CD55 results in phosphorylation of fyn and lck, which are strongly associated with TCR activation.[14,15] This is also achieved by CD97 engagement of CD55 (authors observations). Activated Lck and/or Fyn phosphorylate immunoreceptor tyrosine-based activation motifs (ITAMs) located within the ε, δ, γ and ζ subunits of the TCR complex and provide a mechanism by which costimulators such as CD55 can enhance TCR stimulation of T cells. These events in turn allow recruitment of the tyrosine kinase ζ-chain-associated protein 70 (ZAP-70), which is then activated by autophosphorylation. Ultimately, ZAP-70 activation leads to three main outcomes: the initiation of serine/threonine kinase pathways, increased intracellular calcium levels and cytoskeletal reorganisation[5,16] while CD28 costimulation is thought to promote NFκB activation via vav.[16] NFκB translocates to the nucleus where it initiates transcription of a number of genes essential for T-cell proliferation, including the IL-2 gene (Fig. 3). Costimulatory molecules, including CD28, 41BB, CD27, CD30 and OX40, can induce NFκB-dependent transcription. How CD97 engagement of CD55 enhances TCR signalling, promoting T-cell activation and is still not known.[17]

CD28 is unique in its ability to stimulate effector T cells independent of TCR. Also, it has the capacity to enhance TCR effects both during and after TCR engagement has ceased. This was a possible contributing factor in the clinical trials of anti-CD28 antibody that resulted in cytokine storm.[18] This is not the case for other costimulatory molecules, including CD97-CD55. In order for CD55 to stimulate T cells, there must be simultaneous engagement of TCR-MHCp and CD55-CD97.

## EFFECT ON DIFFERENT T-CELL POPULATIONS

Certain receptor-ligand pairs appear to favour particular T-cell subsets. CD28 is a potent costimulator of CD4 T cells but it may only be expressed by 40% of CD8 T cells. However, 4-1BB expression predominates on CD8 T cells and is thought to be the dominant costimulator of these cells.[19] The effects of CD97 engagement of CD55 on CD8 T cells is similar to that of CD4 cells, in that both cell types show signs of activation, secrete cytokines and proliferate in response to stimulation. While CD97-CD55 engagement alone does not have any noticeable effect on the T cells costimulation via CD3 and CD55 appears to have a great capacity to stimulate both populations of T cells to levels greater than by maximal TCR stimulation alone (Fig. 2).

As far as cytokine production by CD4 cells is concerned, cells costimulated via CD97-CD55 show a significant increase in IL-10 production when compared to CD28 costimulated cells, which predominantly produce IFNγ. When CD55 costimulation of naïve or differentiated CD4 cells was studied, it was observed that only naïve cells had the capacity to differentiate into IL-10 producing cells. Analysis of these cells by ELISPOT and IL-10 capture assays showed that small population of cells (1-3%) was responsible for the high IL-10 production. These cells produced very little IFNγ and no IL4 or IL-2 (ref. 12, and unpublished observations), which is a characteristic of inducible regulatory T cells (Tr1). This might imply that within the naïve cells there are a small sub-population that can be stimulated to become IL-10 producing Tr1 cells. What makes this subset of cells different and why they are preferentially stimulated by CD55 remains to be addressed.

### CD55 Structure and Complement Regulation

In order to consider the consequence of CD55 signalling and its effects on T cells we must consider the role for which CD55 was characterised, that of regulating complement.

CD55 (decay accelerating factor; DAF) is a ~70 kDa GPI-anchored protein comprised of four domains.[20,21] It was originally identified for its role in inhibiting complement activation by binding to the C3 and C5 convertases of the complement pathway, accelerating their decay and thus regulating all three activation pathways.[21] It is comprised of four short consensus repeat (SCR) domains, each approximately 60 amino acids, constrained by

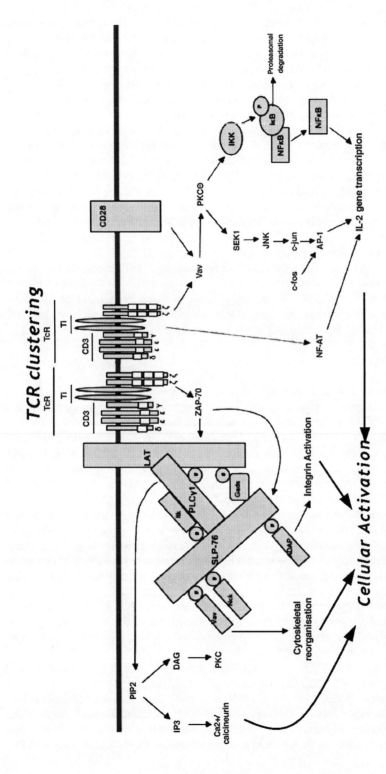

**Figure 3.** Diagram of the signalling events that are initiated in T cells upon TCR engagement leading to cellular activation (left) and the proposed mechanisms of how CD28 engagement integrates with TCR siganls to promote important transcriptional events leading to proliferation and cell survival (right) (Acuto 2003, Jordan 2003 and Rudd 2003).

disulphide bonds in a 1-3, 2-4 manner. These domains are attached to an O-glycosylated, proline-rich spacer region and attached to the outer leaflet of the plasma membrane by a glycosyl-phosphatidylinositol anchor.[22] Complement and their receptors show little structural homology across species making them largely species restricted.[23] The action of CD55 is to accelerate the decay of the main regulatory molecules of complement, the C3 and C5 (in parentheses) convertases C4b2a(C3b) and C3bBb(C3b) into their component parts, after which they are no longer able to re-associate.

A number of domain switching and mutagenesis studies have shown this activity to involve SCR domains 2, 3 and 4 with domain 1 having little or no apparent involvement in convertase dissociation.[24-26] Studies on the precise interaction of CD55 with the convertases and the kinetics of this interaction have been hampered by the very rapid dissociation rates of the convertases upon CD55 binding. Recent biophysical studies have revealed that the kinetics of association/dissociation of C3bBb with CD55 were too rapid for affinity measurements to be made by surface plasmon resonance.[25] This rapid dissociation of the convertases and dampening of complement activation by CD55 protects cells from collateral damage during complement activation, for example, by infectious agents.

The GPI-anchor serves to tether CD55 to the outer leaflet of the plasma membrane. A common feature of gpi-anchored proteins is their association with areas of plasma membrane that are rich in cholesterol, called cholesterol-rich microdomains or lipid rafts. These have been associated with enhanced signalling in a number of cell types, possibly due to the association of a range of kinases to these lipid rafts.[27]

## CD55 Signalling

Despite the lack of an intracellular domain, a number of studies have reported that CD55 is able to mediate signalling events that result in cellular activation. Original studies showed that cross-linking CD55 on human leukocyte populations with anti-CD55 (IgM) antibodies resulted in an increase in release of calcium from intracellular stores, a common feature of cellular activation. Similarly, an increase in oxidative burst was also observed. However, CD55 stimulation alone, without its cross-linking, was insufficient to achieve these activation events. Only when an anti-CD55 IgM was used, which caused CD55 cross-linking, was activation triggered.[28]

Shibuya et al carried out similar studies on human monocytes with an antibody that binds to SCR 3 of CD55 and completely blocks its decay accelerating activity. Antibody binding appeared to activate the monocytes, increasing glucose consumption and enhancing phagocytosis of latex beads, but it failed to stimulate production of cytokines by the cells. Similar experiments were conducted using CD55, engineered to contain a trans-membrane domain that replaced its gpi-anchor. These studies demonstrated the requirement of the gpi-anchor for cellular activation, measured in terms of calcium release and inositol triphosphate production. These also showed that a number of tyrosine kinases, notably $p56^{lck}$ and $p59^{fyn}$, could be coprecipitated with the gpi-anchored form of CD55 but not its trans-membrane form.[29,30]

Antibody mediated CD55 cross-linking has also been shown to cause proliferation of T cells. There is a low level of CD55 expressed on a human T-cell line, but this is upregulated following T-cell activation by mitogens such as calcium ionophores and phorbol esters. Using sub-mitogenic doses of these, antibodies to CD55 resulted in proliferation of the cells and crosslinking of the anti-CD55 antibody enhanced this activation.[31,32] It was also reported that antibodies to SCR3 could induce activation but not those against SCR4, suggesting that not all of the domains are involved in stimulation.[29] Interestingly, CD55 cross-linking alone does not appear to cause these effects, but only in a combination with other signalling events is activation seen. The precise relationship of CD55 with lipid rafts and its influence on signalling is yet to be determined. Also, the role of lipid rafts in a range of cellular events, including that in the immunological synapse of T cells, is still to be fully explained. While CD97 engagement of CD55 has a profound and positive effect on T cells, other possible roles of CD55 need to be considered.

## EFFECTS OF COMPLEMENT ON CD55 AND T CELLS

One of the interesting aspects of any molecule with multiple functions is how those functions relate to each other. With CD55, this concerns its interaction with either CD97 or complement factors or both. A number of lines of evidence suggest that CD55 can engage with CD97 or the C3 convertases without affecting the other's activity. The structural model based on crystallographic and NMR studies (Fig. 1) shows distinct binding sites for CD97 and the C3 convertases on opposite faces of CD55. Other studies have addressed this question more directly by looking at the ability of CD55 to inhibit complement deposition and lysis of cells in the presence or absence of functional CD97. Sheep red blood cells were incubated in the presence of anti-sheep-red-blood-cell (rbc) antisera and fresh serum, causing the complete lysis of the rbcs. When soluble CD55 was introduced into the assays the cells were protected from lysis due to the complement inhibiting activity of CD55. Under these conditions the subsequent addition of soluble and functional CD97 had no effect in reducing the ability of CD55 to inhibit rbc lysis. These indicated that despite the presence of excess CD97, CD55 was able to control complement mediated lysis of red blood cells.[4]

Another investigation examined the ability of CD55 to enhance CD3 mediated stimulation of T cells. T cells were sub-optimally stimulated with plate bound anti-CD3 antibody (OKT3). CD97 engagement of CD55 resulted in increased cellular activation, entry into cell cycle followed by proliferation and cytokine secretion, above that achieved with OKT3 alone. However, this was shown to occur only with plate bound CD97. Soluble CD97 failed to enhance the effects of CD3 stimulation. This is a common feature of costimulatory molecules, with the exception of CD28, which can be activated by soluble 'superagonist' antibodies.

In order to assess any contribution of complement to these effects the level of active C3 components was measured on the T cells. Complement has a slow and natural turnover via the alternative pathway, which can be monitored by examining the level of C3 degradation products deposited on cells cultured with fresh serum. T cells stimulated via CD3, in the presence of soluble or plate bound CD97, showed similar levels of C3c/d on their surface. However, only cells stimulated with plate bound CD97 demonstrated enhanced cellular activation and proliferation. Also, similar results were obtained using serum deficient in a range of complement components and also by increasing the level of both serum and complement components in the cultures. As an answer to the above question, all this data suggests that increased complement activation leading to more C3 fragment deposition on T cells was not the influencing factor in the enhancement of CD3 mediated stimulation of human T cells. It follows that the activation of the T cells was mediated by the CD97-CD55 costimulatory interaction.

In mice, the story may be different, as CD97 shows a reduction during activation. Similarly reports from mice lacking the CD97 ligand (CD55 knockout) show increased T-cell activation and altered cytokine secretion, although these effects may be mediated via increased complement effects on cells and via other cell types.[33-37] This is supported by the observations in both human and mouse systems that the C3 breakdown products (anaphylatoxins) have a potent effect on APCs, which selectively bear their C3a and C5a receptors. The result is an enhanced APC activation and increased ability to stimulate T cells. Hence in the CD55 knockout models, APCs demonstrate an heightened state activation, due to increased anaphylatoxins from the complement cascade resulting in an increased activation of T cells.[35,38] However, in CD97 knockout mice leukocyte migration studies would also suggest that the major role for CD97-CD55 interaction is one of adhesion and migration.[39]

## CONCLUSION

There are many questions that remain unanswered regarding the role of CD97 in T-cell differentiation and function. With most costimulatory receptor-ligand pairs, the signal is invariably through both partners, providing modifying behaviour to both the interacting cells, usually antigen presenting cells and T cells. Like CD55-CD97, many costimulatory pairs can be simultaneously present on APCs and T cells, which suggest some auto-regulatory mechanism. Also, the activity of these costimulators is often regulated by changes in their expression. The questions yet to be addressed are, whether CD97 signals in T cells and how it would do so and whether CD55 interacts with CD97 expressed on the same cell. It is also important to characterise the similarities and the differences in the role of CD55-CD97 interactions in both the human and murine models, as there seem to be a number of apparent differences.

In humans, CD97 appears to have a role in modulating T-cell function via its interaction with CD55 on T cells and may also have an important role in regulating the induction of IL-10 producing Tr1-like regulatory cells. This important axis of immune stimulation and regulation mediated by CD55-CD97 may be a key component of immune homeostasis.

## REFERENCES

1. Eichler W, Aust G, Hamann D. Characterization of an early activation-dependent antigen on lymphocytes defined by the monoclonal antibody BL-Ac(F2). Scand J Immunol 1994; 39(1):111-115.
2. Kop EN, Matmati M, Pouwels W et al. Differential expression of CD97 on human lymphocyte subsets and limited effect of CD97 antibodies on allogeneic T-cell stimulation. Immunology Letters 2009; 123(2):160-168.
3. Lukacik P, Roversi P, White J et al. Complement regulation at the molecular level: the structure of decay-accelerating factor. Proc Natl Acad Sci USA 2004; 101(5):1279-1284.
4. Abbott RJ, Spendlove I, Roversi P et al. Structural and functional characterization of a novel T-cell receptor coregulatory protein complex, CD97-CD55. J Biol Chem 2007; 282(30):22023-22032.
5. Lin HH, Stacey M, Saxby C et al. Molecular analysis of the epidermal growth factor-like short consensus repeat domain-mediated protein-protein interactions: dissection of the CD97-CD55 complex. J Biol Chem 2001; 276(26):24160-24169.
6. Hamann J, van Zeventer C, Bijl A et al. Molecular cloning and characterization of mouse CD97. Int Immunol 2000; 12(4):439-448.
7. Qian YM, Haino M, Kelly K et al. Structural characterization of mouse CD97 and study of its specific interaction with the murine decay-accelerating factor (DAF, CD55). Immunology 1999; 98(2):303-311.

8. Borowski AB, Boesteanu AC, Mueller YM et al. Memory CD8+ T-cells require CD28 costimulation. J Immunol 2007; 179(10):6494-6503.
9. Croft M. Co-stimulatory members of the TNFR family: keys to effective T-cell immunity? Nat Rev Immunol 2003; 3(8):609-620.
10. Shahinian A, Pfeffer K, Lee KP et al. Differential T-cell costimulatory requirements in CD28-deficient mice. Science 1993; 261(5121):609-612.
11. Yashiro Y, Tai XG, Toyo-oka K et al. A fundamental difference in the capacity to induce proliferation of naive T-cells between CD28 and other costimulatory molecules. Eur J Immunol 1998; 28(3):926-935.
12. Capasso M, Durrant LG, Stacey M et al. Costimulation via CD55 on Human CD4+ T-Cells Mediated by CD97. J Immunol 2006; 177(2):1070-1077.
13. Spendlove I, Ramage JM, Bradley R et al. Complement decay accelerating factor (DAF)/CD55 in cancer. Cancer Immunol Immunother 2006; 55(8):987-995.
14. Acuto O, Michel F. CD28-mediated costimulation: a quantitative support for TCR signalling. Nat Rev Immunol 2003; 3(12):939-951.
15. Jordan MS, Singer AL, Koretzky GA. Adaptors as central mediators of signal transduction in immune cells. Nat Immunol 2003; 4(2):110-116.
16. Rudd CE, Raab M. Independent CD28 signaling via VAV and SLP-76: a model for in trans costimulation. Immunol Rev 2003; 192:32-41.
17. Kane LP, Lin J, Weiss A. It's all Rel-ative: NF-kappaB and CD28 costimulation of T-cell activation. Trends Immunol 2002; 23(8):413-420.
18. Suntharalingam G, Perry MR, Ward S et al. Cytokine storm in a phase 1 trial of the anti-CD28 monoclonal antibody TGN1412. N Engl J Med 2006; 355(10):1018-1028.
19. Shuford WW, Klussman K, Tritchler DD et al. 4-1BB costimulatory signals preferentially induce CD8+ T-cell proliferation and lead to the amplification in vivo of cytotoxic T-cell responses. J Exp Med 1997; 186(1):47-55.
20. Hoffman EM. Inhibition of complement by a substance isolated from human erythrocytes. I. Extraction from human erythrocyte stromata. Immunochemistry 1969; 6(3):391-403.
21. Medof ME, Kinoshita T, Nussenzweig V. Inhibition of complement activation on the surface of cells after incorporation of decay-accelerating factor (DAF) into their membranes. J Exp Med 1984; 160(5):1558-1578.
22. Nicholson-weller A, Wang C. Structure and function of decay-accelerating factor Cd55. J Lab Clin Med 1994; 123(4):485-491.
23. Harris CL, Spiller OB, Morgan BP. Human and rodent decay-accelerating factors (CD55) are not species restricted in their complement-inhibiting activities [In Process Citation]. Immunology 2000; 100(4):462-470.
24. Brodbeck WG, Kuttner-Kondo L, Mold C et al. Structure/function studies of human decay-accelerating factor. Immunology 2000; 101(1):104-111.
25. Harris CL, Abbott RJ, Smith RA et al. Molecular dissection of interactions between components of the alternative pathway of complement and decay accelerating factor (CD55). J Biol Chem 2005; 280(4):2569-2578.
26. Kuttner-Kondo L, Hourcade DE, Anderson VE et al. Structure-based mapping of DAF's active site residues that decay accelerate the C3 convertases. J Biol Chem 2007.
27. Sangiorgio V, Pitto M, Palestini P et al. GPI-anchored proteins and lipid rafts. Ital J Biochem 2004; 53(2):98-111.
28. Lund-Johansen F, Olweus J, Symington FW et al. Activation of human monocytes and granulocytes by monoclonal antibodies to glycosylphosphatidylinositol-anchored antigens. Eur J Immunol 1993; 23(11):2782-2791.
29. Shenoy-Scaria AM, Kwong J, Fujita T et al. Signal transduction through decay-accelerating factor. Interaction of glycosyl-phosphatidylinositol anchor and protein tyrosine kinases p56[lck] and p59[fyn] 1. J Immunol 1992; 149(11):3535-3541.
30. Shibuya K, Abe T, Fujita T. Decay-accelerating factor functions as a signal transducing molecule for human monocytes. J Immunol 1992; 149(5):1758-1762.
31. Davis LS, Patel SS, Atkinson JP et al. Decay-accelerating factor functions as a signal transducing molecule for human T-cells. J Immunol 1988; 141(7):2246-2252.
32. Tosello AC, Mary F, Amiot M et al. Activation of T-cells via CD55: recruitment of early components of the CD3-TCR pathway is required for IL-2 secretion. J Inflamm 1998; 48(1):13-27.
33. Heeger PS, Lalli PN, Lin F et al. Decay-accelerating factor modulates induction of T-cell immunity. J Exp Med 2005; 201(10):1523-1530.
34. Lalli PN, Strainic MG, Lin F et al. Decay accelerating factor can control T-cell differentiation into IFN-gamma-producing effector cells via regulating local C5a-induced IL-12 production. J Immunol 2007; 179(9):5793-5802.

35. Lalli PN, Strainic MG, Yang M et al. Locally produced C5a binds to T-cell-expressed C5aR to enhance effector T-cell expansion by limiting antigen-induced apoptosis. Blood 2008; 112(5):1759-1766.
36. Liu J, Miwa T, Hilliard B et al. The complement inhibitory protein DAF (CD55) suppresses T-cell immunity in vivo. J Exp Med 2005; 201(4):567-577.
37. Toomey CB, Cauvi DM, Song W-C et al. Decay-accelerating factor 1 (Daf1) deficiency exacerbates xenobiotic-induced autoimmunity. Immunology 2010; Epub ahead of Print.
38. Strainic MG, Liu J, Huang D et al. Locally produced complement fragments C5a and C3a provide both costimulatory and survival signals to naive CD4+ T-cells. Immunity 2008; 28(3):425-435.
39. Veninga H, Becker S, Hoek RM et al. Analysis of CD97 expression and manipulation: antibody treatment but not gene targeting curtails granulocyte migration. J Immunol 2008; 181(9):6574-6583.

# CHAPTER 13

# F4/80:
# The Macrophage-Specific Adhesion-GPCR and its Role in Immunoregulation

Hsi-Hsien Lin,* Martin Stacey, Joan Stein-Streilein and Siamon Gordon

**Abstract:** As a macrophage-restricted reagent, the generation and application of the F4/80 mAb has greatly benefited the phenotypic characterization of mouse tissue macrophages for three decades. Following the molecular identification of the F4/80 antigen as an EGF-TM7 member of the adhesion-GPCR family, great interest was ignited to understand its cell type-specific expression pattern as well as its functional role in macrophage biology. Recent studies have shown that the F4/80 gene is regulated by a novel set of transcription factors that recognized a unique promoter sequence. Gene targeting experiments have produced two F4/80 knock out animal models and showed that F4/80 is not required for normal macrophage development. Nevertheless, the F4/80 receptor was found to be necessary for the induction of efferent CD8+ regulatory T cells responsible for peripheral immune tolerance. The identification of cellular ligands for F4/80 and delineation of its signaling pathway remain elusive but are critical to understand the in vivo role of this macrophage-specific adhesion-GPCR.

## INTRODUCTION

As a central component of both innate and adaptive immune systems, the macrophage executes many of its cellular functions via cell surface receptors.[1] Studying macrophage-specific cell membrane proteins therefore provides an ideal strategy to investigate macrophage biology.[2,3] The F4/80 mAb was generated for this purpose and

*Corresponding Author: Hsi-Hsien Lin—Department of Microbiology and Immunology, College of Medicine, Chang Gung University, 259 Wen-Hwa 1st Road, Kwei-San, Tao-Yuan, Taiwan.
Email: hhlin@mail.cgu.edu.tw

*Adhesion-GPCRs: Structure to Function*, edited by Simon Yona and Martin Stacey.
©2010 Landes Bioscience and Springer Science+Business Media.

is a well-known reagent for the analysis of mouse tissue macrophages.[4] Its extensive application has been invaluable for the phenotypic and functional characterization of tissue macrophage subpopulations.[5]

Similarly, molecular identification of the F4/80 Ag as one of the founding members of the EGF-TM7 family has helped to understand of this group of novel receptors.[6,7] Most notable is its chimeric structure of a tandem epidermal growth factor (EGF)-like motifs and the class B GPCR-related 7TM domain. Its restricted expression in macrophages also prefigured the leukocyte-restricted expression profiles of other EGF-TM7 family members. The F4/80 gene was the first EGF-TM7 receptor to be inactivated in the germ line and its biological function investigated in vivo.[8,9] The unexpected finding that the F4/80 molecule is critically involved in the generation of efferent $CD8^+$ regulatory cells necessary for peripheral tolerance induced by local exposure to antigen further highlighted the importance of cell surface receptors in the role of macrophages in adaptive immune responses.[8,10] In this chapter, we will review the generation and application of the F4/80 mAb as a tool for macrophage study as well as the molecular characterization of the F4/80 receptor. Finally, the role of F4/80 in immunological tolerance is discussed.

## GENERATION AND APPLICATION OF F4/80 mAb: THE PHENOTYPIC AND FUNCTIONAL CHARACTERIZATION OF MOUSE MACROPHAGE SUBPOPULATIONS

After the "proof of principle" experiment was first established by Milstein's group in the mid 1970s, the monoclonal hybridoma technology was quickly adopted by biologists of all disciplines when its far reaching utility in scientific research were realized.[11] This was especially so in the field of immunology where mAbs specific to the cell surface Ags of leukocytes were generated to analyze the phenotype and function of leukocyte subpopulations as well as the surface Ags themselves. With a specific interest in macrophage biology, a similar approach was applied to generate mouse macrophage-specific mAbs in our laboratory over 30 years ago. One of the first successful hybridoma clones generated and analyzed was F4/80.[4]

The F4/80 mAb was derived from rats that had been immunized with thioglycollate-elicited mouse peritoneal macrophages.[4] Subsequent analysis showed that the F4/80 mAb reactivity was restricted to mouse tissue macrophage subpopulations and macrophage cell lines.[4,5,12,13] No reactivity was observed for other hematopoietic cells of myeloid lineage including neutrophils and monocyte-derived osteoclasts. Cells of lymphoid lineage and other nonhematopoietic cells were also shown to be F4/80 negative. However, it was subsequently demonstrated that the F4/80 Ag is also expressed by murine eosinophils.[14] Likewise, it is interesting to note that Langerhans cells, a type of dendritic cells (DC) in the epidermis, express abundant F4/80 Ag. Immature myeloid-type DC express low levels of F4/80, which are downregulated upon subsequent maturation and migration to draining lymph nodes.[15] Similarly, plasmacytoid DC express low levels of F4/80. The highly restricted F4/80 reactivity patterns have allowed the detailed analysis of the distribution, heterogeneity and ontology of mouse tissue macrophages. The F4/80 mAb remains the tool of choice in mouse macrophage study 30 years after its initial generation.

The F4/80 mAb is a rat IgG2b subtype with no detectable cytotoxic activity. It can be applied in various assay systems, including flow cytometry, Western blotting, immunoprecipitation and immunohistochemistry. Especially useful is its robust reactivity on tissue sections (frozen, paraffin-embedded as well as paraformaldehyde/ glutaraldehyde-perfusion fixation) so the in situ distribution patterns and morphology of tissue macrophage populations can be appreciated. As such, F4/80 positive staining is found on Kupffer cells in liver, red pulp macrophages in spleen, microglia cells in brain as well as other resident macrophages in bone marrow stroma, gut lamina propria, kidney, lymph nodes and peritoneum.[16,17] Indeed, by using a quantitative indirect F4/80 mAb binding assay, it is possible to determine the relative pool sizes of resident macrophages in various mouse tissues.[18]

Not only can the F4/80 molecule be used as a pan-macrophage marker to detect the distribution and localization of tissue macrophages, it can also be applied to examine the phenotypic heterogeneity and the activation state of resident macrophages. Thus, it was known that blood monocytes expressed less F4/80 than their tissue counterparts.[19] In addition, the expression levels of F4/80 are down-regulated in certain activated macrophages such as those isolated from bacilli Calmette-Guerin infected animals or treated with IFN-$\gamma$.[12,20,21] Taken together, F4/80 is widely expressed in macrophages and its expression is tightly regulated, suggesting a unique function.

## MOLECULAR CLONING AND CHARACTERIZATION OF THE F4/80 (Emr1) GENE

Consistent with the flow cytometry and immunohistochemical analysis, the F4/80 Ag was shown to be a plasma membrane protein by radio-iodination labeling and immunoprecipitation.[4] Subsequent biochemical analysis determined that the F4/80 molecule is a 160 kDa cell surface glycoprotein heavily decorated with N- and O-linked glycans, containing a major sialic acid modification through $\alpha$2-6 linkage to galactose.[4,5,22] The posttranslational modification of F4/80 also included a possible attachment of chondroitin sulfate glycosaminoglycans, suggesting the molecule is a potential "part-time" proteoglycan.[17,22]

The molecular identity of the F4/80 Ag was eventually revealed by an expression cloning strategy using rabbit polyclonal antisera.[7] A second independent study also identified the F4/80 cDNA through the differential display analysis of c-myb knock-out animals.[6] Both reports showed that the full-length F4/80 cDNA is a ~3.2 kbp transcript with an open reading frame of 2,796 bp encoding a polypeptide of 931 residues. The deduced amino acid sequence predicted a 27-residue signal peptide and a mature protein of 904 residues with an estimated molecular weight of 99 kDa. Upon analyzing the primary sequence in depth, it was realized that the F4/80 molecule is composed of protein domains homologous to two distinct protein superfamilies, namely the EGF-like and the G protein-coupled receptor (GPCR).[6,7] Thus, the F4/80 protein contained a total of 7 tandem EGF-like motifs at the N-terminus followed by a Ser/Thr-rich stalk region linked to the C-terminal seven-transmembrane (7TM) GPCR domain. This structure identified F4/80 as a member of the EGF-TM7 subfamily of adhesion-GPCR.[16,17,23,24] In fact, nucleotide and amino acid sequence comparison has indicated that F4/80 is most likely the murine ortholog of human EMR1, a previously identified EGF-TM7 receptor.[25] Interestingly, EMR1 protein expression was found to be highly restricted to human eosinophilic granulocytes and was not expressed in human monocytic phagocytes.[26] It seems that F4/80 (mouse Emr1) and human EMR1 are structural orthologs that might possess cell type- and species-specific functions.[24]

Along with other EGF-TM7 receptors, the majority of EGF-like motifs of F4/80 contained a consensus calcium-binding sequence often found in extracellular matrix proteins such as fibrillin-1, fibulin-1 and thrombomodulin.[23,24] Calcium binding therefore might play a role in the conformational arrangement and cellular adhesion function of F4/80. In agreement with earlier data, the long extracellular domain of F4/80 possessed 10 potential N-glycosylation sites, numerous potential O-glycosylation sites and a conserved glycosaminoglycan attachment sequence within the fourth EGF-like motif.[6,7] In addition, an Arg-Gly-Asp (RGD) motif was found in the stalk region, suggesting a possible interaction with integrin molecules. The 7TM region showed a relatively strong homology to the class B/Secretin-like GPCRs, but now has been independently grouped as an adhesion-GPCR.[27] However, unlike the vast majority of adhesion-GPCRs, F4/80 does not have the typical GPCR proteolytic site (GPS) sequence. Indeed, F4/80 and its human counterpart, EMR1, are one of very few known adhesion-GPCRs not to undergo GPS proteolytic modification, in agreement with the earlier result of a ~160 kDa single chain polypeptide.

The F4/80 gene is located at the distal locus of mouse chromosome 17.[6,28] Based on the mouse genome sequence, the assembled F4/80 (Emr1) gene contig encompassed ~124 kbp of DNA and contained a total of 22 exons. Likewise, the human EMR1 gene is mapped to a syntenic region on chromosome 19p13.3 with 21 exons, indicating a close evolutionary link.[24] Indeed, it is likely that the EGF-TM7 genes were derived from an ancestor gene through gene duplication and conversion.[29] Among the human EGF-TM7 family members, EMR4[30] was mapped near to EMR1 on the same locus whereas CD97, EMR2 and EMR3 formed another cluster on chromosome 19p13.1.[24]

Reflecting the macrophage-restricted expression patterns, the F4/80 RNA transcripts, as shown by Northern blot and RT-PCR analysis, were detected predominantly in cells of myeloid lineages as well as macrophage-rich tissues such as bone marrow, spleen, brain, kidney, thymus and fetal liver.[6,7] Functional analysis has identified an enhancer element within the proximal promoter region of the F4/80 gene that confers macrophage-restricted expression pattern.[31] It was found that the Ets transcription factor PU.1 bound to a stretch of purine-rich sequence within the enhancer both in vitro and in vivo. This novel transcription factor-promoter association was considered to be a prerequisite for macrophage-specific expression of F4/80.[31] More recently, the transcription factors MafB and c-Maf were shown to be essential for the expression of F4/80 in macrophages, because the MafB and c-Maf deficient animals both displayed dramatically reduced levels of F4/80 expression in otherwise normal macrophage populations.[32,33] It is interesting to note that MafB and c-Maf both bound to the half-Maf recognition element (MARE) site located within the same enhancer region described above and interacted directly with transcription factors Ets-1. Thus, the restricted expression of F4/80 in macrophages was likely coordinated by a selective set of transcription factors.

In addition to the restricted expression patterns, expression analysis has also revealed extensive alternative splicing of F4/80 transcripts, a well-recognized characteristic of the EGF-TM7 molecules. Thus, multiple F4/80 protein isoforms with different combinations of the EGF-like domains were predicted from differentially spliced RNA transcripts.[6,17] However, little is known about the relative abundance of the protein variants in vivo and their potential functional differences.

## GENERATION AND ANALYSIS OF F4/80-DEFICIENT ANIMALS

In order to investigate the biological functions of F4/80, two different approaches were employed to generate animals deficient in F4/80 gene expression. The first approach involved the production of F4/80-Cre knock-in mice as an attempt to produce a founder mouse strain that would allow analysis of the macrophage-specific functions of genes.[9] The strategy was to replace the first coding exon of the F4/80 gene with the coding sequence of Cre gene, so F4/80 gene targeting can be achieved in F4/80$^{cre/cre}$ homozygous animals. Meanwhile, a mouse founder line that expressed Cre recombinase under the control of the F4/80 promoter (F4/80-Cre knock-in) was also generated.[9] The second F4/80 knock-out animal model used a more conventional strategy by replacing the first coding exon with a β-galactosidase/pGK-Neo cassette.[8]

F4/80 homozygous mutant mice were successfully generated in both cases and found to be true F4/80 null by expression analysis.[8,9] The F4/80 null mutants were healthy and fertile with normal development of tissue macrophages, indicating that F4/80 is not required for the differentiation and development of the macrophage lineage. Likewise, the development and physiological functions of other immune cell populations including B, T and natural killer cells appeared normal. Finally, the anti-inflammatory and anti-infectious functions of the F4/80-deficient macrophages were found to be similar to those of wild-type cells when tested in various in vitro and in vivo conditions. In conclusion, the F4/80 molecule is not necessary for the development and distribution of mouse tissue macrophage populations and the inactivation of F4/80 does not affect the general cellular functions of macrophages analyzed by conventional assays.[8,9]

## THE ROLE OF F4/80 IN IMMUNOREGULATION

Despite its extensive application as macrophage-restricted surface marker, little is known of the cellular functions of F4/80. A limited number of earlier reports suggested a potential immunoregulatory function for F4/80. For example, using whole spleen cell culture from SCID mice, Warschkau et al showed a role for F4/80 in IFN-γ production by these cells when exposed to heat-killed *Listeria monocytogenes* (HKL).[34] IFN-γ is crucial in the control of bacterial growth during infection of facultative intracellular pathogens such as *L. monocytogenes* in vivo. In this in vitro experimental model, HKL-stimulated macrophages secreted TNF-α and IL-12, which in turn activated NK cells for IFN-γ release. It was demonstrated that in addition to cytokine stimulation, cell-cell contact between macrophage and NK provided an additional signal required for optimal NK cellular response. F4/80 was found to be critical in the macrophage-NK interaction in this model because F4/80 mAb specifically down-modulated TNF-α, IL-12 and IFN-γ production.[34] It was thought that the interaction of F4/80 with a potential cellular ligand on NK cell might deliver an important stimulatory signal for cytokine release. Unfortunately, a clear role of F4/80 in host defense against *Listeria* infection in vivo can not be established as the F4/80 deficient mice did not show any gross alteration of immune responses in comparison to wild type control animals.[9]

F4/80 was also implicated subsequently in the induction of immune deviation in the eye. Introduction of exogenous antigens into the anterior chamber of the eye, an immune privileged site, is known to induce a systemic immune tolerance that suppresses T effector cell responses.[35,36] This type of immune tolerance is best studied in a model of immune privilege called anterior chamber-associated immune deviation (ACAID).[35,37] During ACAID induction, Ag injected into the anterior chamber is thought to be captured and processed by indigenous F4/80+ antigen presenting cells (APC), which leave the eye through the venous circulation and migrate to the marginal zone (MZ) of spleen. Following the recruitment of other immune effector cells to the MZ, cellular clusters involving the F4/80+ APC, CD1d+ B cells, CD4+ NKT cells and CD8+ T cells are formed.[35,36,38] As a result of these cellular interactions, Ag-specific efferent CD8+ regulatory T (Treg) cells are produced that actively suppress both T helper Type 1 (Th1)- and Th2-mediated immune responses. It was found that the adoptive transfer of as few as 20 in vitro-generated tolegenic F4/80+ APC into naïve recipients can induce ACAID.[39] Most importantly, systemic administration of mice with F4/80 mAb prevents induction of ACAID, suggesting that not only the F4/80+ APC, but the F4/80 molecule itself are crucial to generate peripheral immune tolerance.[40,41]

The involvement of the F4/80 molecule in the induction of peripheral immune tolerance was subsequently confirmed using F4/80-deficient animals.[8] We found that ACAID can not be induced in the absence of the F4/80 molecule due to failure of the F4/80 knock-out mice to generate functional Ag-specific efferent CD8+ Treg cells. Furthermore, by using F4/80 mAb and F4/80 null APC in an in vitro ACAID model, it was shown that the production of efferent Treg cells only occurred when the APCs in the culture system were able to express the F4/80 Ag. In addition to the ACAID model, we also found that only the wild type but not the F4/80 mutant animals produced an efficient immune tolerance response in a low-dose oral tolerance model. Reconstitution of the F4/80 mutant mice with the F4/80+ APCs restored their ability to induce ACAID as well as low-dose oral tolerance.[8] More recently we found further evidence to support that the F4/80 molecule is required for the generation of ACAID-induced CD4+ CD25- Treg cells, but not ACAID-induced CD4+ CD25+ Treg cells. (J. Stein-Streilein, unpublished results). Therefore, the F4/80 receptor is necessary for the generation of antigen-specific afferent and efferent Treg cells. The molecular mechanism whereby the F4/80 receptor might contribute to the induction of peripheral immune tolerance is currently unknown. Nevertheless, its unique structure suggests that F4/80 might interact with an unidentified cellular ligand via the EGF-like motifs, leading to receptor activation through the 7TM region. Indeed, using the soluble extracellular domain of F4/80 as a probe in an overlay experiment on tissue sections, our recent preliminary data have indicated the presence of a potential cellular ligand on the MZ of spleen (M. Stacey and H-H. Lin, unpublished results). This cellular interaction is F4/80 isoform-specific. Thus, it is highly likely that F4/80+ APC execute part of their tolerogenic ability through the interaction of the F4/80 receptor with an MZ ligand. Thus far, strategies successfully used to identify cellular ligands for other EGF-TM7 receptors have not proved efficient to isolate the putative F4/80 ligands[42].

## CONCLUSION

Among adhesion-GPCRs, the F4/80 molecule is unique in several ways. 1, it is highly restricted in mouse tissue macrophages. On the other hand, its human ortholog, EMR1, is eosinophil-specific. 2, it is one of the very few adhesion-GPCRs that does not contain the conserved GPS motif. As a result, F4/80 does not undergo GPS proteolytic modification. 3, F4/80 is found to be involved in the production of the efferent CD8+

Treg cells necessary for the induction of peripheral tolerance. For the future studies, the top priority will be the linking of the receptor structure to its cellular function. Specific questions include the identification of the potential cellular ligand of F4/80, the signaling pathway mediated by the receptor molecule and the effector molecule(s) produced by macrophages as a result of F4/80 activation.

## ACKNOWLEDGEMENTS

The authors would like to thank the grant support from Chang Gung Memorial Hospital (CMRPD160063 and CMRPD170013) and National Science Council, Taiwan (NSC97-2628-B-182-030 and NSC98-2320-B-182-028).

## REFERENCES

1. Taylor PR, Martinez-Pomares L, Stacey M et al. Macrophage receptors and immune recognition. Annu Rev Immunol 2005; 23:901-944.
2. Martinez-Pomares L, Platt N, McKnight AJ et al. Macrophage membrane molecules: markers of tissue differentiation and heterogeneity. Immunobiology 1996; 195(4-5):407-416.
3. McKnight AJ, Gordon S. Membrane molecules as differentiation antigens of murine macrophages. Adv Immunol 1998; 68:271-314.
4. Austyn JM, Gordon S. F4/80, a monoclonal antibody directed specifically against the mouse macrophage. Eur J Immunol 1981; 11(10):805-815.
5. Starkey PM, Turley L, Gordon S. The mouse macrophage-specific glycoprotein defined by monoclonal antibody F4/80: characterization, biosynthesis and demonstration of a rat analogue. Immunology 1987; 60(1):117-122.
6. Lin HH, Stubbs LJ, Mucenski ML. Identification and characterization of a seven transmembrane hormone receptor using differential display. Genomics 1997; 41(3):301-308.
7. McKnight AJ, Macfarlane AJ, Dri P et al. Molecular cloning of F4/80, a murine macrophage-restricted cell surface glycoprotein with homology to the G-protein-linked transmembrane 7 hormone receptor family. J Biol Chem 1996;271(1):486-489.
8. Lin HH, Faunce DE, Stacey M et al. The macrophage F4/80 receptor is required for the induction of antigen-specific efferent regulatory T-cells in peripheral tolerance. J Exp Med 2005; 201(10):1615-1625.
9. Schaller E, Macfarlane AJ, Rupec RA et al. Inactivation of the F4/80 glycoprotein in the mouse germ line. Mol Cell Biol 2002; 22(22):8035-8043.
10. van den Berg TK, Kraal G. A function for the macrophage F4/80 molecule in tolerance induction. Trends Immunol 2005; 26(10):506-509.
11. Gordon S. The Legacy of cell fusion. Oxford ; New York: Oxford University Press; 1994.
12. Ezekowitz RA, Austyn J, Stahl PD et al. Surface properties of bacillus Calmette-Guerin-activated mouse macrophages. Reduced expression of mannose-specific endocytosis, Fc receptors and antigen F4/80 accompanies induction of Ia. J Exp Med 1981; 154(1):60-76.
13. Hirsch S, Austyn JM, Gordon S. Expression of the macrophage-specific antigen F4/80 during differentiation of mouse bone marrow cells in culture. J Exp Med 1981; 154(3):713-725.
14. McGarry MP, Stewart CC. Murine eosinophil granulocytes bind the murine macrophage-monocyte specific monoclonal antibody F4/80. J Leukoc Biol 1991; 50(5):471-478.
15. Hume DA, Robinson AP, MacPherson GG et al. The mononuclear phagocyte system of the mouse defined by immunohistochemical localization of antigen F4/80. Relationship between macrophages, Langerhans cells, reticular cells and dendritic cells in lymphoid and hematopoietic organs. J Exp Med 1983; 158(5):1522-1536.
16. McKnight AJ, Gordon S. EGF-TM7: a novel subfamily of seven-transmembrane-region leukocyte cell-surface molecules. Immunol Today 1996; 17(6):283-287.
17. McKnight AJ, Gordon S. The EGF-TM7 family: unusual structures at the leukocyte surface. J Leukoc Biol 1998; 63(3):271-280.
18. Lee SH, Starkey PM, Gordon S. Quantitative analysis of total macrophage content in adult mouse tissues. Immunochemical studies with monoclonal antibody F4/80. J Exp Med 1985; 161(3):475-489.

19. Gordon S, Lawson L, Rabinowitz S et al. Antigen markers of macrophage differentiation in murine tissues. Curr Top Microbiol Immunol 1992; 181:1-37.
20. Ezekowitz RA, Gordon S. Down-regulation of mannosyl receptor-mediated endocytosis and antigen F4/80 in bacillus Calmette-Guerin-activated mouse macrophages. Role of T-lymphocytes and lymphokines. J Exp Med 1982; 155(6):1623-1637.
21. Ezekowitz RA, Gordon S. Surface properties of activated macrophages: sensitized lymphocytes, specific antigen and lymphokines reduce expression of antigen F4/80 and FC and mannose/fucosyl receptors, but induce Ia. Adv Exp Med Biol 1982; 155:401-407.
22. Haidl ID, Jefferies WA. The macrophage cell surface glycoprotein F4/80 is a highly glycosylated proteoglycan. Eur J Immunol 1996; 26(5):1139-1146.
23. Stacey M, Lin HH, Gordon S et al. LNB-TM7, a group of seven-transmembrane proteins related to family-B G-protein-coupled receptors. Trends Biochem Sci 2000; 25(6):284-289.
24. Kwakkenbos MJ, Kop EN, Stacey M et al. The EGF-TM7 family: a postgenomic view. Immunogenetics 2004; 55(10):655-666.
25. Baud V, Chissoe SL, Viegas-Pequignot E et al. EMR1, an unusual member in the family of hormone receptors with seven transmembrane segments. Genomics 1995; 26(2):334-344.
26. Hamann J, Koning N, Pouwels W et al. EMR1, the human homolog of F4/80, is an eosinophil-specific receptor. Eur J Immunol 2007; 37(10):2797-2802.
27. Yona S, Lin HH, Siu WO et al. Adhesion-GPCRs: emerging roles for novel receptors. Trends Biochem Sci 2008; 33(10):491-500.
28. McKnight AJ, Macfarlane AJ, Seldin MF et al. Chromosome mapping of the Emr1 gene. Mamm Genome 1997; 8(12):946.
29. Kwakkenbos MJ, Matmati M, Madsen O et al. An unusual mode of concerted evolution of the EGF-TM7 receptor chimera EMR2. FASEB J 2006; 20(14):2582-2584.
30. Stacey M, Chang GW, Sanos SL et al. EMR4, a novel epidermal growth factor (EGF)-TM7 molecule up-regulated in activated mouse macrophages, binds to a putative cellular ligand on B lymphoma cell line A20. J Biol Chem 2002; 277(32):29283-29293.
31. O'Reilly D, Addley M, Quinn C et al. Functional analysis of the murine Emr1 promoter identifies a novel purine-rich regulatory motif required for high-level gene expression in macrophages. Genomics 2004; 84(6):1030-1040.
32. Moriguchi T, Hamada M, Morito N et al. MafB is essential for renal development and F4/80 expression in macrophages. Mol Cell Biol 2006; 26(15):5715-5727.
33. Nakamura M, Hamada M, Hasegawa K et al. c-Maf is essential for the F4/80 expression in macrophages in vivo. Gene 2009; 445(1-2):66-72.
34. Warschkau H, Kiderlen AF. A monoclonal antibody directed against the murine macrophage surface molecule F4/80 modulates natural immune response to Listeria monocytogenes. J Immunol 1999; 163(6):3409-3416.
35. Stein-Streilein J. Immune regulation and the eye. Trends Immunol 2008; 29(11):548-554.
36. Stein-Streilein J, Taylor AW. An eye's view of T regulatory cells. J Leukoc Biol 2007; 81(3):593-598.
37. Niederkorn JY. See no evil, hear no evil, do no evil: the lessons of immune privilege. Nat Immunol 2006; 7(4):354-359.
38. Faunce DE, Stein-Streilein J. NKT cell-derived RANTES recruits APCs and CD8+ T-cells to the spleen during the generation of regulatory T-cells in tolerance. J Immunol 2002; 169(1):31-38.
39. Hara Y, Caspi RR, Wiggert B et al. Analysis of an in vitro-generated signal that induces systemic immune deviation similar to that elicited by antigen injected into the anterior chamber of the eye. J Immunol 1992; 149(5):1531-1538.
40. Wilbanks GA, Mammolenti M, Streilein JW. Studies on the induction of anterior chamber-associated immune deviation (ACAID). II. Eye-derived cells participate in generating blood-borne signals that induce ACAID. J Immunol 1991; 146(9):3018-3024.
41. Wilbanks GA, Streilein JW. Studies on the induction of anterior chamber-associated immune deviation (ACAID). 1. Evidence that an antigen-specific, ACAID-inducing, cell-associated signal exists in the peripheral blood. J Immunol 1991; 146(8):2610-2617.
42. Lin HH, Chang GW, Huang YS et al. Multivalent protein probes for the identification and characterization of cognate cellular ligands for myeloid cell surface receptors. Methods Mol Biol 2009; 531:89-101.

# CHAPTER 14

# SIGNAL TRANSDUCTION MEDIATED THROUGH ADHESION-GPCRs

Norikazu Mizuno and Hiroshi Itoh*

**Abstract:** The signaling cascade of most adhesion-GPCRs remains uncharacterized, as the majority are still orphan receptors and further complicated by their unique structure containing a cleaved long extracellular domain (ECD) and a seven-transmembrane domain (7TM). In this chapter, we review previous reports which suggest G protein-dependent and -independent signaling pathways of adhesion-GPCRs and present our approach to investigate the signal transduction of the adhesion-GPCR, GPR56.

## DIFFICULTIES IN STUDYING THE SIGNAL TRANSDUCTION OF ADHESION-GPCRs

Adhesion-GPCRs have a large extracellular domain (ECD) linked to a seven-transmembrane domain (7TM) via a GPCR proteolytic site (GPS) domain.[1] Even though adhesion-GPCRs belong to the class B secretin family of GPCRs, there is little evidence of G protein-dependent signaling.[2] One reason is that most adhesion-GPCRs are orphan receptors. Only a few ligands of adhesion-GPCRs have been reported. Glycosaminoglycan chondroitin sulfate has been shown to interact with an epidermal growth factor (EGF)-like module containing mucin-like receptor protein 2 (EMR2) through the EGF domain, which mediates cell attachment.[3] EMR3 and EMR4 are suggested to interact with their ligands expressed on the surface of macrophages and activated neutrophils[4] and A20 B-lymphoma cells,[5] respectively. The leukocyte activation antigen CD97, which is a member of the EGF-TM7 protein family, has been shown to bind to the decay accelerating factor (CD55/DAF).[6] The binding site and affinity between CD97 and CD55 have been investigated, demonstrating that the shortest splice variant of CD97 has

*Corresponding Author: Hiroshi Itoh—Department of Cell Biology, Nara Institute of Science and Technology, Takayama, Ikoma, Nara 630-0192, Japan. Email: hitoh@bs.naist.jp

*Adhesion-GPCRs: Structure to Function*, edited by Simon Yona and Martin Stacey.
©2010 Landes Bioscience and Springer Science+Business Media.

the highest CD55 binding capacity.[7] A difference in only three amino acids within the EGF domains of the EMR2 results in an order of magnitude weaker binding to CD55.[8] Structural and functional analysis of the complex of EMR2, a very close homolog of CD97, with CD55 has been performed using X-ray crystallography and NMR-based chemical shift mapping and the findings have suggested that CD97-CD55 complex can regulate the complement system.[9] GPR56 has been shown to contribute to the suppression of melanoma metastasis and tumor growth through binding with tissue transglutaminase (TG2), which is expressed in the extracellular matrix.[10] However, the agonistic activity of TG2 has not been reported yet. Latrophilin was isolated as a target molecule of α-latrotoxin (LTX) from black widow spider venom.[11] The cyclooctadepsipeptide, emodepside, which belongs to an anthelmintic drug for use in cats causes paralysis in nematodes with an LTX-like action. Genetic studies have indicated that the predominant target site of emodepside is presynaptic nematoda latrophilin, LAT-1.[12] LTX and emodespside are useful tools to explore the signal transduction through latrophilin, although they are not the endogenous ligands for latrophilin. BAI1, an adhesion-GPCR expressed on macrophages, specifically binds to phosphatidylserine on the outer plasma membrane of cells undergoing apoptosis as an "eat me" signal.[13]

Another reason for the lack of evidence for G protein-dependent signaling is the complicated structure of adhesion-GPCRs possessing both ECD and 7TM domains. BAI1 was initially identified as a p53-regulated gene whose protein product is able to inhibit angiogenesis.[14] The proteolytically cleaved BAI1 extracellular domain inhibits endothelial cell proliferation by binding αvβ5 integrin via its thrombospondin type 1 repeats.[15] Furthermore, the BAI1 extracellular domain can also inhibit in vivo angiogenesis and tumor xenograft growth in mice. Recently, it has been reported that the 7TM of latrophilin interacts not only with its ECD but also with other ECDs of distinct adhesion-GPCRs, such as EMR2 and GPR56.[16] Therefore, distinct adhesion-GPCRs may cross-interact to induce multiple signaling pathways. This complex signaling network may contribute to elicit a variety of biological phenomena. To complicate analysis of signaling further a common structural feature of adhesion-GPCRs, including CD97 and EMRs, is the existence of alternative mRNA splicing.[17] The splice variants results in receptors with different numbers and arrangements of the EGF repeats with differential ligand binding activities.

## G PROTEIN-DEPENDENT SIGNALING PATHWAY OF ADHESION-GPCRs

Latrophilin, which is activated by LTX, is the most characterized adhesion-GPCR. LTX induces calcium signaling and transmitter release. However, the target of LTX is not only latrophilin but also other receptors, such as neurexin and receptor-like protein tyrosine phosphatase σ. Moreover, LTX forms a channel that is permeable to cations, including $Ca^{2+}$ and small molecules. U73122 (an inhibitor of phospholipase C), thapsigargin (a drug depleting intracellular $Ca^{2+}$ stores) and 2-aminoethoxydiphenyl borate (a blocker of inositol(1,4,5)-trisphosphate-induced $Ca^{2+}$ release) block LTX action, suggesting that G protein signaling may regulate LTX-induced transmitter release. Davletov et al found that LTX initiates extracellular $Ca^{2+}$-dependent and -independent transmitter release.[18] Now, it is widely accepted that extracellular $Ca^{2+}$-dependent release is mediated via neurexin, whereas extracellular $Ca^{2+}$-independent release is mediated via latrophilin. Moreover, Ashoton et al classified the LTX-evoked $Ca^{2+}$-dependent glutamate or GABA release from synaptosomes into a fast phase and a delayed phase.[19] Delayed

Ca²⁺-dependent LTX-induced release can be discriminated by sensitivity to U73112 or thapsigargin. LTX pores are blocked by $La^{3+}$ to disrupt LTX tetramers reversibly. Furthermore, $LTX^{N4C}$, which is a nonpore-forming mutant but can activate latrophilin, causes the release of $Ca^{2+}$ from intracellular stores. Capogna et al used CA3 pyramidal neurons in hippocampal slice cultures to reveal the mechanism of the $LTX^{N4C}$ action on central synaptic transmission and the mechanisms of the receptor-dependent action of native LTX.[20] It has been reported that copurification studies with latrophlin show receptor binding to Gαo[21] and Gαq.[22] These results suggest that latrophilin induces Gαq- or Gβγ-mediated PLCβ activation, leading to $Ca^{2+}$ mobilization. The EGF-TM7 family is also well-characterized, but the G protein-dependent signal pathway through EGF-TM7 remains obscure. We recently indicated that the G protein-coupled receptor GPR56 has the ability to transmit the signal via $Gα_{12/13}$ in neural progenitor cells.[23] GPR56 has been identified as a cortical development-associated gene. Mutations of *GPR56* cause bilateral frontoparietal polymicrogyria (BFPP).[24] In this disorder, the organization of the frontal cortex is disrupted and shows thinner cortical layers and numerous small folds. Several mutations of *GPR56* with BFPP cause impairment of cell surface expression of the receptor.[25] It has also been reported that the expression level of GPR56 is involved in cancer cell adhesion and metastasis.[10,26,27] Coupling of GPR56 with the tetraspanins CD81 and CD9, which are small membrane proteins involved in the regulation of cell migration and mitotic activity, is associated with Gαq. However, pharmacological verification was necessary to determine the coupling to specific G proteins. Therefore to investigate the G protein signaling through GPR56, we used two approaches.

## OLIGOMERIZATION OF ADHESION-GPCR FOR ACTIVATION

First, we examined the effect of GPR56 overexpression on the activation of transcription factors using luciferase reporter genes. The overexpression of GPCR can stimulate ligand-independent signal activation by increasing the active form in a multistate conversion model of GPCR.[28] The classic function of GPCRs is to couple the binding of ligands to the activation of heterotrimeric G proteins, leading to the regulation of their effector proteins. However, many GPCRs have more complex signaling behavior. For example, β2-adrenergic receptor (β2AR) exhibits significant constitutive activity, which can be blocked by inverse agonists, such as carazolol and timolol. Full agonists, such as isoproterenol, are capable of maximal receptor stimulation, whereas partial agonists, such as clenbuterol and dobutamine, are unable to elicit full activity even at saturating concentration.[29] GPCRs can couple with distinct isoforms of the G protein, e.g., β2AR couples to both Gs and Gi in cardiac myocytes and can also activate MAP kinase pathways in a G protein-independent manner through arrestin. Similarly, the process of GPCR desensitization involves multiple pathways, including receptor phosphorylation and arrestin-mediated internalization. Moreover, the activation mechanism of GPCR is complicated by oligomerization and localization to specific membrane compartments and different circumstances in the lipid-bilayer composition. Such multifaceted functional behavior has been observed in many different GPCRs. Ligand-induced oligomerization or clustering of cell surface receptors is found in cytokine receptors, receptor tyrosine kinases, tumor necrosis factor (TNF) receptors, T-cell receptors, Toll-like receptors and GPCRs (but not at all) and is an important mechanism to induce signaling in several biological processes. It has now become accepted that some GPCRs exist as homo- and

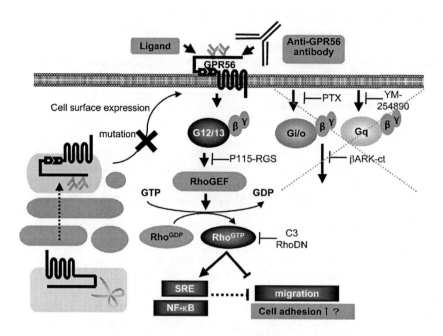

**Figure 1.** Schematic model for the GPR56 signaling pathway. The GPR56 signaling is inhibited by p115-RGS (which blocks $G_{12/13}$ signaling), C3 and RhoDN (which block the Rho signaling) but not PTX (a Gi/o inhibitor), YM254890 (a Gq inhibitor) and βARK-ct (a Gβγ inhibitor). Therefore, GPR56 transmits the signaling through $G\alpha_{12/13}$ and Rho, leading to the stimulation of actin reorganization and SRE- and NF-κB-mediated transcription. Neural progenitor cell migration is negatively regulated through this signaling pathway. The proteolysis in the GPS domain and glycosylation of GPR56 are crucial for the translocation of GPR56 to the plasma membrane.

heterodimers, a finding that has important consequences for receptor pharmacology, signaling and regulation. So far, the oligomerization of many GPCRs with known ligands has been studied, but, recently, dimerization of orphan GPCRs was also reported.[30] Using cell surface cross-linking, co-immunoprecipitation and fluorescence resonance energy transfer analysis of EMR2, it was reported that receptor homo- and hetero-oligomerization plays a regulatory role in modulating the expression and function of leukocyte adhesion-GPCRs.[31] We have shown that GPR56 induces $G\alpha_{12/13}$ and Rho-dependent activation of transcription mediated through the SRE and NF-κB-responsive element in HEK293T cells.[22] Previously, Shashidhar et al reported that GPR56 overexpression in 293 cells activated the TCF-, the PAI-1- and, to a lesser extent, the NF-κB-responsive elements, using a β-galactosidase reporter assay.[24] They also indicated a slight activation of SRE in GPR56-overexpressing cells. More recently, Kim et al reported the differential effects of human GPR56 splice variants on SRE, NFAT, E2F mediated transcriptional activity.[32] However, these activation mechanisms have not been described. Gαi/o, Gαq/11, $G\alpha_{12/13}$ and Gβγ can induce SRE-dependent transcriptional activity under GPCR stimulation. To investigate which Gα isoforms or Gβγ subunits act downstream of GPR56, we examined the effect of the inhibitory molecules of these subunits (Fig. 1). The RGS domain of p115RhoGEF (p115-RGS) selectively binds to $G\alpha_{12/13}$ and prevents the interaction with their downstream effectors.[33] The C-terminal peptide of β–adrenergic receptor kinase (βARK-ct), containing the Gβγ binding domain, inhibits the cell signaling mediated by Gβγ.[34] Pertussis toxin (PTX) and YM-254890 are specific inhibitors of Gαi/o and Gαq/11, respectively.[35-37] p115-RGS completely suppressed the GPR56-induced SRE-luciferase activity, suggesting that GPR56 couples with Gα$_{12/13}$. However, PTX and YM-254890 failed to suppress it. It is known that Gα$_{12/13}$ activates Rho through p115-GEF, which is a Rho-specific guanine nucleotide exchange factor and that Rho induces SRE-mediated transcriptional activation. We examined the effect of the botulinum C3 exoenzyme and the dominant negative mutants of RhoA, Rac1 and Cdc42. The C3 exoenzyme and the RhoA dominant negative mutant inhibited SRE-mediated transcriptional activation by GPR56, whereas the dominant negative mutants of Rac1 and Cdc42 did not. The activation of NF-κB-responsive element-mediated luciferase activity was also inhibited by p115-RGS, the C3 exoenzyme and the RhoA dominant negative mutant. These results suggested that GPR56 couples with Gα$_{12/13}$ and activates SRE- and NF-κB-mediated transcription in a Rho-dependent manner.

## FUNCTIONAL ANTIBODIES AGAINST ADHESION-GPCRs

Secondly, we were able to prepare an agonistic antibody which functions as a ligand for GPR56. As surrogate ligands, some antibodies against the extracellular region of GPCR can stimulate receptors. Autoantibodies against GPCR are known in several types of diseases and the agonistic activity of these autoantibodies could play a pathogenic role in some of the symptoms of these diseases. For example, anti-β1-adrenoceptor autoantibodies, first described in sera of patients with Chagas disease, are now well-documented in patients with idiopathic dilated cardiomyopathy.[38] The antibodies directed to the second extracellular loop of the M2 muscarinic acetylcholine receptor are known to work as a partial agonist.[39,40] The monoclonal antibody, which recognized the second extracellular loop of the human β2AR, could stabilize the receptor in its active formation.[41] Constitutive activation of the thyroid-stimulating hormone receptor (TSHR) in Graves' disease is widely acknowledged to be caused by an antibody-mediated autoimmune reaction.[42] TSHR, also known as the thyrotropin receptor, is a class A rhodopsin family of GPCR. TSHR has a leucine repeat domain in ECD, which is an epitope for the autoantibody in the sera of patients with Graves' disease.[43] We showed that the anti-GPR56 antibody-induced SRE-mediated transcription with Rho activation and inhibited neuronal progenitor cell (NPC) migration. Neutralization with antigen GPR56ECD completely blocked anti-GPR56 antibody-induced SRE activation and inhibition of NPC migration. GPR56 knockdown also attenuated the inhibitory effect of the anti-GPR56 antibody on NPC migration.[23] These results supported the idea that the anti-GPR56 antibody specifically stimulated GPR56 activity as the agonist. An alternative activation mechanism of GPCR is demonstrated by the PARs (protease-activated receptors) and TSH receptors. PARs are activated by a unique proteolytic mechanism whereby thrombin, a serine protease, cleaves the N-terminal exodomain of PARs. The unmasked N-terminus of PAR1 acts as a tethered ligand binding to the receptor to trigger signaling.[44] Synthetic peptides that mimic the tethered ligand domain can activate PAR1 independently of proteolysis. TSHR does not have the GPS domain, but the proteolytic cleavage of ECD is important for ligand-dependent receptor activation. The full-length TSHR is posttranslationally cleaved into the extracellular α-subunit and the 7TM β-subunit. Both subunits are linked by disulfide bonds and, most likely, additional contact points. The α-subunit of the TSHR has a silencing effect on the β-subunit, keeping basal activity low. Without the α-subunit, the β-subunit has high basal activity. Upon binding of TSH, the ECD of TSHR switches from an inverse agonist to a full agonist. A similar mechanism might be triggered by a stimulatory TSHR autoantibody.[45] Inter-molecular interactions between TSHRs resulting in dimerization and multimerization have been proposed and recently confirmed. The leucine-rich repeat domain in the ECD of TSHR is important in the formation of an oligomer, TSH binding and signal transduction. Recently, the crystal structure of the α-subunit of TSHR in a complex with a thyroid-stimulating autoantibody was reported.[46] To investigate whether GPR56 without its ECD is active and whether the cleaved GPR56ECD acts as an agonist or an antagonist itself similar to PARs and TSHR, we constructed the truncated mutant lacking its ECD. This mutant also activated the SRE reporter gene; however, the GPR56ECD failed to affect this activity (unpublished data). A previous report indicated that some mutations of GPR56 in BFPP patients affected the glycosylation of the N-terminal extracellular region and that mutations in the GPS domain inhibited the cleavage at GPS.[25] Moreover, these mutations also caused the impairment of GPR56 trafficking and cell surface expression. Thus, the posttranslational processing of GPR56 in the N-terminal region is also important for cell surface expression and signal transduction. The details of the activation mechanism of GPR56 remain to be clarified. To further understand the regulatory mechanism of GPR56, it will be necessary to elucidate the epitopes of an agonistic antibody against GPR56 and find its blocking antibody.

There are few reports of functional antibodies against adhesion-GPCR. The blocking antibody against CD97 inhibits interleukin-8 (IL-8)-dependent mobilization of hematopoietic and progenitor cells, but it does not inhibit granulocyte-colony stimulating factor (GM-CSF)-dependent mobilization. The blocking antibody, which recognizes the EGF1,2 regions of CD97, affects the adhesion, although the antibody that recognizes the EGF3 region has no effect.[47] It was recently reported that the antibody against CD97 blocked granulocyte trafficking after thioglycollate-induced peritonitis in wild-type but not in CD97 knock-out mice.[48] These results indicate that the CD97 antibody actively induces an inhibitory effect that disturbs normal granulocyte trafficking. Another report showed that incubation of a specific anti-EMR2 antibody targeted towards the stalk region of the receptor enhanced the activation and migration of human neutrophils.[49] Taken together, these observations suggested that the functional antibodies certainly help to clarify the signal transduction pathway of adhesion-GPCRs.

## OTHER STUDIES THAT SUGGEST THE SIGNALING PATHWAY OF ADHESION-GPCRs

Signal transduction of EGF-TM7 receptors, including EMR2 and CD97, remains obscure. Several approaches to demonstrate the signaling of EGF-TM7 receptors by either overexpression of receptors in cell lines, stimulation with purified ligand molecules, such as CD55, or mutations that could give rise to constitutive activation in class B peptide hormone GPCRs have as yet failed.[46] One reason is that adhesion-GPCR might engage in the G protein-independent signaling pathway. It has been reported that the C-terminal region of 7TM was important in migration and invasion through EMR2 and CD97.[49,50] BAI1, GPR124 and GPR125 have an xTxV motif in the C-terminus of their 7TM, which interacts with PDZ.[51,52] However, other adhesion-GPCRs, such as latrophilin and GPR56, do not contain this motif. These observations suggest that adhesion-GPCRs transmit a

# SIGNAL TRANSDUCTION MEDIATED THROUGH ADHESION-GPCRs

signal through both G protein-dependent and G protein-independent mechanisms. BAI1 was isolated as a binding partner of the N-terminus of ELMO1 (engulfment and cell motility 1), the complex of which with Dock180 (dedicator of cytokinesis protein1) and small GTPase, Rac, is a conserved signaling module for promoting the internalization of apoptotic cell debris.[13] BAI1 binds phosphatidylserine exposed on the cell surface undergoing apoptosis. This binding with extracellular thrombospondin repeats mediates 7TM-dependent signaling through the interaction of the guanine nucleotide exchange factor ELMO-DOCK180 complex and Rac.

## COMPLEXITY OF SIGNAL TRANSDUCTION VIA ADHESION-GPCRs

The G protein-dependent signaling through adhesion-GPCR may be complicated. The activation mechanism of GPR56 postulated by previous reports is indicated in Figure 2. GPR56 is coupled with $G_{12/13}$, leading to the activation of Rho and the inhibition of neural progenitor cell migration (A). The agonistic antibody against GPR56ECD can stimulate $G_{12/13}$ signaling. This activation might be caused by the conformational change of 7TM induced by binding of the antibody and its ligand to the ECD (B) or by the dimerization/oligomerization of receptors (C). The ECD of one adhesion-GPCR can interact with the 7TM of other adhesion-GPCRs. GPR56ECD could interact with 7TM of latrophilin

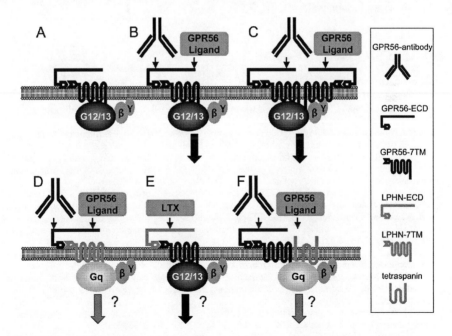

**Figure 2.** Possible G protein-coupled activation mechanisms of GPR56. GPR56 coupled with $G_{12/13}$ (A) is activated by unknown ligands or an agonistic antibody against GPR56ECD. The activation should be mediated through the conformational change of 7TM (B) and/or dimerization/oligomerization, which is also induced by overexpression (C). GPR56 and latrophilin (LPHN) may be cross-activated by each other with cross-assembly of the ECD and 7TM fragments (D,E). Furthermore, the coupling of GPR56 to $G_{12/13}$ might be changed to Gq by the association with tetraspanin.

coupled with Gq.[16] Therefore, the stimulation of the GPR56 antibody may transactivate Gq signaling through 7TM of latrophilin (D). Conversely, the ECD of latrophilin may be able to interact with the 7TM of GPR56, leading to $G_{12/13}$ signaling (E). Moreover, the heterodimer of GPR56 with tetraspanin, such as CD81 or CD9, could convert the interaction with $G_{12/13}$ to Gq (F). Despite the vast and long-standing efforts of research to pair GPCRs to potential ligands, more than 140 GPCRs, except the odorant GPCRs, remain orphan receptors. Standard deorphanization strategies seem to have reached their limit and new strategies are urgently required.[30] To further understand adhesion-GPCR signaling, it will be necessary to clarify the presence of functional agonists and to elucidate the structural and functional relationship between the ECD and 7TM.

## CONCLUSION

In the past decade, the number of adhesion-GPCRs increased and several members were relatively well-characterized. However, understanding of adhesion-GPCR function has been hampered by the lack of identified ligand(s) and by the unique structure of the receptors. Further investigation of adhesion-GPCR signaling will provide the basis for a full understanding of the physiological and pathophysiological role of adhesion-GPCRs. Functional and specific antibody against each adhesion-GPCR should serve as an invaluable probe for not only the ligand hunting, but also the developing therapy for tumors or genetic disorders in which adhesion-GPCRs are involved.

## REFERENCES

1. Yona S, Lin HH, Siu WO et al. Adhesion-GPCRs: emerging roles for novel receptors. Trends Biochem Sci 2008; 33:491-500.
2. Lagerström MC, Schiöth HB. Structural diversity of G protein-coupled receptors and significance for drug discovery. Nat Rev Drug Discov 2008; 7:339-357.
3. Stacey M, Chang GW, Davies JQ et al. The epidermal growth factor-like domains of the human EMR2 receptor mediate cell attachment through chondroitin sulfate glycosaminoglycans. Blood 2003; 102:2916-2924.
4. Stacey M, Lin HH, Hilyard KL et al. Human epidermal growth factor (EGF) module-containing mucin-like hormone receptor 3 is a new member of the EGF-TM7 family that recognizes a ligand on human macrophages and activated neutrophils. J Biol Chem 2001; 276:18863-18870.
5. Stacey M, Chang GW, Sanos SL et al. EMR4, a novel epidermal growth factor (EGF)-TM7 molecule up-regulated in activated mouse macrophages, binds to a putative cellular ligand on B lymphoma cell line A20. J Biol Chem 2002; 277:29283-29293.
6. Hamann J, Vogel B, van Schijndel GM et al. The seven-span transmembrane receptor CD97 has a cellular ligand (CD55, DAF). J Exp Med 1996; 184:1185-1189.
7. Hamann J, Stortelers C, Kiss-Toth E et al. Characterization of the CD55 (DAF)-binding site on the seven-span transmembrane receptor CD97. Eur J Immunol 1998; 28:1701-1707.
8. Lin HH, Stacey M, Saxby C et al. Molecular analysis of the epidermal growth factor-like short consensus repeat domain-mediated protein-protein interactions: dissection of the CD97-CD55 complex. J Biol Chem 2001; 276:24160-24169.
9. Abbott RJ, Spendlove I, Roversi P et al. Structural and functional characterization of a novel T-cell receptor coregulatory protein complex, CD97-CD55. J Biol Chem 2007; 282:22023-22032.
10. Xu L, Begum S, Hearn JD et al. GPR56, an atypical G protein-coupled receptor, binds tissue transglutaminase, TG2 and inhibits melanoma tumor growth and metastasis. Proc Natl Acad Sci USA 2006; 103:9023-9028.
11. Davletov BA, Shamotienko OG, Lelianova VG et al. Isolation and biochemical characterization of a $Ca^{2+}$-independent α-latrotoxin-binding protein. J Biol Chem 1996; 271:23239-23245.
12. Willson J, Amliwala K, Davis A et al. Latrotoxin receptor signaling engages the UNC-13-dependent vesicle-priming pathway in C. elegans. Curr Biol 2004; 14:1374-1379.

13. Park D, Tosello-Trampont AC, Elliott MR et al. BAI1 is an engulfment receptor for apoptotic cells upstream of the ELMO/Dock180/Rac module. Nature 2007; 450:430-434.
14. Nishimori H, Shiratsuchi T, Urano T et al. A novel brain-specific p53-target gene, BAI1, containing thrombospondin type 1 repeats inhibits experimental angiogenesis. Oncogene 1997; 15:2145-2150.
15. Koh JT, Kook H, Kee HJ et al. Extracellular fragment of brain-specific angiogenesis inhibitor 1 suppresses endothelial cell proliferation by blocking $\alpha v\beta 5$ integrin. Exp Cell Res 2004; 294:172-184.
16. Silva JP, Lelianova V, Hopkins C et al. Functional cross-interaction of the fragments produced by the cleavage of distinct adhesion G-protein-coupled receptors. J Biol Chem 2009; 284:6495-506.
17. Kwakkenbos MJ, Kop EN, Stacey M et al. The EGF-TM7 family: a postgenomic view. Immunogenetics 2004; 55:655-666.
18. Davletov BA, Meunier FA, Ashton AC et al. Vesicle exocytosis stimulated by alpha-latrotoxin is mediated by latrophilin and requires both external and stored $Ca^{2+}$. EMBO J 1998; 17:3909-3920.
19. Ashton AC, Volynski KE, Lelianova VG et al. $\alpha$-Latrotoxin, acting via two $Ca^{2+}$-dependent pathways, triggers exocytosis of two pools of synaptic vesicles. J Biol Chem 2001; 276:44695-44703.
20. Capogna M, Volynski KE, Emptage NJ et al. The $\alpha$-latrotoxin mutant $LTX^{N4C}$ enhances spontaneous and evoked transmitter release in CA3 pyramidal neurons. J Neurosci 2003; 23:4044-4053.
21. Lelianova VG, Davletov BA, Sterling A et al. $\alpha$-latrotoxin receptor, latrophilin, is a novel member of the secretin family of G protein-coupled receptors. J Biol Chem 1997; 272:21504-21508.
22. Rahman MA, Ashton AC, Meunier FA et al. Norepinephrine exocytosis stimulated by $\alpha$-latrotoxin requires both external and stored $Ca^{2+}$ and is mediated by latrophilin, G proteins and phospholipase C. Philos Trans R Soc Lond B Biol Sci 1999; 354:379-386.
23. Iguchi T, Sakata K, Yoshizaki K et al. Orphan G protein-coupled receptor GPR56 regulates neural progenitor cell migration via a $G\alpha_{12/13}$ and Rho pathway. J Biol Chem 2008; 283:14469-14478.
24. Piao X, Hill RS, Bodell A et al. G protein-coupled receptor-dependent development of human frontal cortex. Science 2004; 303:2033-2036.
25. Jin Z, Tietjen I, Bu L et al. Disease-associated mutations affect GPR56 protein trafficking and cell surface expression. Hum Mol Genet 2007; 16:1972-1985.
26. Shashidhar S, Lorente G, Nagavarapu U et al. GPR56 is a GPCR that is overexpressed in gliomas and functions in tumor cell adhesion. Oncogene 2005; 24:1673-1682.
27. Ke N, Sundaram R, Liu G et al. Orphan G protein-coupled receptor GPR56 plays a role in cell transformation and tumorigenesis involving the cell adhesion pathway. Mol Cancer Ther 2007; 6:1840-1850.
28. Seifert, R, Wenzel-Seifert K. Constitutive activity of G-protein-coupled receptors: cause of disease and common property of wild-type receptors. Naunyn Schmiedebergs Arch Pharmacol 2002; 366:381-416.
29. Rosenbaum DM, Rasmussen SG, Kobilka BK. The structure and function of G-protein-coupled receptors. Nature 2009; 459:356-363.
30. Levoye A, Dam J, Ayoub MA et al. Do orphan G-protein-coupled receptors have ligand-independent functions? New insights from receptor heterodimers. EMBO Rep 2006; 7:1094-1098.
31. Davies JQ, Chang GW, Yona S et al. The role of receptor oligomerization in modulating the expression and function of leukocyte adhesion-G protein-coupled receptors. J Biol Chem 2007; 282:27343-27353.
32. Kim JE, Han JM, Park CR et al. Splicing variants of the orphan G-protein-coupled receptor GPR56 regulate the activity of transcription factors associated with tumorigenesis. J Cancer Res Clin Oncol 2010; 136:47-53.
33. Shi CS, Sinnarajah S, Cho H et al. G13$\alpha$-mediated PYK2 activation. PYK2 is a mediator of $G_{13\alpha}$-induced serum response element-dependent transcription. J Biol Chem 2000; 275:24470-24476.
34. Koch WJ, Hawes BE, Inglese J et al. Cellular expression of the carboxyl terminus of a G protein-coupled receptor kinase attenuates G$\beta\gamma$-mediated signaling. J Biol Chem 1994; 269:6193–6197.
35. Katada T, Ui M. Direct modification of the membrane adenylate cyclase system by islet-activating protein due to ADP-ribosylation of a membrane protein. Proc Natl Acad Sci USA 1982; 79:3129-3133.
36. Kurose H, Katada T, Amano T et al. Specific uncoupling by islet-activating protein, pertussis toxin, of negative signal transduction via $\alpha$-adrenergic, cholinergic and opiate receptors in neuroblastoma x glioma hybrid cells. J Biol Chem 1983; 258:4870-4875.
37. Takasaki J, Saito T, Taniguchi M et al. A novel $G\alpha q/11$-selective inhibitor. J Biol Chem 2004; 279:47438-47445.
38. Magnusson Y, Wallukat G, Waagstein F et al. Autoimmunity in idiopathic dilated cardiomyopathy. Characterization of antibodies against the $\beta$ 1-adrenoceptor with positive chronotropic effect. Circulation 1994; 89:2760-2767.
39. Elies R, Fu LX, Eftekhari P et al. Immunochemical and functional characterization of an agonist-like monoclonal antibody against the M2 acetylcholine receptor. Eur J Biochem 1998; 251:659-666.
40. Peter JC, Wallukat G, Tugler J et al. Modulation of the M2 muscarinic acetylcholine receptor activity with monoclonal anti-M2 receptor antibody fragments. J Biol Chem 2004; 279:55697-55706.
41. Lebesgue D, Wallukat G, Mijares A et al. An agonist-like monoclonal antibody against the human $\beta$2-adrenoceptor. Eur J Pharmacol 1998; 348:123-133.

42. Rees Smith B, McLachlan SM, Furmaniak J. Autoantibodies to the thyrotropin receptor. Endocr Rev 1988; 9:106-121.
43. Smith BR, Sanders J, Furmaniak J. TSH receptor antibodies. Thyroid 2007; 17:923-938.
44. Arora P, Ricks TK, Trejo J. Protease-activated receptor signalling, endocytic sorting and dysregulation in cancer. J Cell Sci 2007; 120:921-928.
45. Schott M, Scherbaum WA, Morgenthaler NG. Thyrotropin receptor autoantibodies in Graves' disease. Trends Endocrinol Metab 2005; 16:243-248.
46. Sanders J, Chirgadze DY, Sanders P et al. Crystal structure of the TSH receptor in complex with a thyroid-stimulating autoantibody. Thyroid 2007; 17:395-410.
47. van Pel M, Hagoort H, Kwakkenbos MJ et al. Differential role of CD97 in interleukin-8-induced and granulocyte-colony stimulating factor-induced hematopoietic stem and progenitor cell mobilization. Haematologica 2008; 93:601-604.
48. Veninga H, Becker S, Hoek RM et al. Analysis of CD97 expression and manipulation: antibody treatment but not gene targeting curtails granulocyte migration. J Immunol 2008; 181:6574-6583.
49. Yona S, Lin HH, Dri P et al. Ligation of the adhesion-GPCR EMR2 regulates human neutrophil function. FASEB J 2008; 22:741-751.
50. Galle J, Sittig D, Hanisch I et al. Individual cell-based models of tumor-environment interactions: Multiple effects of CD97 on tumor invasion. Am J Pathol 2006; 169:1802-1811.
51. Shiratsuchi T, Futamura M, Oda K et al. Cloning and characterization of BAI-associated protein 1: a PDZ domain-containing protein that interacts with BAI1. Biochem Biophys Res Commun 1998; 247:597-604.
52. Yamamoto Y, Irie K, Asada M et al. Direct binding of the human homologue of the Drosophila disc large tumor suppressor gene to seven-pass transmembrane proteins, tumor endothelial marker 5 (TEM5) and a novel TEM5-like protein. Oncogene 2004; 23:3889-3897.

# CHAPTER 15

# EMERGING ROLES OF BRAIN-SPECIFIC ANGIOGENESIS INHIBITOR 1

Daeho Park and Kodi S. Ravichandran*

**Abstract:** Brain-specific angiogenesis inhibitor 1 (BAI1) encodes a seven-transmembrane protein that belongs to the adhesion-GPCR family.[1-7] Although BAI1 was named for the ability of its extracellular region to inhibit angiogenesis in tumor models, its function in physiological contexts was elusive and remained an orphan receptor until recently.[5,6,8-14] BAI1 is now considered a phagocytic receptor that can recognize phosphatidylserine exposed on apoptotic cells. Moreover, BAI1 has been shown to function upstream of the signaling module comprised of ELMO/Dock180/Rac proteins, thereby facilitating the cytoskeletal reorganization necessary to mediate the phagocytic clearance of apoptotic cells.[15,16] Here, we review the phylogeny, structure, associating proteins, as well as the known and proposed functions of BAI1.

## INTRODUCTION

The seven-transmembrane G protein-coupled receptors (GPCRs) are the most extensively studied family of protein.[17,18] In fact, the genes encoding GPCRs comprise one of the largest families in the human genome.[18,19] GPCR family is classified into five major sub-families. Among them, the adhesion-GPCR subfamily is unique in that the family members possess a long extracellular region, followed by a seven-transmembrane (7TM) domain and a long cytoplasmic tail.[20-23] The adhesion-GPCRs have recently attracted significant attention due to the many unique domains or motifs that have been recognized within their N-termini, linked to protein-protein, cell-cell and cell-matrix interaction and possible biological functions. This chapter specifically focuses on brain-specific angiogenesis inhibitor 1 (BAI1), one of the adhesion-family GPCR proteins that has recently garnered

*Corresponding Author: Kodi S. Ravichandran—Beirne Carter Center for Immunology Research and Department of Microbiology, University of Virginia, Charlottesville, Virginia, USA Email: ravi@virginia.edu

*Adhesion-GPCRs: Structure to Function*, edited by Simon Yona and Martin Stacey.
©2010 Landes Bioscience and Springer Science+Business Media.

significant attention due to its role in clearance of apoptotic cells in the body and its possible role in regulating angiogenesis in the context of glioblastomas. Here, we review the initial identification of BAI1, the evidences linking BAI1 to specific functions, the intracellular binding partners, as well as the potential of BAI1 as a target in glioblastomas.

## INITIAL IDENTIFICATION OF BAI1

The gene encoding brain-specific angiogenesis inhibitor 1 (BAI1) was initially identified as a target whose expression was regulated by the tumor suppressor gene p53. In the original studies, the expression of BAI1 was found to be downmodulated in glioblastomas, compared to high levels of BAI1 expression in normal brain. The authors noted that besides the 7TM region, BAI1 contains several thrombospondin type 1 repeats (TSRs) in its extracellular region; since TSRs have previously been shown to be capable of inhibiting angiogenesis, the role of BAI1 TSR motifs in a rat model of ocular angiogenesis was tested and found to inhibit angiogenesis. This property, along with the predominant expression of BAI1 mRNA in the brain led to its naming as 'brain-specific angiogenesis inhibitor 1'. However, more recent analyses of BAI1 expression in different cell types/tissues or microarray studies suggest that BAI1 is expressed at some level in all tissues, with *Bai1* mRNA detectable in bone marrow, spleen, peritoneal exudate cells and testis. Moreover, the regulation of BAI1 downstream of p53 has also been debated, but the loss of BAI1 expression in multiple glioblastoma lines suggest that BAI1 may play a critical role in normal brain (see below). Two other homologues of BAI1, named BAI2 and BAI3, with similar overall architecture and belonging to the same subfamily of adhesion-GPCRs have also been identified (Fig. 1).[6,7,21] The structural features of BAI1 in comparison to BAI2 and BAI3 are detailed below.

## STRUCTURAL AND FUNCTIONAL DOMAINS OF BAI1

The adhesion-GPCRs can be further classified based on the unique domains or motifs within the extracellular region. BAI1, along with BAI2 and BAI3 constitute a separate subgroup VII among the adhesion-GPCR proteins.[20] One distinction between the subgroup VII and other subgroups of adhesion family GPCRs is presence of thrombospondin type 1 repeats (TSR) in BAI1, BAI2 and BAI3, which are not found in other subgroup adhesion-GPCRs.[20] BAI1 has five TSRs whereas BAI2 and BAI3 have four TSRs. A hormone binding domain (HBD) is also found all three members of subgroup VII adhesion-GPCRs, although HBD is also present in subgroup I, IV and VI. In terms of domains/motifs present, BAI1 is very similar to the other two homologues BAI2 and BAI3. All three proteins contain TSR, HBD and 7TM but only BAI1 has the RGD (arginine-glycine-aspartate) motif upstream of the TSRs (Fig. 1).

BAI1 (1584 residues) contains a long extracellular region (943 residues) followed by a seven-transmembrane heptahelical body and a relatively long 392 residue cytoplasmic tail. The extracellular region of BAI1 contains four recognizable motifs/domains that promote cell-cell and cell-matrix interaction (Fig. 1). The first motif from the N-terminus is the RGD (Arginine-Glycine-Aspartic acid) integrin binding motif. The experiments to date have not ascribed a role for the RGD motif although the possibility that this motif may allow BAI1

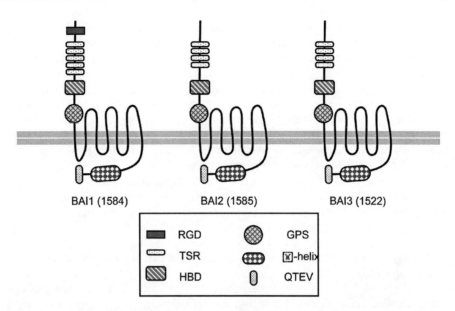

**Figure 1.** Schematic diagram of the subgroup VII of the adhesion-GPCR family. RGD, integrin binding motif; TSR, thrombospondin type 1 repeats; HBD, hormone binding domain; GPS, G protein-coupled receptor proteolytic site; QTEV, PDZ domain binding motif.

expressing cell to 'communicate' with integrins on other cell types is an exciting possibility. In the context of apoptotic cell clearance, the RGD motif appears dispensable.[16]

The thrombospondin type 1 repeats (TSRs) represent the second distinguishable domain/motif within the BAI1. These repeats were originally described by Lawler and Hynes.[24] Thrombospondin-1, a matrix protein, regulates cell proliferation, migration and apoptosis in a variety of physiological conditions.[25] Among the three type of repeats found in thrombospondin (Type 1, 2 and 3), BAI1 contains only repeats that are homologous to Type 1. The TSRs in thrombospondin-1 have been functionally linked to cell attachment, TGF-β activation, inhibition of angiogenesis and cell migration. Many extracellular matrix proteins such as mindin, F-spondin and SCO-spondin contain one or more of the TSR repeats.[24-28] Crystallographic studies of the repeats suggest that TSRs have an elongated structure with a large exposed surface area rather than a spherical structure. In fact, this supports the notion that TSRs are involved in several interactions such as protein-protein, cell-cell and cell-matrix interaction as the protruded structure of TSR could possibly function as a docking site.[29,30] BAI1 has five Type 1 TSRs and initial studies suggest that the TSRs have a role in binding to phosphatidylserine exposed on apoptotic cells (see below).

A hormone binding domain (HBD) follows the TSRs in BAI1. This HBD domain is found in some secretin-like GPCRs, but the specific role of this domain has remained unclear.[20] BAI1 also has a G protein-coupled receptor proteolytic site (GPS) right before the seven-transmembrane domain like other adhesion-GPCRs. The conserved region of GPS domains is about 50 residues long and contain either 2 or 4 cysteine residues that likely form disulphide bridges.[20,22] Previous studies have shown that the extracellular region of BAI1 can be cleaved at the GPS and the cleaved extracellular region of BAI1 was denoted as vasculostatin. The cleaved vasculostatin product could function as an anti-angiogenic and anti-tumorigenic factor.[8,11] However, this form of truncated product may be tissue or cell type specific, as this is not detectable in other conditions (our unpublished observations).

The long extracellular region of BAI1 is followed by the heptahelical seven-transmembrane region of BAI1. The amino acid sequence with the TM of BAI1 contain similar hydrophobic residues in positions analogous to the secretin-family GPCRs.[31] Thus, it was initially classified as a secretin-family GPCR,[31] but more detailed phylogenetic studies have placed BAI1 within the adhesion-GPCR family.[20] To date, there is no direct evidence that BAI1 is dependent on G protein signaling even though GPR56, a member of the subgroup VIII, signals through G proteins.[22] BAI1 might associate with G proteins either through 7-TM or the cytoplasmic tail of BAI1. The canonical DRY motif is not apparent in BAI1 and mutation of one DRY motif does not appear to confer a function in the context of apoptotic cell clearance (unpublished observations). However, the linkage of BAI1 to G proteins needs to be carefully examined. It is also possible that BAI1 uses association with G proteins as well as with other intracellular signaling intermediates to transduce signals into cells.

Although BAI1 has a long cytoplasmic tail, it does not have distinctive domains or motifs except for a proline rich region and QTEV motif at the extreme carboxyl terminus. The QTEV motifs have been shown to interact with proteins containing PDZ domains. Consistent with this hypothesis, a PDZ domain containing protein (BAP1) was identified in a yeast two-hybrid screen with BAI1 as bait and was shown to bind to the QTEV motif of BAI1.[32] However, the function of this interaction and the role of the QTEV motif remain to be determined. Our analysis of the cytoplasmic tail of BAI1 identified α-helix region immediately after the proline rich region, which was found to be important for binding to another protein ELMO1 (see below). The long extracellular region of BAI1 with multiple domains/motifs and its long cytoplasmic tail suggested that BAI1 might play a role in direct signaling from outside of the cell to the inside.

## BAI1 AS AN ENGULFMENT RECEPTOR FOR APOPTOTIC CELLS

Although initial studies suggested BAI1 as a possible regulator of angiogenesis, its physiological role remained elusive for nearly 10 years. Recently, work from our group has identified an important role of BAI1 in phagocytes that recognize and clear apoptotic cells.[16] As a way of background on apoptotic cell clearance, our bodies turnover roughly 1 million cells every second. These cells that are turned over include excess cells generated as part of normal development, aged or nonfunctional cells, or cells that die due to other causes. These dying cells expose 'eat-me' signals on their surface that are in turn recognized by receptors on the phagocytes.[33] The specific recognition of eat-me signals by phagocytic receptors help recognize the dying cells among healthy cells in a tissue and to specifically remove the dying cells. Failure to promptly clear apoptotic cells has been linked to autoimmunity and other disease states in humans and in mouse models.[34,35] BAI1 has recently been identified as an engulfment receptor that recognizes a ligand on apoptotic cells and thereby facilitates apoptotic cell clearance.[15,36] The section below details the key experimental evidence supporting the concept that BAI1 is an engulfment receptor.

Table 1. Summary of BAI1 interacting proteins

| Interacting Protein | Method | Interacting Region within BAI1 | Previously Known Function of the BAI1 Binding Partner | Reference |
|---|---|---|---|---|
| BAP1 | Y2H, GST-pulldown, Colocalization | QTEV motif in the cytoplasmic tail | Unknown | 32 |
| BAP2 (IRSp53) | Y2H, GST-pulldown, Colocalization | Proline rich region in the cytoplasmic tail | Activation of Rac/Cdc42 | 52 |
| BAP3 | Y2H, Immunoprecipitation | Cytoplasmic tail | Unknown | 32 |
| BAP4 (PAHX-AP1) | Y2H, GST-pulldown | Cytoplasmic tail | Unknown | 54 |
| ELMO1 | Y2H, Immunoprecipitation, GST-pulldown | α-helix region in the cytoplasmic tail | Engulfment of apoptotic cells, cell migration | 16 |

One of the highly evolutionarily conserved signaling pathways in clearance of apoptotic corpses (from worm to man) involves the proteins ELMO1/Dock180 and Rac1.[37-42] The two proteins, ELMO1 and Dock180, associate with each other and together activate the small GTPase Rac1. Activated Rac1 then promotes actin polymerization and cytoskeletal rearrangements, which in turn facilitate the phagocyte to wrap the phagocyte membranes around an apoptotic cell, leading to internalization. However, the receptor/membrane protein that function upstream of ELMO/Dock180/Rac module had remained elusive. We performed a yeast two-hybrid screen with ELMO1 as bait and identified the cytoplasmic tail of BAI1 as an interacting partner for the N-terminal region of ELMO1. It is notable that previous studies have showed that the N-terminal region of ELMO1 is important for the localization of the ELMO1/Dock180 complex to the membrane (note: ELMO1 binds via its C-terminal region to Dock180).[39,40,42] Further studies revealed that ELMO1 specifically binds to small α-helix region within the cytoplasmic tail of BAI1.[16] Mutation of this region within BAI1 or mutations within the N-terminal region of ELMO1 abrogated the interaction between BAI1 and ELMO1. It is interesting to note that this α-helix region bound by ELMO1 is not overlapping with regions where other BAI1-associate proteins bind within BAI1 (Table 1).

Several lines of evidence suggested that BAI1 could function as an engulfment receptor for apoptotic cells. First, overexpression of BAI1 in different cell types (macrophages or fibroblasts) enhanced the uptake of apoptotic cells compared to untransfected controls. Second, BAI1 appeared to specifically promote uptake of apoptotic cells compared to uptake of live cells or necrotic cells, suggesting that BAI1 might recognize a specific eat-me signal exposed on apoptotic cells. Third, knockdown of BAI1 expression in macrophages and primary astrocytes inhibited the engulfment of apoptotic targets, commensurate with the extent of siRNA-mediated knockdown that was achievable in these cell types. Fourth, BAI1 localized to the phagocytic cup that forms around the apoptotic target being engulfed and with polymerized actin. Lastly, high expression of BAI1 on the surface of a phagocyte increased both the binding of apoptotic targets as well as the number of targets internalized per phagocytes. Collectively, these data suggested that BAI1 can function as a receptor that can promote uptake of apoptotic cells.

When we addressed what ligands on apoptotic cells might be recognized by BAI1, several observations pointed to the specific recognition by BAI1 of phosphatidylserine, a key eat-me signal that is exposed universally on apoptotic cells. These observations included the following: first, when the thrombospondin repeats on the extracellular domain of BAI1 were expressed as a soluble protein, this acted as a competitive inhibitor and potently inhibited the BAI1-dependent uptake of apoptotic cells via BAI1. Second, the TSR repeats of BAI1 competed with annexin V, a protein that is known to bind phosphatidylserine on apoptotic cells. Third, the mixing of the soluble TSRs derived from BAI1 to apoptotic cells resulted in the decoration of the cell surface of apoptotic cells, but not live cells. Fourth and perhaps more conclusively, the soluble TSRs of BAI1 directly bound to PtdSer on the lipid membrane strips and the binding to phosphatidylserine was stereo-specific. Lastly, BAI1 overexpressing cells specifically promoted the increased the uptake of phosphatidylserine lipid vesicles compared to phosphatidylcholine lipid vesicles. These data suggested that BAI1 can bind to phosphatidylserine on apoptotic cells and that the binding is mediated via the thrombospondin repeats on the extracellular surface of TSRs of BAI1 is the region, which binds phosphatidylserine. Lastly, the TSR repeats of BAI1 also showed a functional role in engulfment of apoptotic cells in vivo in a mouse model of cell clearance.

With respect to intracellular signaling via BAI1 during recognition of apoptotic cells, the interaction of ELMO1 with BAI1 was essential for BAI1-mediated uptake. Knockdown of ELMO1 or mutation of the ELMO1 binding site within the cytoplasmic tail of BAI1 abolished ability of BAI1 to promote engulfment of apoptotic cells. We also observed that BAI1 formed a trimeric complex with ELMO1/Dock180. Furthermore, coexpression of BAI1, ELMO1 and Dock180 showed the maximal uptake of apoptotic targets; conversely, dominant negative forms of either ELMO1 or Dock180 inhibited BAI1-dependent engulfment.

Collectively, these data identified a new physiological role for BAI1 as an engulfment receptor for recognition and uptake of apoptotic cells by phagocytes. Although several other engulfment receptors have been previously known, BAI1 represents the first adhesion-GPCR involved apoptotic cell clearance. The current model suggests that the recognition of phosphatidylserine exposed on apoptotic cells via the TSRs of BAI1 leads to intracellular signaling via the ELMO/Dock180/Rac module leading to cytoskeletal reorganization and the internalization of apoptotic corpses (Fig. 2).

Our most recent studies suggest that BAI1 is the most abundantly expressed phosphatidylserine recognition receptor in Sertoli cells of the testes.[35] These Sertoli cells play a crucial role in clearance of apoptotic germ cells during development and as part of normal testicular homeostasis (see Davies and Kirchhoff in this volume for adhesion-GPCRs in the male reproductive tract). Studies where BAI1 function was disrupted or its signaling through ELMO1 was affected, these mice show a severe defect in clearance of apoptotic germ cells and sperm output. Thus, BAI1 may play a crucial role in regulating spermatogenesis.

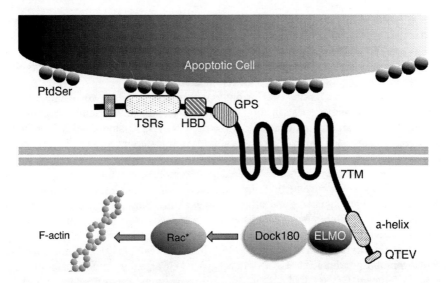

**Figure 2.** BAI1-mediated phosphatidylserine recognition and signaling. Phosphatidylserine on apoptotic cells is recognized by the TSRs of BAI1. The apoptotic recognition signal is then transduced into cells, which results in association of the cytoplasmic tail of BAI1 with the ELMO/Dock180/Rac signal module and causes Rac activation at the site of apoptotic cell recognition. Cytoskeleton rearrangement results from Rac activation initiates engulfment of apoptotic cells. PtdSer, phosphatidylserine; RGD, integrin binding motif; TSR, thrombospondin type 1 repeats; HBD, hormone binding domain; GPS, G protein-coupled receptor proteolytic site; 7TM, 7 transmembrane domain; QTEV, PDZ domain binding motif.

## ROLE OF BAI1 IN GLIOBLASTOMAS

When tumors reach a certain size (1-2 mm), they have to overcome limitations in oxygen and nutrient supply for further growth and/or metastasis.[43] This problem is overcome by the tumor through generation of new vasculature, termed 'angiogenesis'. Glioblastoma multiforme (GBMs), (or malignant diffuse gliomas, WHO Grade IV) are one of the most highly vascularized tumors.[44] Approaches to block angiogenesis is a major focus of recent therapies toward this tumor.[43] Remarkably, BAI1 is expressed at high levels in normal astrocytes, but BAI1expression is reduced or lost in many glioblastomas.[6,10,11] Because the TSRs of BAI1 can potently inhibit angiogenesis (see below), it is thought that loss of BAI1 expression may remove a natural 'block' in angiogenesis and thereby promote vascularization of the tumor.[10,11,16] Based on this premise, re-expression of BAI1 or the extracellular fragment of BAI1 is being considered an attractive therapeutic target.

In the initial cloning of BAI1, it was recognized that the TSRs of BAI1 may play a role in angiogenesis; when recombinant GST-BAI1 fusion proteins were introduced into rat cornea, the recombinant protein containing the TSRs of BAI1 inhibited neovascularization induced by bFGF.[6] In subsequent studies, overexpression of BAI1 in Panc-1, a human pancreatic adenocarcinoma cell line, was used to test the effect of BAI1 on angiogenesis. While there was no obvious difference in growth in vitro between BAI1 transfected and control LacZ transfected Panc-1 cells, the BAI1 overexpressing cells showed retarded tumor growth and suppression of angiogenesis in immunodeficient mice manifested by diminished vascularity of the tumor.[45] Furthermore, impaired angiogenesis was observed when U373MG cells transduced with adenovirus expressing BAI1 were transplanted into transparent skin-fold chambers of SCID mice, compared to control U373MG cells not expressing BAI1. Moreover, In vivo inoculation of U373MG cells in to the brain of mice BAI1 expressing U373MG cells showed reduced intratumoral vascular density and more necrosis compared to control U373MG injected cells.[10]

Despite these studies linking BAI1 to angiogenesis and tumor growth, it unclear how BAI1 might regulate angiogenesis in normal brain. Early studies used soluble recombinant fragments of BAI1 and the latter studies have used overexpression studies. Van Meir and colleagues suggested the existence of a soluble extracellular region of BAI1, called vasculostatin.[11] They found a 120 kDa fragment in the conditioned medium of BAI1 transfected 293T cells and detected an analogous extracellular fragment of BAI1 in brain lysates. Moreover, introduction of vasculostatin inhibited migration of endothelial cells and reduced angiogenesis and tumor growth. Thus, they proposed a mechanism that generation of a soluble factor by cleavage of a pre-existing transmembrane protein might inhibits angiogenesis and tumorigenesis. Interestingly, the anti-angiogenic activity of vasculostatin was reported to require CD36, which has been previously known to bind TSR of thromobospondin-1. The CLESH domain of CD36 appeared important for binding to vasculostatin and CD36 knockout mice did not show inhibition of neovascularization when micropellets containing vasculostatin and bFGF were implanted into the mice cornea.[8] Overall, these studies suggest that the extracellular region of BAI1 might be proteolytically cleaved at the GPCR proteolytic site. However, others have not reported on a similar soluble fragment of BAI1 and we have not observed a similar fragment under the conditions of our expression (unpublished observations). This raises an interesting possibility that the cleavage and processing BAI1 may differ between cell types and could be highly relevant for deciphering its normal function in angiogenesis.

Besides the possible role of BAI1 in angiogenesis, another possibility needs to be entertained in the context of its role in development of glioblastomas. Our recent studies have identified a fundamentally new role for BAI1, as a membrane receptor involved in recognition and clearance of dying cells.[16] Remarkably, Glioblastomas often have a necrotic core within the tumors and the presence of necrotic cells is often linked to poor prognosis for GBMs.[43,46-49] Intriguingly, the extracellular region of BAI1 capable of inhibiting angiogenesis in model systems is also the same region that mediates recognition and clearance of dying cells.[16] Thus, it becomes critical to fully understand BAI1 function in the context of GBMs, whether it solely relates to the function of BAI1 in inhibiting angiogenesis, its role as an engulfment receptor or both, prior to its potential use as a therapeutic target.

## OTHER KNOWN INTERACTING PARTNERS OF BAI1

Until two years ago, the role of BAI1 in a physiological context or its ligands remained unclear. Therefore, to deduce the function of BAI1 through its interacting proteins, search for BAI1 associated proteins has been extensively performed. This approach primarily involved yeast two-hybrid screens using the cytoplasmic tail of BAI1. So far, four BAI1-associated proteins (BAP1, 2, 3, 4) have been identified (Table 1). These screens did not pick up ELMO1 which was identified independently.[16] BAP1, 2 and 3 were identified with yeast two-hybrid screens using the cytoplasmic tail of BAI1 as bait,[32,50-53] whereas the cytoplasmic tail of BAI1 was fished out by BAP4 (PAHX-AP1).[54]

BAP1 is a novel member of the MAGUK (membrane-associated guanylate kinase homologue) family, comprising a guanylate kinase domain, WW domain and multiple PDZ domains. BAI1 has the QTEV motif at the extreme end of C-termini and this motif has previously shown to bind proteins containing PDZ domains. BAP1 possesses PDZ domain and associates with the cytoplasmic tail of BAI1 via interaction between the PDZ domain of BAP1 and the QTEV motif of BAI1.[32] The transcript of BAP1 was detected in several tissues such as heart, lung, kidney and pancreas as well as brain by northernblotting.[32] The role of BAP1, also called membrane-associated guanylate kinase 1 (MAGI-1) is unclear, but MAGUK family proteins are involved in organization of receptors, ion-channels and signaling molecules. Members of MAGUK family usually localize at tight junctions, septate junctions and synaptic junctions.[55,56] Thus, the interaction between BAI1 and BAP1 might play a potential role of BAI1 in cell adhesion and signal transduction, although no studies to date have addressed this possibility. However, the BAI1:BAP1 interaction appears dispensable for engulfment of apoptotic cells, since BAI1 can be tagged at the C-terminus with FLAG or GFP tag (thereby destroying the C-terminal QTEV motif) and these tagged proteins behave normally in promoting clearance of apoptotic cells.

The second BAI1 interacting protein, BAP2, is known previously as IRSp53.[52] The SH3 domain of BAP2 bound the proline rich region of the cytoplasmic tail of BAI1, which stretches from 1389 to 1437. BAP2 consists of an I-BAR domain, a partial-CRIB motif interrupted by an SH3-binding site, an SH3 domain and a potential WW domain binding site. It might be involved in linking Rac1/Cdc42 to WAVE to form lamellipodia at the site of Rac activation.[57-59] The expression profile of BAP2 by northern blotting is similar to that of BAI1 in the brain and they are also colocalized in COS-7 cells at overexpression conditions. Enrichment of BAI1 in growth cone and colocalization of BAI1 with BAP1 suggest that BAI1 might be involved in growth cone guidance. However, systematic functional studies have not been undertaken to address the role of the BAI1:IRSp53 interaction. In our studies to date, we are unable to detect a function for BAP2/IRSp53 in clearance of apoptotic cells (unpublished observations).

BAP3 is a novel C2 domain-containing protein with homology to Munc13 and synaptotagmin. The mRNA expression pattern of BAP3 is very similar to BAI1. The homology of BAP3 with Munc13 and synaptotagmin suggests that BAI1 may have neuronal functions.[53] However, so far the role of BAP3 has not been studied.

BAP4 interaction with BAI1 was identified differently from the other BAP proteins.[54] The cDNA fragment encoding the cytoplasmic tail of BAI1 was identified in a two-hybrid screen performed using PAHX-AP1 (which stands for PHAX-associated protein 1) and has now been renamed as BAP4. PAHX is a protein related to an autosomal recessive disorder of the lipid metabolism.[54] So far, the role of BAP4 is unclear but it is possible that BAP4 functions as an adaptor protein to link BAI1 to PAHX for regulation of the lipid metabolism—perhaps in digesting the contents of apoptotic cells.

To summarize, four BAP proteins have been identified, besides ELMO1 and the interaction between BAI1 and BAI1-associated proteins have been tested using yeast two-hybrid assays and immunoprecipitation/colocalization studies after overexpression of these proteins (Table 1). However, physiological role of BAI1 association with these BAPs remains unknown. Clearly, more investigations are necessary to better define a physiological role for BAPs as well as BAI1.

## CONCLUSION

Although BAI1 remained an orphan receptor for nearly 10 years, recent studies have begun to shed new light on BAI1 function under physiological conditions. However, a number of unanswered questions remain. First, the specific roles of BAI1 in physiological settings need to be better defined. This would hopefully be achieved through mouse knockout studies targeting *Bai1*. This would help define the role of BAI1 in engulfment of apoptotic cells in the different tissue contexts as well as its role in normal brain functions. With the availability of testable glioblastoma mouse models, the BAI1 knockout mice may also prove useful for these studies. Second, at the molecular level, the two known functions of BAI1 are anti-angiogenesis and recognition of phosphatidylserine, both being performed by the same domain containing TSRs. It would be interesting to define the mechanism by which the unrelated functions can be executed by the same domain. It is also unclear whether particular TSRs may be involved in phosphatidylserine recognition versus anti-angiogenesis. Along the same lines, the cleaved versus noncleaved versions of BAI1 as well as the role of the other domains/motifs in BAI1 need to be better defined. Third, although several interaction partners of BAI1 have been previously identified based on yeast two-hybrid screens, their biological roles have not been defined, except for ELMO1 (in process in nature).[16] Defining these molecules and their respective pathways may shed insights on BAI1 function as well as potentially other adhesion-GPCR family members. Fourth, how engagement of BAI1 on the extracellular side, such as the recognition of phosphatidylserine on apoptotic cells, is communicated through the 7-TM stalk to the intracellular signaling pathways need to be precisely defined. This is likely to be challenging, but highly rewarding. Fifth, one of the hallmarks of apoptotic cell clearance process in vivo is that it is anti-inflammatory and it is not known whether BAI1 mediated recognition of phosphatidylserine on apoptotic cells also leads to elicitation of anti-inflammatory mediators. Defining this would help in a better understanding of the role of BAI1 in basic clean up of dead cells, versus other larger functions. Lastly, although BAI1 is considered a member of the GPCR family, whether BAI1 links to G proteins for mediating some of its signals has not been adequately addressed. The recent identification of BAI1 function in apoptotic cell recognition and clearance should prove useful to test these possibilities. It is likely that future studies on BAI1 in the coming years may yield new, exciting and therapeutically relevant information on this exciting adhesion-GPCR.

## ACKNOWLEDGMENTS

This work was supported by grants from the National Institutes of Health/NIGMS (USA) to K.S.R., as well as a grant from The Goldhirsh Foundation (to K.S.R.).

## REFERENCES

1. Kee HJ, Ahn KY, Choi KC et al. Expression of brain-specific angiogenesis inhibitor 3 (BAI3) in normal brain and implications for BAI3 in ischemia-induced brain angiogenesis and malignant glioma. FEBS Lett 2004; 569(1-3):307-316.
2. Kaur B, Brat DJ, Calkins CC et al. Brain angiogenesis inhibitor 1 is differentially expressed in normal brain and glioblastoma independently of p53 expression. Am J Pathol 2003; 162(1):19-27.

3. Mori K, Kanemura Y, Fujikawa H et al. Brain-specific angiogenesis inhibitor 1 (BAI1) is expressed in human cerebral neuronal cells. Neurosci Res 2002; 43(1):69-74.
4. Yoshida Y, Oshika Y, Fukushima Y et al. Expression of angiostatic factors in colorectal cancer. Int J Oncol 1999; 15(6):1221-1225.
5. Fukushima Y, Oshika Y, Tsuchida T et al. Brain-specific angiogenesis inhibitor 1 expression is inversely correlated with vascularity and distant metastasis of colorectal cancer. Int J Oncol 1998; 13(5):967-970.
6. Nishimori H, Shiratsuchi T, Urano T et al. A novel brain-specific p53-target gene, BAI1, containing thrombospondin type 1 repeats inhibits experimental angiogenesis. Oncogene 1997; 15(18):2145-2150.
7. Shiratsuchi T, Nishimori H, Ichlse H et al. Cloning and characterization of BAI2 and BAI3, novel genes homologous to brain-specific angiogenesis inhibitor 1 (BAI1). Cytogenet Cell Genet 1997; 79(1-2):103-108.
8. Kaur B, Cork SM, Sandberg EM et al. Vasculostatin inhibits intracranial glioma growth and negatively regulates in vivo angiogenesis through a CD36-dependent mechanism. Cancer Res 2009; 69(3):1212-1220.
9. Kudo S, Konda R, Obara W et al. Inhibition of tumor growth through suppression of angiogenesis by brain-specific angiogenesis inhibitor 1 gene transfer in murine renal cell carcinoma. Oncol Rep 2007; 18(4):785-791.
10. Kang X, Xiao X, Harata M et al. Antiangiogenic activity of BAI1 in vivo: implications for gene therapy of human glioblastomas. Cancer Gene Ther 2006; 13(4):385-392.
11. Kaur B, Brat DJ, Devi NS et al. Vasculostatin, a proteolytic fragment of brain angiogenesis inhibitor 1, is an antiangiogenic and antitumorigenic factor. Oncogene 2005; 24(22):3632-3642.
12. Yoon KC, Ahn KY, Lee JH et al. Lipid-mediated delivery of brain-specific angiogenesis inhibitor 1 gene reduces corneal neovascularization in an in vivo rabbit model. Gene Ther 2005; 12(7):617-624.
13. Koh JT, Kook H, Kee HJ et al. Extracellular fragment of brain-specific angiogenesis inhibitor 1 suppresses endothelial cell proliferation by blocking alphavbeta5 integrin. Exp Cell Res 2004; 294(1):172-184.
14. Hatanaka H, Oshika Y, Abe Y et al. Vascularization is decreased in pulmonary adenocarcinoma expressing brain-specific angiogenesis inhibitor 1 (BAI1). Int J Mol Med 2000; 5(2):181-183.
15. Park D, Hochreiter-Hufford A, Ravichandran KS. The phosphatidylserine receptor TIM-4 does not mediate direct signaling. Curr Biol 2009; 19(4):346-351.
16. Park D, Tosello-Trampont AC, Elliott MR et al. BAI1 is an engulfment receptor for apoptotic cells upstream of the ELMO/Dock180/Rac module. Nature 2007; 450(7168):430-434.
17. Rosenbaum DM, Rasmussen SG, Kobilka BK. The structure and function of G-protein-coupled receptors. Nature 2009; 459(7245):356-363.
18. Fredriksson R, Lagerstrom MC, Lundin LG et al. The G-protein-coupled receptors in the human genome form five main families. Phylogenetic analysis, paralogon groups and fingerprints. Mol Pharmacol 2003; 63(6):1256-1272.
19. Klabunde T, Hessler G. Drug design strategies for targeting G-protein-coupled receptors. Chembiochem 2002; 3(10):928-944.
20. Bjarnadottir TK, Fredriksson R, Hoglund PJ et al. The human and mouse repertoire of the adhesion family of G-protein-coupled receptors. Genomics 2004; 84(1):23-33.
21. Haitina T, Olsson F, Stephansson O et al. Expression profile of the entire family of Adhesion G protein-coupled receptors in mouse and rat. BMC Neurosci 2008; 9:43.
22. Yona S, Lin HH, Siu WO et al. Adhesion-GPCRs: emerging roles for novel receptors. Trends Biochem Sci 2008; 33(10):491-500.
23. Bjarnadottir TK, Fredriksson R, Schioth HB. The adhesion GPCRs: a unique family of G protein-coupled receptors with important roles in both central and peripheral tissues. Cell Mol Life Sci 2007; 64(16):2104-2119.
24. Lawler J, Hynes RO. The structure of human thrombospondin, an adhesive glycoprotein with multiple calcium-binding sites and homologies with several different proteins. J Cell Biol 1986; 103(5):1635-1648.
25. Chen H, Herndon ME, Lawler J. The cell biology of thrombospondin-1. Matrix Biol 2000; 19(7):597-614.
26. Silverstein RL. The face of TSR revealed: an extracellular signaling domain is exposed. J Cell Biol 2002; 159(2):203-206.
27. Manodori AB, Barabino GA, Lubin BH et al. Adherence of phosphatidylserine-exposing erythrocytes to endothelial matrix thrombospondin. Blood 2000; 95(4):1293-1300.
28. Adams JC, Tucker RP. The thrombospondin type 1 repeat (TSR) superfamily: diverse proteins with related roles in neuronal development. Dev Dyn 2000; 218(2):280-299.
29. Huwiler KG, Vestling MM, Annis DS et al. Biophysical characterization, including disulfide bond assignments, of the anti-angiogenic type 1 domains of human thrombospondin-1. Biochemistry 2002; 41(48):14329-14339.
30. Smith KF, Nolan KF, Reid KB et al. Neutron and X-ray scattering studies on the human complement protein properdin provide an analysis of the thrombospondin repeat. Biochemistry 1991; 30(32):8000-8008.
31. Harmar AJ. Family-B G-protein-coupled receptors. Genome Biol 2001; 2(12):REVIEWS3013.

32. Shiratsuchi T, Futamura M, Oda K et al. Cloning and characterization of BAI-associated protein 1: a PDZ domain-containing protein that interacts with BAI1. Biochem Biophys Res Commun 1998; 247(3):597-604.
33. Ravichandran KS, Lorenz U. Engulfment of apoptotic cells: signals for a good meal. Nat Rev Immunol 2007; 7(12):964-974.
34. Elliott MR, Ravichandran KS. Clearance of apoptotic cells: implications in health and disease. J Cell Biol 1059; 189(7):1059-1070.
35. Elliott MR, Zheng S, Park D et al. Unexpected requirement for ELMO1 in apoptotic germ cell clearance in vivo. Nature 2010; In Press.
36. Park D, Tosello-Trampont AC, Elliott MR et al. BAI1 is an engulfment receptor for apoptotic cells upstream of the ELMO/Dock180/Rac module. Nature 2007; 450:430-434.
37. Lu M, Ravichandran KS. Dock180-ELMO cooperation in Rac activation. Methods Enzymol 2006; 406:388-402.
38. Lu M, Kinchen JM, Rossman KL et al. A Steric-inhibition model for regulation of nucleotide exchange via the Dock180 family of GEFs. Curr Biol 2005; 15(4):371-377.
39. Lu M, Kinchen JM, Rossman KL et al. PH domain of ELMO functions in trans to regulate Rac activation via Dock180. Nat Struct Mol Biol 2004; 11(8):756-762.
40. Grimsley CM, Kinchen JM, Tosello-Trampont AC et al. Dock180 and ELMO1 proteins cooperate to promote evolutionarily conserved Rac-dependent cell migration. J Biol Chem 2004; 279(7):6087-6097.
41. Brugnera E, Haney L, Grimsley C et al. Unconventional Rac-GEF activity is mediated through the Dock180-ELMO complex. Nat Cell Biol 2002; 4(8):574-582.
42. Gumienny TL, Brugnera E, Tosello-Trampont AC et al. CED-12/ELMO, a novel member of the CrkII/Dock180/Rac pathway, is required for phagocytosis and cell migration. Cell 2001; 107(1):27-41.
43. Anderson JC, McFarland BC, Gladson CL. New molecular targets in angiogenic vessels of glioblastoma tumours. Expert Rev Mol Med 2008; 10:e23.
44. Kleihues P, Louis DN, Scheithauer BW et al. The WHO classification of tumors of the nervous system. J Neuropathol Exp Neurol 2002; 61(3):215-225; discussion 226-219.
45. Duda DG, Sunamura M, Lozonschi L et al. Overexpression of the p53-inducible brain-specific angiogenesis inhibitor 1 suppresses efficiently tumour angiogenesis. Br J Cancer 2002; 86(3):490-496.
46. Cao Y, Nagesh V, Hamstra D et al. The extent and severity of vascular leakage as evidence of tumor aggressiveness in high-grade gliomas. Cancer Res 2006; 66(17):8912-8917.
47. Muccio CF, Esposito G, Bartolini A et al. Cerebral abscesses and necrotic cerebral tumours: differential diagnosis by perfusion-weighted magnetic resonance imaging. Radiol Med 2008; 113(5):747-757.
48. Ray SK, Patel SJ, Welsh CT et al. Molecular evidence of apoptotic death in malignant brain tumors including glioblastoma multiforme: upregulation of calpain and caspase-3. J Neurosci Res 2002; 69(2):197-206.
49. Sinha S, Bastin ME, Whittle IR et al. Diffusion tensor MR imaging of high-grade cerebral gliomas. AJNR Am J Neuroradiol 2002; 23(4):520-527.
50. Fujiwara T, Mammoto A, Kim Y et al. Rho small G-protein-dependent binding of mDia to an Src homology 3 domain-containing IRSp53/BAIAP2. Biochem Biophys Res Commun 2000; 271(3):626-629.
51. Kim MY, Ahn KY, Lee SM et al. The promoter of brain-specific angiogenesis inhibitor 1-associated protein 4 drives developmentally targeted transgene expression mainly in adult cerebral cortex and hippocampus. FEBS Lett 2004; 566(1-3):87-94.
52. Oda K, Shiratsuchi T, Nishimori H et al. Identification of BAIAP2 (BAI-associated protein 2), a novel human homologue of hamster IRSp53, whose SH3 domain interacts with the cytoplasmic domain of BAI1. Cytogenet Cell Genet 1999; 84(1-2):75-82.
53. Shiratsuchi T, Oda K, Nishimori H et al. Cloning and characterization of BAP3 (BAI-associated protein 3), a C2 domain-containing protein that interacts with BAI1. Biochem Biophys Res Commun 1998; 251(1):158-165.
54. Koh JT, Lee ZH, Ahn KY et al. Characterization of mouse brain-specific angiogenesis inhibitor 1 (BAI1) and phytanoyl-CoA alpha-hydroxylase-associated protein 1, a novel BAI1-binding protein. Brain Res Mol Brain Res 2001; 87(2):223-237.
55. Funke L, Dakoji S, Bredt DS. Membrane-associated guanylate kinases regulate adhesion and plasticity at cell junctions. Annu Rev Biochem 2005; 74:219-245.
56. Gonzalez-Mariscal L, Betanzos A, Avila-Flores A. MAGUK proteins: structure and role in the tight junction. Semin Cell Dev Biol 2000; 11(4):315-324.
57. Krugmann S, Jordens I, Gevaert K et al. Cdc42 induces filopodia by promoting the formation of an IRSp53:Mena complex. Curr Biol 2001; 11(21):1645-1655.
58. Miki H, Yamaguchi H, Suetsugu S et al. IRSp53 is an essential intermediate between Rac and WAVE in the regulation of membrane ruffling. Nature 2000; 408(6813):732-735.
59. Takenawa T, Miki H. WASP and WAVE family proteins: key molecules for rapid rearrangement of cortical actin filaments and cell movement. J Cell Sci 2001; 114(Pt 10):1801-1809.

CHAPTER 16

# ADHESION-GPCRs IN THE MALE REPRODUCTIVE TRACT

Ben Davies* and Christiane Kirchhoff

**Abstract**  The male reproductive tract expresses a diverse array of adhesion-GPCRs, many in a highly specific and regulated manner. Despite this specificity of expression, little is known about the function of this receptor family in male reproductive physiology. Insights into function are beginning to emerge with the increasing availability of genetically modified mice harbouring mutations in these genes. Gpr64 is the best characterised of the adhesion-GPCRs in the male reproductive system and the phenotype of Gpr64 knock-out mice implicates this receptor in the regulation of fluid absorption in the efferent ducts and proximal epididymis. This chapter summarizes recent data concerning this receptor and other family members in the male reproductive system.

## INTRODUCTION

Within the male reproductive tract a diverse array of cell types contribute to the reproductive competence of the organism. In mammals, the sertoli and leydig cells of the testis support and regulate the proliferation and subsequent differentiation of germ cells into spermatozoa. Spermatozoa exit the testis via the efferent ductules into the epididymis, where the spermatozoa mature and gain their reproductive competence. Mirroring these diverse functions, the cell types of the male reproductive tract display a vast complexity of gene expression[1,2] and consistent with the complexity, it appears that the male reproductive tract express a large number of adhesion-GPCRs, the topic of this chapter.

An initial survey of mouse and human expressed sequence tags (ESTs) revealed the expression of at least 13 different family members within the male reproductive tract.[3] A more comprehensive RT-PCR expression analysis of rodent adhesion-GPCRs also detected above background expression for many of the known family members,[4] although only testis was

*Corresponding Author: Ben Davies—Wellcome Trust Centre for Human Genetics, University of Oxford, Roosevelt Drive, OX3 7BN, UK. Email: ben.davies@well.ox.ac.uk

*Adhesion-GPCRs: Structure to Function*, edited by Simon Yona and Martin Stacey.
©2010 Landes Bioscience and Springer Science+Business Media.

**Table 1.** Reported expression of adhesion-GPCRs within the male reproductive tract

| Adhesion-GPCR | Expression Site | Detection Method |
|---|---|---|
| Gpr64 | Nonciliated principal cells of efferent ducts and caput epididymis | Northern,[5] ISH,[5,6] Knock-in reporter,[11] IHC[10] |
| Gpr124 | Testis | RT-PCR[16] |
| Gpr125 | Seminiferous tubules, germ cells | RT-PCR,[16] Knock-in reporter,[17] IHC[17] |
| Celsr1 | Testis | RT-PCR[23] |
| Celsr2 | Sertoli cells, germ cells | IHC,[23] RT-PCR[23] |
| Celsr3 | Elongated spermatids | IHC,[23] RT-PCR[23] |
| Vlgr1 | Testis | RT-PCR[4,25] |
| Gpr56 | Testis | Northern,[30] RT-PCR[4] |
| Gpr97 | Testis | RT-PCR[4] |
| Gpr110 | Testis | RT-PCR[4] |
| Gpr113 | Testis | Northern,[33] RT-PCR[4] |
| Emr1 | Testis | RT-PCR[4] |
| Cd97 | Testis | IHC[35] |

taken for analysis. Table 1 summarizes what is known in the literature about adhesion-GPCR expression within the male reproductive tract. It is clearly premature to conclude that a weak RT-PCR expression equates to a functional significance within the male reproductive system, yet there is considerable more in depth evidence to suggest that members of this diverse receptor family do play an important role in the function and regulation of male fertility.

## GPR64

GPR64, originally reported as HE6, was identified as part of a differential expression screen for human epididymis specific transcripts.[5] Northern and in situ hybridization analysis revealed the expression to be highly specific and strong within the proximal epididymis, representing approximately 0.01% of all cDNA clones. Mammalian orthologues from mouse and rat were subsequently cloned by homology and similar proximal epididymal specific expression was demonstrated in these species by Northern and immunohistochemical analysis.[6] GPR64 has the classic features of an adhesion-GPCR—a secretin-like seven transmembrane domain together with an adjacent cysteine rich GPS (GPCR proteolysis site) motif and a long N-terminal extracellular domain. The N-terminal domain lacks any significant functional domains and is solely characterised by serine/threonine/proline rich sequence which is predicted to be extensively glycosylated. Deglycosylation Western analysis using an N-terminal specific antibody has confirmed this to be the case.[6]

In terms of family homology, GPR64 is thus most similar in overall structure to group II and IV adhesion-GPCR family members.[4,7] Little functional information has been attributed to the majority of these receptors; however the two best characterised, GPR56 and CD97, both share a common function in cell motility and both have been implicated in the regulation of tumorogenesis. Consistent with their proposed function, extracellular matrix or cell surface ligands have been characterised for these receptors. GPR56 has been demonstrated to bind transglutaminase-2[8] and CD97 has been shown to interact with glycosaminoglycans and integrins.[9] Although no ligand or binding partner

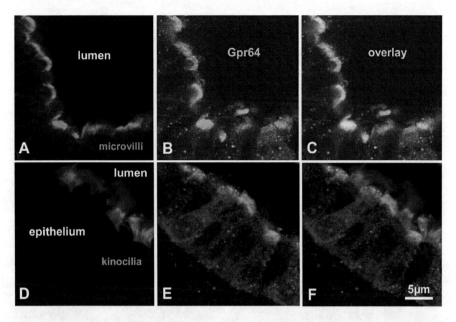

**Figure 1.** Colocalization study of GPR64 receptor protein and apical differentiations of epithelial cells in efferent ducts of the mouse. A-C) Confocal microscopy after dual labeling of GPR64 and actin shows that the receptor is present on the microvilli of nonciliated cells. Cy2-conjugated second antibody was used in combination with GPR64 antibody (green fluorescence); F-actin was labeled by phalloidin staining (red fluorescence). Colocalization (yellow) is most prominent in microvillar shafts and a zone immediately below the brush border (yellow fluorescence). D-E) Dual labeling of GPR64 and acetylated α-tubulin labeling stabile microtubules shows that the receptor is absent from kinocilia of ciliated cells. Cy2-conjugated second antibody was used in combination with GPR64 antibody (green fluorescence) and Cy3-conjugated second antibody in combination with acetylated α-tubulin (red fluorescence). Note that GPR64-related and tubulin-related fluorescence are associated with alternating apical structures of epithelial cells. A color version of this figure is available at www.landesbioscience.com/curie.

has been ascribed to GPR64, interactions with components of the extracellular matrix or other membrane proteins seem likely.

Antibodies raised to both the extracellular and transmembrane region have shown GPR64 to be present as a two subunit heterodimer in epididymis preparations, suggesting activate cleavage of the GPS site occurs.[6] Colocalisation of these two subunits was seen in all expression studies performed, suggesting that the two subunits remain associated—no evidence for a soluble ectodomain could be found, although this cannot be ruled out. Both mRNA and protein expression studies have revealed a gradient of *Gpr64* expression, strongest within the efferent ducts and the proximal epididymis, decreasingly rapidly towards more distal epididymal regions.[5,6]

Within the efferent ducts, confocal immunohistochemical studies have revealed GPR64 to be specifically expressed in the apical compartment of nonciliated principal cells.[10] These cells are primarily responsible for the absorbance of testicular fluid and, consistent with this function, the apical luminal compartment is organised into microvilli of variable length. Using phalloidin staining of F-actin which highlights the microvillar brush border, colocalisation with GPR64 could be seen (Fig. 1A-C). In contrast, no GPR64 expression was detected in the tufts of (kino-) cilia of the ciliated cells[10] (Fig. 1,D-F) which are thought to be primarily involved in back and forth movement of luminal contents via long kinocilia.

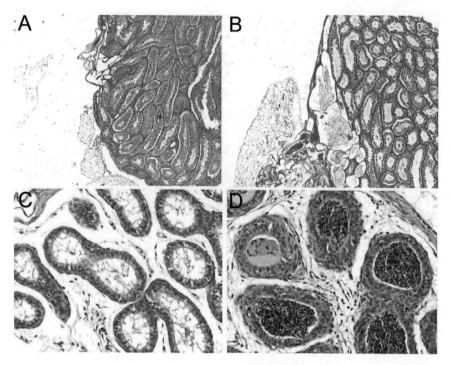

**Figure 2.** Histology of testis and efferent ducts in wild-type and knock-out *Gpr64* mice. A-B) Representative photomicrographs (magnification, × 100) of hematoxylin- and eosin-stained sections through the rete testis of wild-type (A) and knock-out (B). Note the dilated rete testis and seminiferous tubules in the knock-out mice. C-D) Representative photomicrographs (magnification, × 100) of hematoxylin- and eosin-stained sections through the efferent ductules of wild-type (C) and knock-out (D) mice. The stasis and accumulation of spermatozoa can clearly be seen in the knock-out tubules. A color version of this figure is available at www.landesbioscience.com/curie. Reproduced from reference 11 with permission from American Society for Microbiology.

This highly specific expression pattern suggests that GPR64 could play a role in the structural and functional organisation of the microvilli responsible for luminal fluid uptake.

To gain an insight into the function of GPR64, knock-out mice lacking the entire seven transmembrane domain of this X-chromosomal located gene were generated.[11] *Gpr64* mRNA and protein expression in the resulting hemizygous knock-out mice was completely absent. The absence of *Gpr64* led to a rapid age-dependent decline in male fertility. Histological investigation revealed the cause of this infertility was a dysregulation of testicular fluid reabsorption. Before the onset of spermatogenesis, a build up of testicular fluid was observed within the efferent ducts and the rete testis of mutant mice (Fig. 2A-B) which, in older males led to a stasis of spermatozoa within the efferent ducts and proximal epididymis (Fig. 2C-D). Morphologically, the deletion of *Gpr64* had no apparent effect on the morphology or cell types present within the efferent ducts and the proximal epididymis, suggesting a specific function in the regulation of water uptake for this receptor. This phenotype agrees with the proposed function of GPR64 as determined by its highly specific expression pattern.

To establish a molecular explanation for the observed phenotype and to investigate the general effects of *Gpr64* disruption on gene expression, differential cDNA library and microarray screening was performed using adult epididymal tissue from wild-type and *Gpr64*

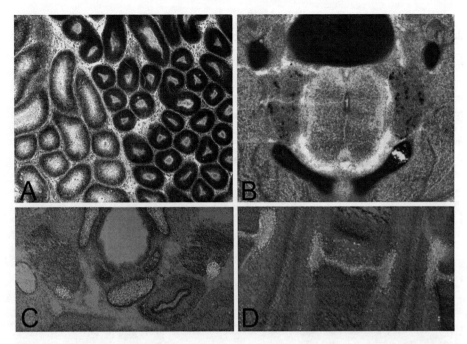

**Figure 3.** β-galactosidase reporter expression in *Gpr64* knock-out mice. Photomicrograph of a BLUO-GAL-stained section through the epididymis of an adult HE6 knockout (A), dorsal root ganglia of a P0 *Gpr64* knock-out (B), parathyroid of an E16.5 *Gpr64* knock-out (C) and digits of a P0 *Gpr64* knock-out, showing the activity of the β-galactosidase gene. A and B are bright field images and C and D are images under a polarizing filter where the BLUO-GAL precipitate is easily detected as white birefringent crystals. A color version of this figure is available at www.landesbioscience.com/curie.

knock-out mice.[12] Both techniques revealed an overlapping set of differentially expressed transcripts, the majority of which were found to be downregulated in the knock-out rather than the wild-type mice. It is interesting that several of the differentially regulated transcripts encode proteins that are implicated in fluid uptake. For example, *Cldn10A* (Claudin 10A), a gene strongly expressed in the proximal epididymis with a role in tight junction formation[13] controlling paracellular fluid movement, was found to be down-regulated over seven-fold in knock-out mice. *Slc1a1* encodes a glutamate transporter which contributes to the regulation of osmolyte balance in the proximal epididymis[14] and was found to be down-regulated four-fold in knock-out epididymis. Since many epididymal genes are regulated by testis derived luminal factors and the stasis of spermatozoa seen in knock-out mice may impede these signals, it is too speculative to conclude a causal link between receptor activation and gene expression changes in a classic GPCR mediated manner and the significance of these expression changes demands further investigation.

In the knock-out model of *Gpr64*, the transmembrane region was replaced by a beta-galactosidase reporter gene under the control of an internal ribosome entry signal. The resulting reporter expression was seen to faithfully recapitulate the endogenous expression pattern in the epdidiymis.[11] Despite its initial characterisation as an epididymis specific transcript, *Gpr64* expression was also seen within a subset of neurons in the dorsal root and trigeminal ganglia, within the synovial membranes of the developing joints and in the developing parathyroid (Fig. 3). Although many of these sites of expression must be confirmed by RNA and protein analysis, it is tempting to speculate that within these tissues there are cells which, similar to those of the proximal epididymis, are also involved in some degree of fluid dynamics—the secretion of synovial fluid, the release of hormones and neurotransmitters. Although the functional significance of this expression is as yet unclear and no nonreproductive phenotype has been reported for the *Gpr64* knock-out mice, it would be interesting to establish whether GPR64 has a conserved function outside of the reproductive system. Recently, *GPR64* has been shown to be specifically expressed in synovial fibroblasts isolated from osteoarthritis patients, suggesting a role for this receptor in adult joint pathophysiology.[15]

Overall it would appear that the specifically expressed adhesion-GPCR has an important role in regulating the exchange of fluid within the efferent ducts and the proximal epididymis. An electron microscopic investigation of the subcellular organisation of the nonciliated principal cells which express this receptor so specifically in the apical compartment is warranted and a further examination of GPR64 function in tissues outside of the reproductive system may provide important insights into its function.

## GPR124 AND GPR125

GPR124 and GPR125, originally reported as TEM5 and TEM5-like protein, were identified as interacting partners for the human homologue of Drosophila disc large (hDlg) which has diverse roles in cell polarity, adhesion and tumorogenesis.[16] As yet no function has been attributed to these receptors whose N-terminal regions contain a Leucine Rich Region, a region of immunoglobulin homology and a putative hormone binding domain. Northern analysis has revealed broad expression of *Gpr124* and *Gpr125* in multiple tissues including strong expression in the testis and the prostate.

Testis expression of *Gpr125* was further explored using a knock-in mouse model, in which a beta-galactosidase reporter gene was inserted under the control of the endogenous *Gpr125* promoter and strong expression was revealed in the first layer of cells adjacent to the basal layer of the seminiferous tubules.[17] Expression was subsequently confirmed by antibody staining which revealed strong expression in spermatogonia. When cultured in vitro on mouse testicular stromal cells, GPR125 expressing cells were found to represent spermatogonial progenitor cells, expressing markers of undifferentiated spermatogonia. Long-term culture of these GPR125+ progenitor cells led to the formation of distinct colonies of multipotent adult stem cells (MASCs) which were propagatable on embryonic fibroblast feeder layers. The ability of these cells to differentiate in vitro to multiple GPR125 negative cell types, to cause teratoma formation in SCID mice and to contribute to all cell lineages following injection into blastocysts demonstrated the multipotency of these cells. GPR125 is thus a marker for these stems cells which are now being considered as an ethical alternative to embryonic stem cells for future stem cell regenerative therapies.[18] Interestingly, the function of the GPR125 in this system is unknown and significantly, progenitor cells propagated from homozygous *Gpr125* knock-out mice are still able to maintain their progenitor stem cell phenotype.[17]

It is thus unclear what direct function GPR125 may have in the development of spermatogonia and certainly the knock-out experiments imply that this receptor may have no causative role in defining cell fate; however the exquisite regulation of its expression suggests an indirect or possibly redundant role in germ cell development. Indeed, in support of potential functional redundancy, testis expression of the highly related *Gpr124*[16] could be compensating for loss of GPR125 function in these mutants.

## CELSR1-3

The three CELSR adhesion-GPCRs are vertebrate homologues of the Drosophila *flamingo* receptor, which all contain seven cadherin domains in their extracellular N termini together with multiple Laminin G and Epidermal Growth Factor binding domains and a hormone binding domain.[19] Similar to *flamingo*'s role in Drosophila wing development, studies with mutant mice have demonstrated a role for CELSR1 in the regulation of planar cell polarity of stereocilia in the inner ear.[20] Furthermore, again mirroring flamingo's role in the fly, all mammalian CELSR receptors have been implicated in aspects of neuronal development.[20-22]

Outside of the brain, all three CELSR receptors are expressed in the rat testis, each showing a unique developmental pattern of expression.[23] *Celsr1* and *Celsr2* expression is highest at postnatal day 7, after which time point only a low level of *Celsr2* expression can be detected. In contrast, *Celsr3* is expressed only weakly at postnatal day 7 but increases strongly throughout postnatal development coincident with increasing germ cell numbers. Further immunocytochemical analysis refined the site of expression to sertoli (and possibly germ cells) for Celsr1 and Celsr2 and to elongated spermatids for Celsr3.[23] Interestingly, no CELSR staining colocalised with cadherin-based cell adherens junctions; CELSR2 staining localised in the Golgi and in the late endocytic vesicles, perhaps indicative of receptor internalization following active signalling. Despite the lack of CELSR protein at adherens junctions, recombinant CELSR2 cadherin domain fragments led to germ cell detachment in Sertoli-germ cell cocultures, implying that the cadherin domains of this receptor subfamily do have some functionality.

Sertoli cell-germ cell adhesion is an important regulatory factor in germ cell proliferation and differentiation[24] and consequently, a role for these receptors in germ cell development may be expected. Little evidence for a functional involvement in fertility can be provided by the available mouse mutants for these three genes. Homozygous *Celsr1* and *Celsr3* mutants are embryonic or neonatal lethal and subsequently no information regarding the functional consequences of loss of these genes in the reproductive system is available.[20,21] Arguing against a direct involvement in germ cell development, *Celsr2* homozygous mutants are viable and no fertility phenotype has been reported.[22] However, an in vivo role for CELSR2 cannot be ruled out due to the degree of functional redundancy that might be expected for these highly homologous genes.

## OTHER ADHESION-GPCRs

An extensive RT-PCR study has revealed expression of other adhesion-GPCR family members within the testis[4] and in particular, GPR98 (VLGR1), GPR56, GPR113 and EMR1 have revealed relatively high or specific expression (in comparison with other tissues) (see Table 1). Adhesion-GPCRs whose expression has been corroborated by at least one further study are discussed below.

Expression of *VLGR1*, the largest known cell surface receptor with an extracellular domain characterised by 35 imperfect calcium binding Calx-β domains, has also been investigated within the human testis by RT-PCR.[25] Interestingly, only a truncated potentially extracellular isoform, known as Vlgr1c, was detected with full length transcripts being absent from this tissue. Several independent mutant mouse studies have revealed a role for this receptor in the organisation of auditory hair bundles, with mutant mice being severely deaf.[26-28] Furthermore, GPS sequence mutations of the human *VLGR1* gene have been found to be the cause of Usher syndrome Type 2 disease, a genetic disease characterised by blindness and deafness.[29] Both *Vlgr1* mutant mice and patients with Usher syndrome Type 2 patients show normal fertility, perhaps implying that this receptor type has no functional role within the male reproductive tract. However, due to the complexity of alternative splicing within the 5' end of the gene, it is not apparent whether the testis specific transcript of *VLGR1* is still expressed in these mutants and thus the functional significance of this expression remains unclear.

Similarly *Gpr56* expression has been confirmed within the testis by Northern analysis.[30] A role for this receptor in the development of the cerebral cortex has been revealed by mapping studies which revealed mutations in *Gpr56* to associate with the human condition bilateral frontoparietal polymicrogyria.[31] Mutant mice have also been generated which display a comparable cortical deficit.[32] Despite the testis expression of *Gpr56*, no fertility phenotype has been reported for the mutant mice.

GPR113, an adhesion-GPCR with a single extracellular hormone binding domain, was found to be strongly expressed in taste receptor cells.[33] Subsequent Northern expression showed a highly restricted pattern of strong expression within the testis although it remains unclear which cell types express this receptor. No further functional analysis of this receptor has been performed, although an ES cell line harbouring a gene trap insertion within the gene has been reported[34] and a closer investigation of reproductive parameters is warranted.

The comprehensive study of adhesion-GPCR expression by expression profiling revealed high *Emr1* expression within the testis,[4] although no testis ESTs have been reported for this receptor. The related CD97 was found to be abundant within the interstitium of the mouse testis and within the caput region of the epididymis,[35] most probably identifying invading macrophages. It is thus possible that the *Emr1* expression reported is also confined to cells of the immune system which are well-known to be important in the normal development of male reproductive tissues.[36]

## CONCLUSION

Expression studies have revealed that the male reproductive tract expresses a number of adhesion-GPCRs. Studies with mutant mice are beginning to unravel the function of many of these receptors in specific tissues; however the functional significance of male reproductive expression for most family members remains unclear. Where mutant mice have been generated, reproductive phenotypes are potentially being masked by co-expression of related receptor family members. A specific role has however been elucidated for GPR64 in the regulation of fluid reabsorption with the efferent ducts and epididymis.

# REFERENCES

1. Divina P, Vlcek C, Strnad P et al. Global transcriptome analysis of the C57BL/6J mouse testis by SAGE: evidence for nonrandom gene order. BMC Genomics 2005; 6(1):29.
2. Johnston DS, Jelinsky SA, Bang HJ et al. The mouse epididymal transcriptome: transcriptional profiling of segmental gene expression in the epididymis. Biol Reprod 2005; 73(3):404-413.
3. Bjarnadottir TK, Fredriksson R, Hoglund PJ et al. The human and mouse repertoire of the adhesion family of G-protein-coupled receptors. Genomics 2004; 84(1):23-33.
4. Haitina T, Olsson F, Stephansson O et al. Expression profile of the entire family of Adhesion G protein-coupled receptors in mouse and rat. BMC Neurosci 2008; 9:43.
5. Osterhoff C, Ivell R, Kirchhoff C. Cloning of a human epididymis-specific mRNA, HE6, encoding a novel member of the seven transmembrane-domain receptor superfamily. DNA Cell Biol 1997; 16(4):379-389.
6. Obermann H, Samalecos A, Osterhoff C et al. HE6, a two-subunit heptahelical receptor associated with apical membranes of efferent and epididymal duct epithelia. Mol Reprod Dev 2003; 64(1):13-26.
7. Bjarnadottir TK, Fredriksson R, Schioth HB. The adhesion GPCRs: a unique family of G protein-coupled receptors with important roles in both central and peripheral tissues. Cell Mol Life Sci 2007; 64(16):2104-2119.
8. Xu L, Begum S, Hearn JD et al. GPR56, an atypical G protein-coupled receptor, binds tissue transglutaminase, TG2 and inhibits melanoma tumor growth and metastasis. Proc Natl Acad Sci USA 2006; 103(24):9023-9028.
9. Wang T, Ward Y, Tian L et al. CD97, an adhesion receptor on inflammatory cells, stimulates angiogenesis through binding integrin counterreceptors on endothelial cells. Blood 2005; 105(7):2836-2844.
10. Kirchhoff C, Osterhoff C, Samalecos A. HE6/GPR64 adhesion receptor colocalizes with apical and subapical F-actin scaffold in male excurrent duct epithelia. Reproduction 2008; 136(2):235-245.
11. Davies B, Baumann C, Kirchhoff C et al. Targeted deletion of the epididymal receptor HE6 results in fluid dysregulation and male infertility. Mol Cell Biol 2004; 24(19):8642-8648.
12. Davies B, Behnen M, Cappallo-Obermann H et al. Novel epididymis-specific mRNAs downregulated by HE6/Gpr64 receptor gene disruption. Mol Reprod Dev 2007; 74(5):539-553.
13. Van Itallie CM, Anderson JM. Claudins and epithelial paracellular transport. Annu Rev Physiol 2006; 68:403-429.
14. Wagenfeld A, Yeung CH, Lehnert W et al. Lack of glutamate transporter EAAC1 in the epididymis of infertile c-ros receptor tyrosine-kinase deficient mice. J Androl 2002; 23(6):772-782.
15. Galligan CL, Baig E, Bykerk V et al. Distinctive gene expression signatures in rheumatoid arthritis synovial tissue fibroblast cells: correlates with disease activity. Genes Immun 2007; 8(6):480-491.
16. Yamamoto Y, Irie K, Asada M et al. Direct binding of the human homologue of the Drosophila disc large tumor suppressor gene to seven-pass transmembrane proteins, tumor endothelial marker 5 (TEM5) and a novel TEM5-like protein. Oncogene 2004; 23(22):3889-3897.
17. Seandel M, James D, Shmelkov SV et al. Generation of functional multipotent adult stem cells from GPR125+ germline progenitors. Nature 2007; 449(7160):346-350.
18. Dym M, He Z, Jiang J et al. Spermatogonial stem cells: unlimited potential. Reprod Fertil Dev 2009; 21(1):15-21.
19. Formstone CJ, Little PF. The flamingo-related mouse Celsr family (Celsr1-3) genes exhibit distinct patterns of expression during embryonic development. Mech Dev 2001; 109(1):91-94.
20. Curtin JA, Quint E, Tsipouri V et al. Mutation of Celsr1 disrupts planar polarity of inner ear hair cells and causes severe neural tube defects in the mouse. Curr Biol 2003; 13(13):1129-1133.
21. Tissir F, Bar I, Jossin Y et al. Protocadherin Celsr3 is crucial in axonal tract development. Nat Neurosci 2005; 8(4):451-457.
22. Deltagen. MGI Direct Data Submission 2005; MGI:3604450.
23. Beall SA, Boekelheide K, Johnson KJ. Hybrid GPCR/cadherin (Celsr) proteins in rat testis are expressed with cell type specificity and exhibit differential Sertoli cell-germ cell adhesion activity. J Androl 2005; 26(4):529-538.
24. Fritz IB. Somatic cell-germ cell relationships in mammalian testes during development and spermatogenesis. Ciba Found Symp 1994; 182:271-274; discussion 274-281.
25. McMillan DR, Kayes-Wandover KM, Richardson JA et al. Very large G protein-coupled receptor-1, the largest known cell surface protein, is highly expressed in the developing central nervous system. J Biol Chem 2002; 277(1):785-792.
26. Skradski SL, Clark AM, Jiang H et al. A novel gene causing a mendelian audiogenic mouse epilepsy. Neuron 2001; 31(4):537-544.

27. McGee J, Goodyear RJ, McMillan DR et al. The very large G-protein-coupled receptor VLGR1: a component of the ankle link complex required for the normal development of auditory hair bundles. J Neurosci 2006; 26(24):6543-6553.
28. Yagi H, Tokano H, Maeda M et al. Vlgr1 is required for proper stereocilia maturation of cochlear hair cells. Genes Cells 2007; 12(2):235-250.
29. Weston MD, Luijendijk MW, Humphrey KD et al. Mutations in the VLGR1 gene implicate G-protein signaling in the pathogenesis of Usher syndrome type II. Am J Hum Genet 2004; 74(2):357-366.
30. Liu M, Parker RM, Darby K et al. GPR56, a novel secretin-like human G-protein-coupled receptor gene. Genomics 1999; 55(3):296-305.
31. Piao X, Hill RS, Bodell A et al. G protein-coupled receptor-dependent development of human frontal cortex. Science 2004; 303(5666):2033-2036.
32. Li S, Jin Z, Koirala S et al. GPR56 regulates pial basement membrane integrity and cortical lamination. J Neurosci 2008; 28(22):5817-5826.
33. LopezJimenez ND, Sainz E, Cavenagh MM et al. Two novel genes, Gpr113, which encodes a family 2 G-protein-coupled receptor and Trcg1, are selectively expressed in taste receptor cells. Genomics 2005; 85(4):472-482.
34. Zambrowicz BP, Abuin A, Ramirez-Solis R et al. Wnk1 kinase deficiency lowers blood pressure in mice: a gene-trap screen to identify potential targets for therapeutic intervention. Proc Natl Acad Sci USA 2003; 100(24):14109-14114.
35. Kiessling AA, Mullen TE, Kiessling RL et al. Detection in Mice and Men of a Novel Class of Leukocyte/Macrophages Essential for Normal Development of Reproductive Tract Tissues. Fertility and sterility 2000; 74(3):S86.
36. Cohen PE, Nishimura K, Zhu L et al. Macrophages: important accessory cells for reproductive function. J Leukoc Biol 1999; 66(5):765-772.

# APPENDIX

# MAMMALIAN* ADHESION-GPCRs

Simon Yona and Martin Stacey

| Family | Receptor Name | Expression | Function | References |
|---|---|---|---|---|
| EGF-like | EMR1 | Eosinophils, mononuclear cells (mouse only) | Peripheral immune tolerance (mouse only) | 1-4 |
| | EMR2 | Myeloid cells | PMN activation and migration, receptor for chondroitin sulphate | 5-7 |
| | EMR3 | Myeloid cells | Mature PMN marker | 8 |
| | EMR4 | Macrophages, DCs | Unknown | 9,10 |
| | CD97 | Leukocytes, smooth muscle | Tumour angiogenesis, CD55 and chondroitin sulphate receptor, cell migration | 11-14 |
| | ETL | Heart, smooth muscle | Unknown | 15 |
| BAI-like | BAI1 | CNS, bone marrow, monocyte, macrophage | Angiogenesis inhibitor, apoptotic cell clearance, phosphatidylserine receptor | 16-18 |
| | BAI2 | CNS, skeletal muscle, heart | Angiogenesis inhibitor | 19-21 |

*continued on next page*

* A further 90 uncharacterised adhesion-GPCRs have been identified in the Sea Urchin.[61]
Putative adhesion-GPCRs have been discovered in protozoans (http://smart.embl-heidelberg.de).

*Adhesion-GPCRs: Structure to Function*, edited by Simon Yona and Martin Stacey.
©2010 Landes Bioscience and Springer Science+Business Media.

**Appendix.** Continued

| Family | Receptor Name | Expression | Function | References |
|---|---|---|---|---|
| | BAI3 | CNS, heart | Angiogenesis inhibitor | 19,20,22 |
| CELSR-like | CELSR1 | CNS, eye, kidney, lung, spleen, testis | Brain development | 23-27 |
| | CELSR2 | CNS, eye, heart, kidney, lung, spleen, testis | Dendritic maintenance, growth and arborisation, germ cell survival | 23,25,27 |
| | CELSR3 | CNS, eye, testis | Brain development | 23,27-29 |
| IgG-like | GPR116 | Lung, heart, kidney | Unknown | 30 |
| | GPR124 | Endothelial cells | Tumour angiogenesis | 31,32 |
| | GPR125 | Stem cells, testis | Stem cell marker | 33,34 |
| Latrophilin-like | Latrophilin-1 | CNS | α-latrotoxin (spider venom) receptor | 35-38 |
| | Latrophilin-2 | CNS, lung, liver | Heart valve development | 36,37,39 |
| | Latrophilin-3 | CNS | Unknown | 36,37 |
| Miscellaneous | GPR113 | CNS, testis, taste receptor cells | Unknown | 40 |
| | GPR64 | Testis, joints | Fluid resorption | 41 |
| | GPR126 | Endothelial cell, placenta | Height variation, myelination of Schwann cells | 42,43-45 |
| | GPR56 | CNS, natural killer cells | Brain development, neuronal cell migration, metastasis inhibitor, TG2 receptor, Natural killer cell function | 46-50 |
| | VLGR1 | CNS | Sensory neuronal development | 51-53 |
| | GPR97 | Bone marrow, mast cells (gnf-altas) | Unknown | 54 |
| | GPR110 | Lung and prostate tumour, cornea (gnf atlas) | Oncogene | 55 |
| | GPR111 | Unknown | Unknown | 54 |
| | GPR112 | Unknown | Unknown | 54 |
| | GPR114 | Unknown | Unknown | 54 |
| | GPR115 | Unknown | Unknown | 54 |
| | GPR123 | CNS | Unknown | 56 |
| | GPR128 | Small/large intestine (gnf, atlas) | Unknown | 54,57 |
| | GPR133 | CNS | Height variation | 58,59 |
| | GPR144 | Unknown | Unknown | 60 |

## ACKNOWLEDGMENT

This appendix was reproduced from Yona S, Lin HH, Siu WO, Gordon S, Stacey M. Adhesion-GPCRs: Emerging roles for novel receptors. Trends Biochem Sci 2008; 33(10):491-500; ©2008 with permission from Elsevier.

## REFERENCES

1. Hamann J, Koning N, Pouwels W et al. EMR1, the human homolog of F4/80, is an eosinophil-specific receptor. Eur J Immunol 2007; 37(10):2797-2802.
2. Lin HH, Faunce D, Stacey M et al. The macrophage receptor F4/80 is involved in the induction of CD8$^+$ regulatory T-cells in peripheral tolerance. J Exp Med 2005; 201(10):1615-1625.
3. Austyn JM, Gordon S. F4/80, a monoclonal antibody directed specifically against the mouse macrophage. Eur J Immunol 1981; 11(10):805-815.
4. Hume DA, Robinson AP, MacPherson GG et al. The mononuclear phagocyte system of the mouse defined by immunohistochemical localization of antigen F4/80. Relationship between macrophages, Langerhans cells, reticular cells and dendritic cells in lymphoid and hematopoietic organs. J Exp Med 1983; 158(5):1522-1536.
5. Yona S, Lin HH, Dri P et al. Ligation of the adhesion-GPCR EMR2 regulates human neutrophil function. FASEB J 2008; 22(3):741-751.
6. Chang GW, Davies JQ, Stacey M et al. CD312, the human adhesion-GPCR EMR2, is differentially expressed during differentiation, maturation and activation of myeloid cells. Biochem Biophys Res Commun 2007; 353(1):133-138.
7. Stacey M, Chang GW, Davies JQ et al. The epidermal growth factor-like domains of the human EMR2 receptor mediate cell attachment through chondroitin sulphate glycosaminoglycans. Blood 2003; 102(8):2916-2924.
8. Matmati M, Pouwels W, van Bruggen R et al. The human EGF-TM7 receptor EMR3 is a marker for mature granulocytes. J Leukoc Biol 2007; 81(2):440-448.
9. Stacey M, Chang GW, Sanos SL et al. EMR4, a novel epidermal growth factor (EGF)-TM7 molecule up-regulated in activated mouse macrophages, binds to a putative cellular ligand on B lymphoma cell line A20. J Biol Chem 2002; 277(32):29283-29293.
10. Caminschi I, Lucas KM, O'Keeffe MA et al. Molecular cloning of F4/80-like-receptor, a seven-span membrane protein expressed differentially by dendritic cell and monocyte-macrophage subpopulations. J Immunol 2001; 167(7):3570-3576.
11. Wang T, Ward Y, Tian L et al. CD97, an adhesion receptor on inflammatory cells, stimulates angiogenesis through binding integrin counterreceptors on endothelial cells. Blood 2005; 105(7):2836-2844.
12. Kwakkenbos MJ, Pouwels W, Matmati M et al. Expression of the largest CD97 and EMR2 isoforms on leukocytes facilitates a specific interaction with chondroitin sulfate on B-cells. J Leukoc Biol 2005; 77(1):112-119.
13. Hamann J, Vogel B, van Schijndel GM et al. The seven-span transmembrane receptor CD97 has a cellular ligand (CD55, DAF). J Exp Med 1996; 184(3):1185-1189.
14. Aust G, Wandel E, Boltze C et al. Diversity of CD97 in smooth muscle cells. Cell Tissue Res 2006; 324(1):139-147.
15. Nechiporuk T, Urness LD, Keating MT. ETL, a novel seven-transmembrane receptor that is developmentally regulated in the heart. ETL is a member of the secretin family and belongs to the epidermal growth factor-seven-transmembrane subfamily. J Biol Chem 2001; 276(6):4150-4157.
16. Park D, Tosello-Trampont AC, Elliott MR et al. BAI1 is an engulfment receptor for apoptotic cells upstream of the ELMO/Dock180/Rac module. Nature 2007; 450(7168):430-434.
17. Nishimori H, Shiratsuchi T, Urano T et al. A novel brain-specific p53-target gene, BAI1, containing thrombospondin type 1 repeats inhibits experimental angiogenesis. Oncogene 1997; 15(18):2145-2150.
18. Koh JT, Lee ZH, Ahn KY et al. Characterization of mouse brain-specific angiogenesis inhibitor 1 (BAI1) and phytanoyl-CoA alpha-hydroxylase-associated protein 1, a novel BAI1-binding protein. Brain Res Mol Brain Res 2001; 87(2):223-237.
19. Shiratsuchi T, Nishimori H, Ichise H et al. Cloning and characterization of BAI2 and BAI3, novel genes homologous to brain-specific angiogenesis inhibitor 1 (BAI1). Cytogenet Cell Genet 1997; 79(1-2):103-108.
20. Kee HJ, Koh JT, Kim MY et al. Expression of brain-specific angiogenesis inhibitor 2 (BAI2) in normal and ischemic brain: involvement of BAI2 in the ischemia-induced brain angiogenesis. J Cereb Blood Flow Metab 2002; 22(9):1054-1067.
21. Jeong BC, Kim MY, Lee JH et al. Brain-specific angiogenesis inhibitor 2 regulates VEGF through GABP that acts as a transcriptional repressor. FEBS Lett 2006; 580(2):669-676.

22. Kee HJ, Ahn KY, Choi KC et al. Expression of brain-specific angiogenesis inhibitor 3 (BAI3) in normal brain and implications for BAI3 in ischemia-induced brain angiogenesis and malignant glioma. FEBS Lett 2004; 569(1-3):307-316.
23. Formstone CJ, Barclay J, Rees M et al. Chromosomal localization of Celsr2 and Celsr3 in the mouse; Celsr3 is a candidate for the tippy (tip) lethal mutant on chromosome 9. Mamm Genome 2000; 11(5):392-394.
24. Curtin JA, Quint E, Tsipouri V et al. Mutation of Celsr1 disrupts planar polarity of inner ear hair cells and causes severe neural tube defects in the mouse. Curr Biol 2003; 13(13):1129-1133.
25. Shima Y, Kengaku M, Hirano T et al. Regulation of dendritic maintenance and growth by a mammalian 7-pass transmembrane cadherin. Dev Cell 2004; 7(2):205-216.
26. Tissir F, De-Backer O, Goffinet AM et al. Developmental expression profiles of Celsr (Flamingo) genes in the mouse. Mech Dev 2002; 112(1-2):157-160.
27. Beall SA, Boekelheide K, Johnson KJ. Hybrid GPCR/cadherin (Celsr) proteins in rat testis are expressed with cell type specificity and exhibit differential Sertoli cell-germ cell adhesion activity. J Androl 2005; 26(4):529-538.
28. Formstone CJ, Little PF. The flamingo-related mouse Celsr family (Celsr1-3) genes exhibit distinct patterns of expression during embryonic development. Mech Dev 2001; 109(1):91-94.
29. Tissir F, Bar I, Jossin Y et al. Protocadherin Celsr3 is crucial in axonal tract development. Nat Neurosci 2005; 8(4):451-457.
30. Abe J, Suzuki H, Notoya M et al. Ig-hepta, a novel member of the G protein-coupled hepta-helical receptor (GPCR) family that has immunoglobulin-like repeats in a long N-terminal extracellular domain and defines a new subfamily of GPCRs. J Biol Chem 1999; 274(28):19957-19964.
31. Vallon M, Essler M. Proteolytically processed soluble tumor endothelial marker (TEM) 5 mediates endothelial cell survival during angiogenesis by linking integrin alpha(v)beta3 to glycosaminoglycans. J Biol Chem 2006; 281(45):34179-34188.
32. Carson-Walter EB, Watkins DN et al. Cell surface tumor endothelial markers are conserved in mice and humans. Cancer Res 2001; 61(18):6649-6655.
33. Seandel M, Falciatori I, Shmelkov SV et al. Niche players: spermatogonial progenitors marked by GPR125. Cell Cycle 2008; 7(2):135-140.
34. Seandel M, James D, Shmelkov SV et al. Generation of functional multipotent adult stem cells from GPR125+ germline progenitors. Nature 2007; 449(7160):346-350.
35. Sugita S, Ichtchenko K, Khvotchev M et al. alpha-Latrotoxin receptor CIRL/latrophilin 1 (CL1) defines an unusual family of ubiquitous G-protein-linked receptors. G-protein coupling not required for triggering exocytosis. J Biol Chem 1998; 273(49):32715-32724.
36. Matsushita H, Lelianova VG, Ushkaryov YA. The latrophilin family: multiply spliced G protein-coupled receptors with differential tissue distribution. FEBS Lett 1999; 443(3):348-352.
37. Ichtchenko K, Bittner MA, Krasnoperov V et al. A novel ubiquitously expressed alpha-latrotoxin receptor is a member of the CIRL family of G-protein-coupled receptors. J Biol Chem 1999; 274(9):5491-5498.
38. Lajus S, Vacher P, Huber D et al. Alpha-latrotoxin induces exocytosis by inhibition of voltage-dependent K+ channels and by stimulation of L-type $Ca^{2+}$ channels via latrophilin in beta-cells. J Biol Chem 2006; 281(9):5522-5531.
39. Doyle SE, Scholz MJ, Greer KA et al. Latrophilin-2 is a novel component of the epithelial-mesenchymal transition within the atrioventricular canal of the embryonic chicken heart. Dev Dyn 2006;235(12):3213-3221.
40. LopezJimenez ND, Sainz E, Cavenagh MM et al. Two novel genes, Gpr113, which encodes a family 2 G-protein-coupled receptor and Trcg1, are selectively expressed in taste receptor cells. Genomics 2005; 85(4):472-482.
41. Davies B, Baumann C, Kirchhoff C et al. Targeted deletion of the epididymal receptor HE6 results in fluid dysregulation and male infertility. Mol Cell Biol 2004; 24(19):8642-8648.
42. Moriguchi T, Haraguchi K, Ueda N et al. DREG, a developmentally regulated G protein-coupled receptor containing two conserved proteolytic cleavage sites. Genes Cells 2004; 9(6):549-560.
43. Stehlik C, Kroismayr R, Dorfleutner A et al. VIGR—a novel inducible adhesion family G-protein coupled receptor in endothelial cells. FEBS Lett 2004; 569(1-3):149-155.
44. Zhao J, Li M, Bradfield JP et al. The role of height-associated loci identified in genome wide association studies in the determination of pediatric stature. BMC Med Genet 2010; 11:96.
45. Liu JZ, Medland SE, Wright MJ et al. Genome-wide association study of height and body mass index in Australian twin families. Twin Res Hum Genet 2010; 13(2):179-193.
46. Jin Z, Tietjen I, Bu L et al. Disease-associated mutations affect GPR56 protein trafficking and cell surface expression. Hum Mol Genet 2007; 16(16):1972-1985.
47. Piao X, Hill RS, Bodell A et al. G protein-coupled receptor-dependent development of human frontal cortex. Science 2004; 303(5666):2033-2036.
48. Xu L, Hynes RO. GPR56 and TG2: possible roles in suppression of tumor growth by the microenvironment. Cell Cycle 2007; 6(2):160-165.

# APPENDIX: MAMMALIAN ADHESION-GPCRs

49. Iguchi T, Sakata K, Yoshizaki K et al. Orphan G protein-coupled receptor GPR56 regulates neural progenitor cell migration via a Galpha 12/13 and Rho pathway. J Biol Chem 2008; 283(2):14469-14478.
50. Della Chiesa M, Falco M et al. GPR56 as a novel marker identifying the CD56dull CD16$^+$ NK cell subset both in blood stream and in inflamed peripheral tissues. Int Immunol 2010; 22(2):91-100.
51. McGee J, Goodyear RJ, McMillan DR et al. The very large G-protein-coupled receptor VLGR1: a component of the ankle link complex required for the normal development of auditory hair bundles. J Neurosci 2006; 26(24):6543-6553.
52. McMillan DR, Kayes-Wandover KM, Richardson JA et al. Very large G protein-coupled receptor-1, the largest known cell surface protein, is highly expressed in the developing central nervous system. J Biol Chem 2002; 277(1):785-792.
53. McMillan DR, White PC. Loss of the transmembrane and cytoplasmic domains of the very large G-protein-coupled receptor-1 (VLGR1 or Mass1) causes audiogenic seizures in mice. Mol Cell Neurosci 2004; 26(2):322-329.
54. Fredriksson R, Lagerstrom MC, Hoglund PJ et al. Novel human G protein-coupled receptors with long N-terminals containing GPS domains and Ser/Thr-rich regions. FEBS Lett 2002; 531(3):407-414.
55. Lum AM, Wang BB, Beck-Engeser GB et al. Orphan receptor GPR110, an oncogene overexpressed in lung and prostate cancer. BMC Cancer10:40.
56. Lagerstrom MC, Rabe N, Haitina T et al. The evolutionary history and tissue mapping of GPR123: specific CNS expression pattern predominantly in thalamic nuclei and regions containing large pyramidal cells. J Neurochem 2007; 100(4):1129-1142.
57. Chase A, Ernst T, Fiebig A et al. TFG, a target of chromosome translocations in lymphoma and soft tissue tumors, fuses to GPR128 in healthy individuals. Haematologica 95(1):20-26.
58. Vanti WB, Nguyen T, Cheng R et al. Novel human G-protein-coupled receptors. Biochem Biophys Res Commun 2003; 305(1):67-71.
59. Tonjes A, Koriath M, Schleinitz D et al. Genetic variation in GPR133 is associated with height: genome wide association study in the self-contained population of Sorbs. Hum Mol Genet 2009; 18(23):4662-4668.
60. Bjarnadottir TK, Fredriksson R, Hoglund PJ et al. The human and mouse repertoire of the adhesion family of G-protein-coupled receptors. Genomics 2004; 84(1):23-33.
61. Whittaker CA, Bergeron KF, Whittle J et al. The echinoderm adhesome. Dev Biol 2006; 300(1):252-266.

# INDEX

## A

Adaptive immunity  128, 130, 132, 133
Adhesion  1-9, 11, 12, 14, 30, 31, 37-40,
    45, 46, 49-57, 59, 60, 62, 64-70, 76,
    79, 80, 83, 87-89, 92, 94, 98-100,
    102-106, 109-114, 116-118, 121-126,
    128, 129, 136, 146, 149, 151, 152,
    154, 157-164, 167-170, 172, 175,
    176, 179, 180, 184-186, 189, 191
Adhesion-GPCR  14, 30, 31, 37-40,
    45, 46, 49-57, 60, 62, 64-69, 76,
    79, 80, 83, 87-89, 94, 98-100, 105,
    109-112, 114, 117, 118, 121-126,
    129, 136, 149, 151, 152, 154,
    157-164, 167-170, 172, 176, 179,
    180, 184-186, 189, 191
Adhesion receptor  4, 98, 102, 103, 106,
    109, 117
Adult stem cell  184
A-latrotoxin (LTX)  38, 55, 56, 158, 159
Alternative splicing  61, 77, 100, 118,
    121, 129, 130, 152, 186
Amino-terminal thrombospondin-like
    (TspN)  79
Angiogenesis  89, 103, 109, 112, 113,
    116, 117, 125, 158, 167-170, 173,
    174, 176, 189, 190
Anterior chamber-associated immune
    deviation (ACAID)  154
Antibody  25, 55, 60-62, 67, 69, 103,
    111, 122, 124, 125, 128, 129,
    131-136, 141, 142, 144, 145,
    161-164, 180, 181, 184
Antibody treatment  128, 129, 132, 134,
    135
Antigen-presenting cell (APC)  138, 140,
    146, 154
Apoptosis  102, 106, 116, 158, 163, 169
Audiogenic seizure (AGS)  76, 77, 81,
    83, 89
Autoimmunity  170
Autosomal-dominant partial epilepsy
    with auditory feature (ADPEAF)  79
Avian ankle link antigen (ALA)  81, 82

## B

Basement membrane (BM)  87, 91-94
Bilateral frontoparietal polymicrogyria
    (BFPP)  55, 87, 90-92, 98, 101, 159,
    162, 186
Blastomere  40-44
Brain angiogenesis inhibitor 1  4, 80, 87,
    89, 115-117, 121, 124, 125, 158, 162,
    163, 167-176, 189
Brain angiogenesis inhibitor 1-3  87, 89
Brain angiogenesis inhibitor (BAI)  3, 5,
    6, 89, 113, 115, 117, 118, 125, 189
BUB/BnJ  80, 81

## C

Cadherin  6, 9, 14-16, 18-20, 23, 24, 26, 28-31, 38, 39, 81, 87, 185
Cadherin 23 (CDH23)  81
Calcium  9, 14, 31, 59, 60, 68, 76, 79, 81, 82, 141, 142, 144, 152, 158, 186
Calx-b  76, 78, 79, 82
Cancer  59, 72, 89, 98, 101-106, 110, 113, 116, 118, 125, 126, 159
Carcinoma  110-112, 116
CD55  111, 112, 123, 129, 130, 132, 134-136, 138-142, 144-146, 157, 158, 162, 189
CD81  94, 104, 105, 159, 164
CD97  5, 50, 54, 55, 57, 60, 79, 80, 110-112, 114, 117, 118, 121-125, 128-136, 138-142, 145, 146, 152, 157, 158, 162, 180, 186, 189
Cdh-6  39
*Caenorhabditis elegans*  8, 19, 37-42, 44-46, 59, 64, 65, 69, 70, 72
Cell adhesion  6, 79, 89, 92, 99, 104, 105, 110, 159, 175, 185
Cell-cell interaction  37, 38, 64
Cell-ECM interaction  103
Cell migration  55, 72, 94, 112, 126, 159, 160, 163, 169, 171, 189, 190
Celsr  5-7, 14-16, 18, 19, 30, 31, 37, 38, 113, 114, 185, 190
Celsr 1-3  113
Central nervous system (CNS)  6, 19, 26, 27, 80, 81, 87-89, 94, 112, 125, 189, 190
Chondroitin sulfate  129, 130, 132, 134, 151, 157
Cleavage site  49, 50-53, 66, 101
Cobblestone lissencephaly  91, 92
Collagen-induced arthritis (CIA)  125, 128, 132-135
Colocalization  55, 132, 171, 175, 181
Colorectal carcinoma  112
Complement  123, 129, 138, 142, 144-146, 158
Cortical development  87, 94, 159
Costimulation  135, 138, 140-142
Crosslinking  93, 103, 104, 106, 116, 144, 145, 160
C-terminal fragment (CTF)  61, 64-71, 101, 104
Cytokine  140-142, 144-146, 153, 159

## D

Delayed type hypersensitivity (DTH)  128, 130-134
Dendritic cell  122, 123, 150
Differentiation  20, 46, 93, 123, 140, 146, 153, 179, 181
Dock180  124, 125, 163, 167, 171-173
Drosophila  11, 14, 15, 18-23, 25-31, 39, 42, 45, 46, 80, 184, 185

## E

Efferent ductule  179, 182
EGF-TM7  5, 121-123, 125, 126, 128, 129, 135, 141, 149-152, 154, 157, 159, 162
ELMO1  163, 170-172, 174-176
Embryonic development  18, 31, 38, 40, 55
EMR1  5, 111, 112, 114, 121-123, 151, 152, 154, 185, 189
EMR2  1, 3-5, 52-56, 99, 112, 114, 121, 123, 124, 139, 152, 157, 158, 160, 162, 189
Engulfment of apoptotic cell  171-173, 175, 176
Eosinophil  122, 123, 150, 154, 189
Epididymis  179, 180-184, 186
Epilepsy  76, 77, 79
Epitempin (EPTP)  9, 76, 79
Evolution  1, 5, 7, 8, 11, 12, 39, 62, 64, 83, 129
Exocytosis  59, 60, 69
Experimental autoimmune encephalomyelitis (EAE)  128, 132-134
Extracellular domain  14, 15, 30, 31, 38, 56, 64, 79, 116, 121, 152, 154, 157, 158, 172, 180, 186
Extracellular matrix  2, 7, 64, 70, 87, 91-93, 98, 106, 109, 116, 123, 129, 152, 158, 169, 180, 181

## F

F4/80 121-123, 149-155
Fertility 180, 182, 185, 186
Flamingo 14, 18-23, 25-31, 37-39, 113, 185
Fluid absorption 179
Frings 77, 81
Frizzled 3, 4, 19-23, 26, 27, 29, 40, 44, 67
Functional domain 168, 180

## G

Gα12/13 94, 104-106, 159-161
Galactose-binding lectin (GBL) 9, 64, 72
Glioblastoma 116, 168, 173, 174, 176
Glycosylation 52, 54, 77, 91, 99, 100, 111, 118, 122, 152, 160, 162
GPCR proteolytic site (GPS) 1, 2, 7, 9, 30, 39, 46, 49-57, 65-67, 69, 78, 79, 81, 87, 90, 100, 101, 104, 121, 122, 125, 152, 154, 157, 160-162, 169, 173, 174, 180, 181, 186
GPR56 54, 55, 87, 89, 90, 92-94, 98-106, 115-117, 121, 125, 126, 157-164, 170, 180, 185, 190
GPR64 115-118, 180-182, 184, 186, 190
GPR98 76, 185
G protein-coupled receptor (GPCR) 1-9, 11, 12, 14, 15, 30, 31, 37-40, 45, 46, 49-57, 59-62, 64-70, 76, 77, 79, 80, 83, 87-89, 94, 98-101, 103, 105, 109-112, 114, 117, 118, 121-126, 128, 129, 135, 136, 149-152, 154, 157-164, 167-170, 172-174, 176, 179, 180, 183-186, 189, 191
G protein signaling 158, 159, 170
G protein α subunit (Gα) 101, 160
GPS motif 39, 52, 53, 65-67, 100, 101, 121, 154
Granule cell adhesion 92
Granulocyte 128-135, 138, 152, 162

## H

HE6 115, 117, 180, 183
Hormone receptor motif (HRM) 15, 31, 46, 65, 69

## I

Immunoregulation 149, 153
Inducible regulatory T cells (Tr1) 138, 142, 146
Innate immunity 133
Inositol 1,4,5-trisphosphate 158
Intracellular trafficking 30, 90

## K

Knockout mouse 87, 135

## L

Laminin G (LamG) 9, 14, 15, 79, 185
Langerhans cell 122, 150
Lasso 59, 70
Lat-1 37, 39, 42-44, 46, 59, 69, 70, 72, 158
Lat-2 69, 70, 72
Latrophilin 5, 37-39, 42, 44, 46, 50, 54-56, 59-73, 87, 89, 100, 110, 114, 117, 118, 158, 159, 162-164, 190
Latrotoxin 38, 55, 56, 59-62, 64-69, 71, 158, 190
Leucine-rich glioma inactivated (LGI) 79
Ligand 2, 3, 19, 22, 27, 31, 38, 40, 46, 56, 59, 60, 64-67, 69, 70, 73, 76, 79, 80, 83, 87, 92-94, 103-106, 111, 112, 118, 123, 128-130, 132, 134-136, 138-142, 146, 149, 153-155, 157-164, 170, 172, 174, 180
Ligand receptor pair 56
Lipopolysaccharide 129
Lung carcinoma 110

## M

Macrophage  122, 123, 125, 128, 130, 132, 133, 135, 149-155, 157, 158, 171, 186, 189
Male reproductive tract  172, 179, 180, 186
Mechanism  27, 30, 31, 37, 38, 40, 42, 44, 46, 49, 52, 60, 67-69, 72, 83, 92, 98, 100, 102-104, 128, 129, 135, 136, 140, 141, 143, 146, 154, 159-163, 174, 176
Melanoma  98, 99, 101-103, 106, 116, 117, 158
Metastasis  98, 101-106, 109, 116, 117, 158, 159, 173, 190
Microenvironment  102-104, 109, 116-118
Monoclonal antibody  111, 122, 128, 161
Monogeneic audiogenic seizure-susceptible (MASS1)  76-78, 80, 115, 117
Mouse model  55, 87, 89, 116, 124, 129, 132, 170, 172, 176, 184
Multipotent  184
Multipotent adult stem cell (MASC)  184
Mutant  19-21, 23-28, 31, 39, 40, 42-44, 46, 52, 54, 55, 67, 68, 70, 71, 76, 80-83, 89-91, 153, 154, 159, 161, 162, 182, 185, 186

## N

$Na^+/Ca^{2+}$ exchanger  79
Neural  6, 14, 16, 17, 19, 20, 24, 25, 27, 29, 31, 64, 80, 89, 93, 94, 105, 159, 160, 163
Neural progenitor cell  89, 94, 105, 159, 160, 163
Neurexin  60, 62, 68, 158
Neuroepithelium-notable (Neurepin)  77
Neurogenesis  64, 79, 80, 83
Neuronal overmigration  87
Neurotransmitter  38, 59, 60, 68, 69, 72, 184
Neurotransmitter release  59, 68
Neutrophil  123, 125, 128, 129, 131-133, 150, 162
N-terminal fragment (NTF)  60, 61, 64-70, 72, 92, 94, 104
N-termini  1, 2, 5-7, 9, 109, 167
Nucleophilic attack  50, 52-54

## O

Olfactomedin  9, 61, 64, 65
Oral tolerance  154
Organ  6, 14, 16, 18, 24, 31, 38, 46, 99, 102, 123
Orthologue  1, 3-5, 8, 59, 61, 62, 64-66, 69, 70, 72, 123, 180

## P

Pentraxin (PTX)  9, 78, 79, 160, 161
Peripheral tolerance  123, 150, 155
Phagocytosis  125, 144
Phospholipase C  31, 68, 72, 158
Phylogeny  1, 167
Pial basement membrane  87, 91, 92
Planar cell polarity  14, 18, 25, 31, 37, 45, 113, 185
Postsynaptic density protein 95/ Drosophila disks large/zona occludens-1 (PDZ)  71, 76, 80, 83, 162, 169, 170, 173, 175
Presynaptic  69, 71, 72, 158
Protease  50, 52, 161
Protein-protein interaction  100
Proteolysis  49-52, 54, 55, 57, 64-66, 79, 100, 160, 161, 180

## R

Rac  125, 163, 167, 171-173, 175
RBL domain  39, 46, 72
Receptor  1-6, 15, 19, 22, 37-40, 42, 45, 46, 49-52, 54-56, 59, 60, 64-71, 73, 76, 79, 80, 87-91, 93, 98-103, 105, 106, 109, 114, 115, 117, 118, 121-126, 128, 129, 135, 138, 140, 142, 144, 146, 149-152, 154, 155, 157-164, 167, 169-176, 179-186, 189-191
Regulation  14, 18, 28, 30, 49, 54, 64, 87, 92, 102, 113, 135, 140, 142, 146, 159, 160, 168, 175, 179, 180, 182-186
Regulatory T cell  142, 149

Relay 42
Rhamnose-binding lectin 64
Rho 94, 160, 161, 163
RhoA 94, 104-106, 161

## S

Self-catalytic reaction 50
Seven-transmembrane domain (7TM) 2, 5, 11, 14-16, 18-20, 23, 24, 26, 28-31, 38, 39, 46, 49, 50, 55, 56, 150-152, 154, 157, 158, 161-164, 167-169, 173
Signaling 3, 18-23, 26, 30, 31, 37, 40, 42, 44, 46, 57, 60, 66-71, 80, 83, 87, 89, 93, 94, 101, 103-106, 113, 116, 125, 135, 141-145, 149, 155, 157-164, 167, 170-173, 175, 176, 185
Signal transduction 60, 67, 68, 98, 112, 128, 157, 158, 162, 163, 175
Spermatozoa 179, 182, 183
Spindle orientation 37, 44, 46
SRE 160-162
Stem cell 7, 99, 102, 104, 106, 128, 129, 134, 184, 190
Stereocilia 25, 81, 82, 185
Suppressor 101, 104, 105, 117, 168

## T

T cell 123, 131-133, 135, 138, 140-146, 149, 154, 159
Testes 78, 172

Tetraspanin 92-94, 105, 159, 163, 164
Thrombospondin repeat 9, 125, 163, 172
Thyroid-stimulating hormone receptor (TSHR) 161, 162
Thyrotropin receptor 161
Tissue polarity 18, 37, 38, 42, 45, 46
Tissue transglutaminase (TG2) 93, 94, 103-106, 116, 158, 190
Tolerance 123, 149, 150, 153-155, 189
Treg cell 154, 155
Tumor cell interaction 109
Tumorigenesis 109-112, 118, 174
Tumor progression 103-105, 113
Tumor suppression 103

## U

Unique 1, 2, 5-7, 14, 16, 31, 49, 50, 52, 65, 92, 121, 138, 142, 149, 151, 154, 157, 161, 164, 167, 168, 185
Usher syndrome 76, 82, 83, 89, 186

## V

Ventricular zone 16, 77, 80, 88, 89, 92
Very large G protein-coupled receptor-1 (VLGR1) 6, 76-83, 89, 115, 117, 185, 186, 190
Vlgr1del7TM 80, 81

## W

Wnts 19